WHEELED MOBILE ROBOTICS

WHEELED MOBILE ROBOTICS

From Fundamentals Towards Autonomous Systems

Edited by

GREGOR KLANČAR

ANDREJ ZDEŠAR

SAŠO BLAŽIČ

IGOR ŠKRJANC

Butterworth-Heinemann
An imprint of Elsevier
elsevier.com

Butterworth-Heinemann is an imprint of Elsevier
The Boulevard, Langford Lane, Kidlington, Oxford OX5 1GB, United Kingdom
50 Hampshire Street, 5th Floor, Cambridge, MA 02139, United States

Notices
Knowledge and best practice in this field are constantly changing. As new research and experience
broaden our understanding, changes in research methods, professional practices, or medical treatment
may become necessary.

Practitioners and researchers must always rely on their own experience and knowledge in evaluating
and using any information, methods, compounds, or experiments described herein. In using such
information or methods they should be mindful of their own safety and the safety of others, including
parties for whom they have a professional responsibility.

To the fullest extent of the law, neither the Publisher nor the authors, contributors, or editors, assume
any liability for any injury and/or damage to persons or property as a matter of products liability,
negligence or otherwise, or from any use or operation of any methods, products, instructions, or ideas
contained in the material herein.

Library of Congress Cataloging-in-Publication Data
A catalog record for this book is available from the Library of Congress

British Library Cataloguing-in-Publication Data
A catalogue record for this book is available from the British Library

ISBN: 978-0-12-804204-5

For information on all Butterworth-Heinemann publications
visit our website at https://www.elsevier.com/

Working together
to grow libraries in
developing countries

www.elsevier.com • www.bookaid.org

Publisher: Joe Hayton
Acquisition Editor: Sonnini R. Yura
Editorial Project Manager: Ana Claudia Abad Garcia
Production Project Manager: Kiruthika Govindaraju
Cover Designer: Mark Rogers

Typeset by SPi Global, India

CONTENTS

Companion site link is: http://booksite.elsevier.com/9780128042045/

PREFACE

This is an introductory book on the subject of wheeled mobile robotics, covering all the essential knowledge and algorithms that are required in order to achieve autonomous capabilities of mobile robots. The book can serve both as a textbook for engineering students and a reference book for professionals in the field. As a textbook, it is suitable for courses in mobile robotics, especially if the courses' main emphasis is on mobile robots with wheels. The book covers topics from mathematical modeling of motion, sensors and measurements, control algorithms, path planning, nondeterministic events, and state estimation. The theory is supported with examples that have solutions and excerpts of Matlab code listings in order to make it simple for the reader to try, evaluate, and modify the algorithms. Furthermore, at the end of the book some interesting practical projects are depicted, which can be used for laboratory practice or to consolidate the theory learned.

The field of autonomous mobile robotics is an extremely popular field of research and development. Within this field the majority of mobile robots use wheels for motion. Wheeled mobile robots leave tracks where no man has ever been (e.g., extraterrestrial explorations), man should not go (e.g., dangerous or contaminated areas), or act alongside humans (e.g., human support and assistance). Wheeled mobile robots have already started to penetrate into our homes in the form of robotic floor cleaners and lawn mowers, automatic guided vehicles can be found inside many industrial facilities, prototype self-driving cars already drive in the normal traffic, and in the nearby future many new applications of autonomous wheeled mobile robots are expected to appear, even in some unforeseen ways. The purpose of this book is to shed some light onto the design and give a deeper insight into the algorithms that are required in order to implement autonomous capabilities of wheeled mobile robots. The book addresses essential problems in wheeled mobile robotics and presents to the reader various approaches that can be used to tackle them. The presented algorithms range from basic and simple solutions to more advanced and state-of-the-art approaches.

The complete Matlab code listings of the algorithms that are presented in this book are also available for download at the book companion website: http://booksite.elsevier.com/9780128042045/. Therefore it should be simple for the readers to try, evaluate, tweak and tune the algorithms by themselves. The website also contains many animated videos from the examples that give an additional insight into the topics and algorithms covered in the book.

G. Klančar, A. Zdešar, S. Blažič, I. Škrjanc
Ljubljana, Slovenia
August 2016

ACKNOWLEDGEMENTS

The authors would like to recognize all the individuals who contributed, both directly and indirectly, to this book. We wish to thank all our colleagues, the current and past members of the Laboratory of Autonomous Mobile Systems, and the Laboratory of Modelling, Simulation and Control, Faculty of Electrical Engineering, University of Ljubljana, Slovenia, who inspired and contributed to this book. We are grateful to our colleague Dr. Matevž Bošnak for providing some insight into the field of assistive mobile robotics in walking rehabilitation therapy. Special acknowledgement goes to our former Head of the Laboratory, Prof. Dr. Drago Matko for introducing our laboratory into the field of mobile robotics. Our gratitude also goes to all graduate students, research associates and technical staff from the Faculty of Electrical Engineering, University of Ljubljana, and researchers from institutions around the world with whom we have been involved in a diverse range of projects that in a way also enabled creation of this book. We would like to thank Asst. Prof. Dr. Fernando A. Auat Cheein, Department of Electronics Engineering, Universidad Técnica Federico Santa María, Chile, for providing us with some material from his field of research, mobile robotics in agriculture. This work would have not been possible without support of the Slovenian Research Agency that backed many of our research and application projects.

CHAPTER 1

Introduction to Mobile Robotics

1.1 INTRODUCTION

1.1.1 Robots

The word *robot* has roots in Slavic languages. The meaning of *"robota"* in Polish is work or labor, while the word is more archaic in Czech or Slovenian language and means statute labor or corvée. The famous Czech writer Karel Čapek coined the word robot and used it in the science fiction play *R.U.R.*, which stands for Rossum's Universal Robots. The robots in the play are a sort of artificial human; in modern terminology cyborgs or androids would be more appropriate. The play witnessed huge success and the word robot was adopted by the majority of world languages. While the word robot is not even 100 years old, the idea of mechanic creatures goes very deep into history.

In Greek mythology we find many creatures that are used for particular tasks. Spartoi are mythical, fierce, and armed men who sprang up from the dragon's teeth sown by Cadmus. They assisted Cadmus in building the Cadmeia or citadel of Thebes. Talos, created by Hephaestus, was a giant automaton made of bronze to protect Europa in Crete from pirates and invaders. The Greek god of blacksmiths and craftsmen, Hephaestus, is also credited for some other mechanical structures. Automata can be also found in ancient Jewish, Chinese, and Indian legends. The idea of mechanical automata that resembled either humans or animals was then present in literature throughout the history. It really became popular in the 19th century and, especially, the 20th century. In the 20th century robots found a new popular media to depict them and bring them to life: film. Some of the ideas in literature and films were attributed as science fiction at the time of creation, and later this fiction became a reality.

The robotic designs were not present only in fiction. Very early inventors tried to construct mechanical automata. The Greek mathematician

Fig. 1.1 Model of a bicycle and a programmable cart built based on Leonardo da Vinci's notes.

Archytas is believed to have designed and built the first artificial, self-propelled flying device in the 4th century BC. A mechanical bird propelled by steam was said to have actually flown some 200 m. In the comprehensive heritage of Leonardo da Vinci several mechanical designs can be found. Among the rough sketches scattered throughout Leonardo's notes, Rosheim [1] has reconstructed a programmable cart (Fig. 1.1) used as a base for Leonardo's inventions, such as a robot lion and a robot knight. The spring of industrial revolution technological advancement resulted in the outburst of automation that gradually led to the mobile robotics we know today.

1.1.2 Mobile

The word *mobile* has its roots in the Latin word of the same meaning, "*mōbilis*." The majority of animal species possess the ability of locomotion. While some animals use passive systems to do so (they can move with water or air motion), others have developed more or less sophisticated mechanisms for active movement. Some animals perform locomotion in the 3D space (swimming in the water, flying in the air, moving through the soil), and others more or less follow the 2D surface of the water or the ground, while some animals are capable of combining different ways of movement. In the context of mobile robots we are concerned with systems that can move using their locomotion apparatus. The latter very often mimics the one of a human or a certain animal. Copying from biological

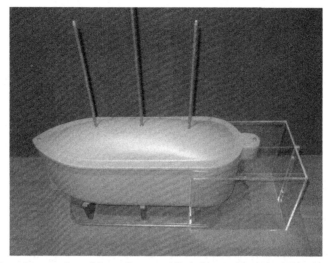

Fig. 1.2 A working replica of the radio-controlled electrical boat built by Nikola Tesla.

systems also very often successfully solves some technical problems that arise during artificial locomotion system design.

The other important aspect of a mobile system is that being mobile also means that the distance to the human operator can become large. This implies that the system needs to either possess a certain level of autonomy, meaning that it has to move across the space without the help of the operator, or accept the commands from a distant operator, meaning that the system is able to move tele-operated. Nikola Tesla was the first to design and build a radio-controlled electrical boat (Fig. 1.2) at the end of the 19th century. In the 20th and the 21st century the level of autonomy continuously rose. Yet, all the existing mobile systems are human operated on a certain level.

1.1.3 Wheels

Although very primitive animal species are able to move, it is often a nontrivial task to design an artificial system that is able to mimic animal locomotion. While wheels or similar structures cannot be found in the animal world, the vehicles with wheels are known to enable energy-efficient motion over the ground. The surface has to be smooth enough although appropriately constructed wheeled vehicles can also move over rugged terrain, such as steps. It is not known where and when the wheel

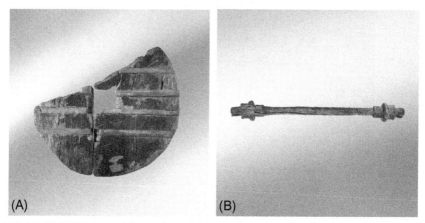

Fig. 1.3 The 5200-year-old wooden wheel (A) with the axle (B) found in the Ljubljana Marshes pile dwelling is one of the world's most significant cultural heritage items in terms of both its age and technological accomplishment. The diameter of the wheel is 70 cm and the axle length is 120 cm. *(Museum and Galleries of Ljubljana, photos by M. Paternoster.)*

was invented but the established belief is that the first wheels were used in Mesopotamia approximately 4000 BC. From there, they were spread around the world. Some experts attribute the wheel's invention to prehistoric Europe. The oldest wooden wheel with an axle, 5200 years old, was discovered in Slovenia in the Ljubljana Marshes (Fig. 1.3).

1.1.4 Autonomous Mobile Systems

Mobile systems can be defined as systems that are not attached to the environment and can move in a certain space. In terms of the environment they move across they can be classified into some principal groups:

Ground mobile systems Various types of mobile platforms can be found here such as mobile vehicles with wheels or caterpillars, legged robots (humanoids or animal mimicking), or robots that mimic some other type of animal locomotion, for example, snakes. Ground mobile systems with wheels or caterpillars that do not carry the operator are often referred to as unmanned ground vehicles.

Aerial mobile systems This group consists of mobile systems that fly in a certain aerial space (airplanes, helicopters, drones, rockets, animal-mimicking flying systems; when used without a pilot they are referred to as unmanned aerial vehicles) or orbit the Earth or some other celestial body (satellites).

Water and underwater mobile systems In this group we find different types of ships, boats, submarines, autonomous underwater vehicles, etc.

In this book only the wheeled mobile vehicles will be discussed although a large portion of the presented material can be used for other types of mobile systems with appropriate modifications.

Mobile systems are regarded as autonomous when they are capable of autonomous motion within their environment. Autonomy has to be guaranteed by the following:

- from the energy point of view: the robot should carry some source of energy; and
- from the decision point of view: the robot should be capable of taking certain decisions and performing appropriate actions.

In reality this means that the mobile system takes the commands from the human operator based on the level of autonomy the system is capable of. The system responds by trying to carry out the commanded task and appropriate subtasks on the lower levels. In case there are no unforeseen circumstances, the task is usually executed in a certain time interval. Based on the level of the robot's autonomy, the following typical commands can be taken from the operator:

Desired wheel velocities The robot takes the commands in terms of its wheel velocities and very basic control algorithms with appropriate sensors (typically rotary encoders) to ensure that the wheels rotate as commanded.

Desired robot longitudinal and angular velocities The computer program running on the robot is aware of the robot's kinematic model, so that it can compute the appropriate wheel velocities to achieve the desired robot velocities.

Desired robot path or robot trajectory The robot is capable of determining and controlling its pose within its environment; the pose is usually defined as joint information about position and orientation relative to some coordinate system. On this level we find robot localization that uses different sensors mounted either on the robot or in the environment to obtain the best possible approximation of the robot's pose. The problems also arise in control due to nonlinear character of the system, erroneous sensor information, wheel slipping, poor models, delays, etc.

Desired operation inside of the known environment with potential obstacles The robot has to perform an operation in the known environment but with some (static or moving) obstacles. On this level the robot is capable of planning its path and replanning it if some obstacles prevent the fulfillment of the operation.

Desired operation inside of the unknown environment The robot does not know its environment. It has to simultaneously perform the actions of determining its pose and building its map of environment; the approach is known as SLAM (simultaneous localization and mapping).

Desired mission The robot starts the mission that it has to accomplish. It operates in the environment where it might cooperate with other robots or agents. It needs to have a certain understanding of its missions. It needs to have certain priorities of its missions, so that it can abort a certain mission and/or take a mission with a higher priority. A lot of decision making has to be built in the robot. A simple example is that the robot needs to inspect the energy level in batteries and has to recharge its batteries if necessary.

The robots are not understood as autonomous on the first two levels shown above. Also, there exist other classifications. It is important that we understand the tasks and intelligence that the robot possesses on a certain level.

The main mechanical and electronic parts of an autonomous mobile robot are the following:

Mechanical parts Rigid and moving parts (body, wheels, tracks, legs, etc.)

Actuators Electrical motors (DC, stepper, servomotor, etc.)

Sensors Rotation encoders, proximity and distance sensors, inertial navigation unit, global navigation satellite system, etc.

Computers Micro-controllers, portable personal computer, embedded systems, etc.

Power unit Batteries, solar panels, etc.

Electronics Actuator drive, sensor measurement, power distribution, and telecommunication electronics

This book is about algorithms that are required to process sensor data and drive the actuators in order to give autonomous capabilities to the mobile robot.

There are several properties that make use of wheeled mobile robots appealing. Mobile robots allow human-free access to hazardous environments (e.g., minefields, radioactive environments, deep-sea explorations, etc.) and access to remote or inaccessible environments (e.g., exploration of extraterrestrial planets, nanoscale robots in medicine, etc.). Robotic systems can do physically demanding tasks instead of humans. The introduction of automation, robotics, and mobile systems also allows for greater productivity, product or service quality, and can reduce labor costs.

Nowadays, mobile systems are used in numerous applications in different fields, which are constantly expanding due to rapid technological development. An incomplete list of applications of wheeled mobile robots includes the following:

- medical services, operation support, laboratory analyses (e.g., in situations with the risk of infections)
- cleaning applications (floor vacuuming, sweeping and mopping in homes or large buildings, window cleaning, etc.)
- applications in agriculture, like automated fruit picking, seed planting, grass mowing, etc.
- forest maintenance and logging
- consumer goods stores
- inspection and surveillance of hazardous areas (detection and deactivation of mines in minefields, review of nuclear reactors, cleaning of sewer pipes)

- space missions (satellites, inspection and servicing of satellites, planetary exploration)
- depths of the sea (robots for cabling and screening the seabed)
- robots for loading and unloading goods or materials from airplanes, ships, or trucks
- military robots (reconnaissance robots, airplanes, and various autonomous projectiles)
- security robots (security guards to control storage, buildings, etc.)
- help for the elderly and disabled (autonomous wheelchairs, rehabilitation robots)
- consumer applications (robotic pets, robot soccer, etc.)
- systems at research institutions aimed at learning and developing new algorithms

Predicting the way the future will look like has always been a hard task. Technologies and applications that are available today, were hard to imagine even a decade ago. However, in the near future autonomous wheeled mobile robots are expected to become an even more integral part of our daily lives: they will work alongside us in modern factories of the future, help us with domestic chores, drive us on the road, save lives in rescue missions and much more. How technological development brought us to current point is depicted in a short history review in the next section.

1.2 HISTORY

This section presents some important milestones in the history of wheeled mobile robotics [2]. Focus is given to the application of wheeled mobile robots; however, some other technological achievements that influenced the field are mentioned.

1898   At an electrical exhibition at Madison Square Garden in New York City, Nikola Tesla demonstrated a radio-controlled unmanned boat [3]. Nikola Tesla holds a patent [4] for the invention.

1939–45 During World War II, cruise missiles V-1 [5] and V-2 [6] were developed in Germany. At the same time an American, Norbert Wiener, was working on automatic aiming of antiaircraft guns [7].

1948–49 W. Grey Walter constructed two autonomous robots, Elmer and Elsie [8], which resembled turtles and were capable of following a light source (photodiode), detecting obstacles (contact switch), and preventing collisions.

1961–63 Johns Hopkins University developed *Beast* [9], a self-surviving mobile robot that was able to wander through white halls, seeking black wall outlets to recharge its batteries.

1966–72 The Stanford Research Institute was developing Shakey the robot [10], which had a camera, sonar range finders, collision detection sensors, and a wireless link. It was the first general-purpose mobile robot capable of sensing and interpreting the environment, and then navigating between the obstacles on its own. Results of the project include development of A* search algorithm, Hough transform, and visibility graph.

1969 A first robotic lawn mower, MowBot, was introduced and patented [11].

1970 The Soviet Union successfully landed the first lunar rover Lunokhod 1 on the Moon's surface, which was remote controlled from the Earth and it carried several cameras and other sensors. In 301 days of operation the rover traveled approximately 10 km, returned more than 25,000 images, and made several soil analyses [12].

1973 The Soviet Union deployed a second lunar rover Lunokhod 2 on the surface of the Moon. During the 4-month mission the rover covered a distance of 39 km, a record for off-Earth driving distance that lasted until 2014 [13].

1976 NASA unmanned spacecrafts Viking 1 and Viking 2 (each consisting of an orbiter and a lander) entered Mars' orbit, and the landers soft-landed on Mars' surface several days later [14].

1977 French Laboratory for Analysis and Architecture of Systems started developing mobile robot Hilare 1 [15], which was equipped with ultrasonic and laser range finders and a camera on a robotic arm.

1979 The Stanford Cart (initial model introduced in 1962) became capable of vision-based navigation through an obstacle course [16].

1982 A first model from the series of commercial HERO robots became available, which were mainly targeted for home and educational use [17].

1986 The team of Ernst Dieter Dickmanns [18] developed the robot car VaMoRs that was able to drive by itself on streets without traffic, reaching speeds up to 90 km/h.

1995 An affordable mobile robot, Pioneer, for education and research appeared on the market [19].

1996   The First robotic soccer tournament was organized, and the Federation of International Robot-soccer Association (FIRA) was established a year later [20].

1996–97   On board of Mars Pathfinder, NASA sent a rover, Sojourner, to Mars [21], which was commanded from Earth, but it was able to autonomously drive along the predefined path and avoid hazardous situations on the way.

2002   A first model of the Roomba floor vacuuming robot for domestic use was launched [22].

2004   Mars twin rovers Spirit and Opportunity landed on Mars [23]. The Spirit rover became stuck in 2009, but the Opportunity rover is still active, and in 2014 it surpassed the record for the longest off-Earth rover distance set by Lunokhod 2.

2004   The first competition of the DARPA Grand Challenge was held in the Mojave desert (United States). None of the autonomous vehicles completed the 240 km route [24].

2005   In the second DARPA Grand Challenge, autonomous vehicle Stanley from Stanford University finished the route first, and four other vehicles of 23 competitors were able to complete the mission [25].

2007   DARPA Urban Grand Challenge was organized, where six autonomous vehicles completed the course in an urban area. Obeying all traffic regulation was required and the vehicles needed to merge into traffic [26].

2009   The initial version of the Robot Operating System (ROS 0.4) was released [27].

2009   Google started testing their self-driving technology with the modified Toyota Prius on highways in California [28].

2010   In the VisLab Intercontinental Autonomous Challenge [29], four autonomous vehicles completed the almost 16,000 km long journey from Parma, Italy to Shanghai, China, with virtually no human intervention.

2012   NASA successfully landed the robotic rover Curiosity on Mars [30]; it is still active.

2014   Google unveiled their new prototype autonomous vehicle that has no steering wheel and no pedals [28].

2016   As of June, the fleet of Google self-driving vehicles had driven a total of 2,777,585 km in autonomous mode [31].

1.3 ABOUT THE BOOK

The book can serve as a primary or secondary source in mobile robotics courses for undergraduate and postgraduate study. It covers the main topics from wide area of mobile robotics. After the reader is introduced to a particular topic and the formulation of the problem is given, various approaches that can be used to solve the particular problem are presented, ranging from basic to more advanced and state-of-the-art algorithms. Several examples are included for better understanding, many of them also include a short Matlab script code so that they can be easily reused in practical work.

The contents of the book are as follows: Chapter 2 is about motion modeling for mobile robots, Chapter 3 covers various control approaches of wheeled mobile systems, Chapter 4 presents path planning approaches, Chapter 5 gives an overview of sensors used in mobile systems, Chapter 6 presents the approaches of dealing with nondeterministic events in mobile systems, and Chapter 7 describes some applications of autonomous guided vehicles. In Chapter 8 at the end of the book some interesting projects are overlayed, which can be used for laboratory practice or to consolidate learned theory.

REFERENCES

[1] M. Rosheim, Leonardo'S Lost Robots, Springer, Berlin, 2006.
[2] Mobile robot, 2016, https://en.wikipedia.org/wiki/Mobile_robot (accessed 18.07.16).
[3] Nikola Tesla, 2016, https://en.wikipedia.org/wiki/Nikola_Tesla (accessed 18.07.16).
[4] N. Tesla, Method of and apparatus for controlling mechanism of moving vessels or vehicles, US Patent 613,809, 1898.
[5] V-1 flying bomb, 2016, https://en.wikipedia.org/wiki/V-1_flying_bomb (accessed 18.07.16).
[6] V-2 rocket, 2016, https://en.wikipedia.org/wiki/V-2_rocket (accessed 18.07.16).
[7] D. Jerison, D. Stroock, Norbert Wiener, Notices of the American Mathematical Society 42 (4) (1995) 430–438.
[8] William Grey Walter, 2016, https://en.wikipedia.org/wiki/William_Grey_Walter (accessed 18.07.16).
[9] D.P. Watson, D.H. Scheidt, Autonomous systems, Johns Hopkins APL Tech. Dig. 26 (4) (2005) 368–376.
[10] N.J. Nilsson, Shakey the robot, Technical report, SRI International, 1984.
[11] S.L. Bellinger, Self-propelled random motion lawnmower, US Patent 3,698,523, 1972.
[12] I. Karachevtseva, J. Oberst, F. Scholten, A. Konopikhin, K. Shingareva, E. Cherepanova, E. Gusakova, I. Haase, O. Peters, J. Plescia, M. Robinson, Cartography of the Lunokhod-1 landing site and traverse from LRO image and stereo-topographic data, Planet. Space Sci. 85 (2013) 175–187.

[13] Lunokhod 2, 2016, https://en.wikipedia.org/wiki/Lunokhod_2 (accessed 18.07.16).

[14] Viking project information, 2016, http://nssdc.gsfc.nasa.gov/planetary/viking.html (accessed 18.07.16).

[15] G. Giralt, R. Chatila, M. Vaisset, An integrated navigation and motion control system for autonomous multisensory mobile robots, in: Autonomous Robot Vehicles, Springer, 1990, pp. 420–443.

[16] Stanford cart, 2012, http://web.stanford.edu/learnest/cart.htm (accessed 18.07.16).

[17] HERO (robot), 2016, https://en.wikipedia.org/wiki/HERO_(robot) (accessed 18.07.16).

[18] E.D. Dickmanns, Dynamic Vision for Perception and Control of Motion, Springer Science & Business Media, Heidelberg, 2007.

[19] MobileRobots Inc drives autonomous robots to market, 2016, http://www.mobilerobots.com/AboutMobileRobots.aspx (accessed 2016.03.08).

[20] FIRA, 2016, http://www.fira.net/contents/sub01/sub01_3.asp (accessed: 2016.03.08).

[21] NASA—Mars Pathfinder and Sojourner, 2013, http://www.nasa.gov/mission_pages/mars-pathfinder/ (accessed 18.07.16).

[22] iRobot history, 2016, http://www.irobot.com/About-iRobot/Company-Information/History.aspx (accessed 18.07.16).

[23] NASA Mars Exploration Rovers—Spirit and Opportunity, 2016, http://www.nasa.gov/mission_pages/mer/index.html (accessed 18.07.16).

[24] DARPA Grand Challenge 2004, 2014, http://archive.darpa.mil/grandchallenge04 (accessed 18.07.16).

[25] DARPA Grand Challenge 2005, 2014, http://archive.darpa.mil/grandchallenge05 (accessed 18.07.16).

[26] DARPA Urban Challenge, 2014, http://archive.darpa.mil/grandchallenge (accessed 18.07.16).

[27] ROS.org, History, 2016, http://www.ros.org/history/ (accessed 18.07.16).

[28] On the road—Google Self-Driving Car Project, 2016, https://www.google.com/selfdrivingcar/where (accessed 18.07.16).

[29] The VisLab Intercontinental Autonomous Challenge, 2016, http://viac.vislab.it (accessed 18.07.16).

[30] NASA Mars Science Laboratory—Curiosity, 2016, https://www.nasa.gov/mission_pages/msl/index.html (accessed 18.07.16).

[31] Google Self-Driving Car Project: Monthly Report, 2016, https://static.googleusercontent.com/media/www.google.com/en//selfdrivingcar/files/reports/report-0616.pdf (accessed 18.07.16).

CHAPTER 2

Motion Modeling for Mobile Robots

2.1 INTRODUCTION

Humans took advantage of wheel drive for thousands of years, and the underlying structure of a prehistoric two-wheeled cart (Fig. 2.1) can be found in modern cars and wheeled robots. This chapter deals with motion modeling of different wheeled mobile systems. The resulting model can be used for various purposes. In this book it will be mainly used for designing locomotion strategies of the system. *Locomotion* is the process of moving an autonomous system from one place to another.

Motion models can describe robot kinematics, and we are interested in mathematics of robot motion without considering its causes, such as forces or torques. *Kinematic model* describes geometric relationships that are present in the system. It describes the relationship between input (control) parameters and behavior of a system given by state-space representation. A kinematic model describes system velocities and is presented by a set of differential first-order equations.

Dynamic models describe a system motion when forces are applied to the system. This models include physics of motion where forces, energies, system mass, inertia, and velocity parameters are used. Descriptions of dynamic models are given by differential equations of the second order.

In wheeled mobile robotics, usually kinematic models are sufficient to design locomotion strategies, while for the other systems such as robots in space, air, or walking robots, dynamic modeling is also needed.

2.2 KINEMATICS OF WHEELED MOBILE ROBOTS

Several types of kinematic models exist:
- *Internal kinematics* explains the relation between system internal variables (e.g., wheel rotation and robot motion).

Fig. 2.1 Two-wheeled cart. *(Museum and Galleries of Ljubljana, painting by I. Rehar.)*

- *External kinematics* describes robot position and orientation according to some reference coordinate frame.
- *Direct kinematics* and *inverse kinematics*. A direct kinematics describes robot states as a function of its inputs (wheel speeds, joints motion, wheel steering, etc.). From inverse kinematics one can design a motion planning, which means that the robot inputs can be calculated for a desired robot state sequence.
- *Motion constraints* appear when a system has less input variables than degrees of freedom (DOFs). Holonomic constraints prohibit certain robot poses while a nonholonomic constraint prohibits certain robot velocities (the robot can drive only in the direction of the wheels' rotation).

In the following, some examples of internal kinematics for wheeled mobile robots (WMR) will be given. The robot pose in a plane is defined by its state vector

$$\boldsymbol{q}(t) = \begin{bmatrix} x(t) \\ y(t) \\ \varphi(t) \end{bmatrix}$$

in a global coordinate frame (X_g, Y_g), as illustrated in Fig. 2.2. A moving frame (X_m, Y_m) is attached to the robot. The relation between the global and moving frame (external kinematics) is defined by the translation vector $[x, y]^T$ and rotation matrix:

$$\boldsymbol{R}(\varphi) = \begin{bmatrix} \cos\varphi & \sin\varphi & 0 \\ -\sin\varphi & \cos\varphi & 0 \\ 0 & 0 & 1 \end{bmatrix}$$

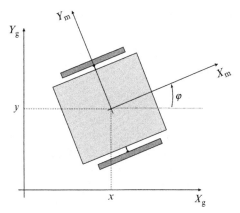

Fig. 2.2 Robot in the plane.

WMR moves on the ground using wheels that rotate on the ground due to the friction contact between wheels and the ground. At moderate WMR speeds an idealized rolling wheel model is usually used where the wheel can move by rotation only and no slip in the driving and lateral direction is considered. Each wheel in WMR can freely rotate around its own axis; therefore, there exists a common point that lies in the intersection of all wheel axes. This point is called *instantaneous center of rotation* (ICR) or *instantaneous center of curvature* and defines a point around which all the wheels follow their circular motion with the same angular velocity ω according to ICR. See Refs. [1–3] for further information.

2.2.1 Differential Drive

Differential drive is a very simple driving mechanism that is quite often used in practice, especially for smaller mobile robots. Robots with this drive usually have one or more castor wheels to support the vehicle and prevent tilting. Both main wheels are placed on a common axis. The velocity of each wheel is controlled by a separate motor. According to Fig. 2.3 the input (control) variables are the velocity of the right wheel $v_R(t)$ and the velocity of the left wheel $v_L(t)$. The meanings of other variables in Fig. 2.3 are as follows: r is the wheel radius; L is the distance between the wheels and $R(t)$ is the instantaneous radius of the vehicle driving trajectory (the distance between the vehicle center (middle point between the wheels) and ICR point). In each instance of time both wheels have the same angular velocity $\omega(t)$ around the ICR,

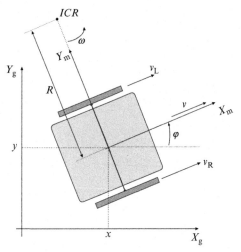

Fig. 2.3 Differential drive kinematics.

$$\omega = \frac{v_L(t)}{R(t) - \frac{L}{2}}$$

$$\omega = \frac{v_R(t)}{R(t) + \frac{L}{2}}$$

from where $\omega(t)$ and $R(t)$ are expressed as follows:

$$\omega(t) = \frac{v_R(t) - v_L(t)}{L}$$

$$R(t) = \frac{L}{2} \frac{v_R(t) + v_L(t)}{v_R(t) - v_L(t)}$$

Tangential vehicle velocity is then calculated as

$$v(t) = \omega(t)R(t) = \frac{v_R(t) + v_L(t)}{2}$$

Wheel tangential velocities are $v_L(t) = r\omega_L(t)$ and $v_R(t) = r\omega_R(t)$, where $\omega_L(t)$ and $\omega_R(t)$ are left and right angular velocities of the wheels around their axes, respectively. Considering the above relations the internal robot kinematics (in local coordinates) can be expressed as

$$\begin{bmatrix} \dot{x}_m(t) \\ \dot{y}_m(t) \\ \dot{\varphi}(t) \end{bmatrix} = \begin{bmatrix} v_{X_m}(t) \\ v_{Y_m}(t) \\ \omega(t) \end{bmatrix} = \begin{bmatrix} \frac{r}{2} & \frac{r}{2} \\ 0 & 0 \\ -\frac{r}{L} & \frac{r}{L} \end{bmatrix} \begin{bmatrix} \omega_L(t) \\ \omega_R(t) \end{bmatrix} \qquad (2.1)$$

Robot external kinematics (in global coordinates) is given by

$$
\begin{bmatrix} \dot{x}(t) \\ \dot{y}(t) \\ \dot{\varphi}(t) \end{bmatrix} = \begin{bmatrix} \cos(\varphi(t)) & 0 \\ \sin(\varphi(t)) & 0 \\ 0 & 1 \end{bmatrix} \begin{bmatrix} v(t) \\ \omega(t) \end{bmatrix} \tag{2.2}
$$

where $v(t)$ and $\omega(t)$ are the control variables. Model (2.2) can be written in discrete form (2.3) using Euler integration and evaluated at discrete time instants $t = kT_s$, $k = 0, 1, 2, \ldots$ where T_s is the following sampling interval:

$$
\begin{aligned}
x(k+1) &= x(k) + v(k)T_s \cos(\varphi(k)) \\
y(k+1) &= y(k) + v(k)T_s \sin(\varphi(k)) \\
\varphi(k+1) &= \varphi(k) + \omega(k)T_s
\end{aligned} \tag{2.3}
$$

Forward and Inverse Kinematics

A robot pose at some time t is obtained by integration of the kinematic model, which is known as *odometry* or dead reckoning. Determination of the robot pose for given control variables is called direct (also forward) kinematics:

$$
\begin{aligned}
x(t) &= \int_0^t v(t) \cos(\varphi(t))\,dt \\
y(t) &= \int_0^t v(t) \sin(\varphi(t))\,dt \\
\varphi(t) &= \int_0^t \omega(t)\,dt
\end{aligned} \tag{2.4}
$$

If constant velocities v and ω are assumed during the sample time the integration of Eq. (2.4) can be done numerically using the Euler method. The direct kinematics is then given by

$$
\begin{aligned}
x(k+1) &= x(k) + v(k)T_s \cos(\varphi(k)) \\
y(k+1) &= y(k) + v(k)T_s \sin(\varphi(k)) \\
\varphi(k+1) &= \varphi(k) + \omega(k)T_s
\end{aligned} \tag{2.5}
$$

If trapezoidal numerical integration is used a better approximation is obtained as follows:

$$
\begin{aligned}
x(k+1) &= x(k) + v(k)T_s \cos\left(\varphi(k) + \frac{\omega(k)T_s}{2}\right) \\
y(k+1) &= y(k) + v(k)T_s \sin\left(\varphi(k) + \frac{\omega(k)T_s}{2}\right) \\
\varphi(k+1) &= \varphi(k) + \omega(k)T_s
\end{aligned} \tag{2.6}
$$

If exact integration is applied the direct kinematics is

$$x(k+1) = x(k) + \frac{v(k)}{\omega(k)} \left(\sin\left(\varphi(k) + \omega(k)T_s\right) - \sin(\varphi(k)) \right)$$

$$y(k+1) = y(k) - \frac{v(k)}{\omega(k)} \left(\cos\left(\varphi(k) + \omega(k)T_s\right) - \cos(\varphi(k)) \right) \qquad (2.7)$$

$$\varphi(k+1) = \varphi(k) + \omega(k)T_s$$

where integration in Eq. (2.7) is done inside the sampling time interval where constant velocities v and ω are assumed to obtain increments:

$$\Delta x(k) = v(k) \int_{kT_s}^{(k+1)T_s} \cos\left(\varphi(t)\right) dt$$

$$= v(k) \int_{kT_s}^{(k+1)T_s} \cos\left(\varphi(k) + \omega(k)(t - kT_s)\right) dt$$

$$\Delta y(k) = v(k) \int_{kT_s}^{(k+1)T_s} \sin\left(\varphi(t)\right) dt$$

$$= v(k) \int_{kT_s}^{(k+1)T_s} \sin\left(\varphi(k) + \omega(k)(t - kT_s)\right) dt$$

The development of the inverse kinematics is a more challenging task than the above cases of direct kinematics. We use inverse kinematics to determine control variables to drive the robot to the desired robot pose or path trajectory. Robots are usually subjected to nonholonomic constraints (Section 2.3), which means that not all driving directions are possible. There are also many possible solutions to arrive to the desired pose.

One simple solution to the inverse kinematics problem would be if we allow a differential robot to drive only forward ($v_R(t) = v_L(t) = v_R \Longrightarrow \omega(t) = 0$, $v(t) = v_R$) or only rotate at the spot ($v_R(t) = -v_L(t) = v_R \Longrightarrow \omega(t) = \frac{2v_R}{L}$, $v(t) = 0$) at constant speeds. For rotation motion equation (2.4) simplifies to

$$x(t) = x(0)$$
$$y(t) = y(0) \qquad (2.8)$$
$$\varphi(t) = \varphi(0) + \frac{2v_R t}{L}$$

and for straight motion equation (2.4) simplifies to

$$x(t) = x(0) + v_R \cos(\varphi(0))t$$
$$y(t) = y(0) + v_R \sin(\varphi(0))t \qquad (2.9)$$
$$\varphi(t) = \varphi(0)$$

Motion strategy could then be to orient the robot to the target position by rotation and then drive the robot to the target position by a straight motion and finally align (with rotation) the robot orientation with the desired orientation in the desired robot pose. The required control variables for each phase (rotation, straight motion, rotation) can easily be calculated from Eqs. (2.8), (2.9).

If we consider a discrete time notation where control speeds $v_R(k)$, $v_L(k)$ are constant during time interval T_s and changes of the control speeds are only possible at time instants $t = kT_s$ then we can write equations for the robot motion. For rotation motion $(v_R(k) = -v_L(k))$ it follows

$$x(k+1) = x(k)$$
$$y(k+1) = y(k) \qquad (2.10)$$
$$\varphi(k+1) = \varphi(k) + \frac{2v_R(k)T_s}{L}$$

and the straight motion equation $(v_R(k) = v_L(k))$ is

$$x(k+1) = x(k) + v_R(k) \cos(\varphi(k))T_s$$
$$y(k+1) = y(k) + v_R(k) \sin(\varphi(k))T_s \qquad (2.11)$$
$$\varphi(k+1) = \varphi(k)$$

So for the desired robot motion inside time interval $t \in [kT_s, (k+1)T_s)$ the inverse kinematics can be calculated for each sample time by expressing control variables from Eqs. (2.10), (2.11).

As already stated, there are many other solutions to drive the robot to the desired pose using smooth changing trajectories. The inverse kinematic problem is easier for the desired smooth target trajectory $(x(t), y(t))$ that the robot should follow so that its orientation is always tangent to the trajectory. Trajectory is defined in time interval $t \in [0, T]$. Supposing the robot's initial pose is on trajectory, and there is a perfect kinematic model and no disturbances, we can calculate required control variables v as follows:

$$v(t) = \pm\sqrt{\dot{x}^2(t) + \dot{y}^2(t)} \qquad (2.12)$$

where the sign depends on the desired driving direction (+ for forward and − for reverse). The tangent angle of each point on the path is defined as

$$\varphi(t) = \arctan2(\dot{y}(t), \dot{x}(t)) + l\pi \tag{2.13}$$

where $l \in \{0, 1\}$ defines the desired driving direction (0 for forward and 1 for reverse) and the function arctan2 is the four-quadrant inverse tangent function. By calculating the time derivative of Eq. (2.13) the robot's angular velocity $\omega(t)$ is obtained:

$$\omega(t) = \frac{\dot{x}(t)\ddot{y}(t) - \dot{y}(t)\ddot{x}(t)}{\dot{x}^2(t) + \dot{y}^2(t)} = v(t)\kappa(t) \tag{2.14}$$

where $\kappa(t)$ is the path curvature. Using relations (2.12), (2.14) and the defined desired robot path $x(t), y(t)$, the robot control inputs $v(t)$ and $\omega(t)$ are calculated. The necessary condition in the path-design procedure is a twice-differentiable path and a nonzero tangential velocity $v(t) \neq 0$. If for some time t the tangential velocity is $v(t) = 0$, the robot rotates at a fixed point with the angular velocity $\omega(t)$. The angle $\varphi(t)$ cannot be determined from Eq. (2.12), and therefore, $\varphi(t)$ must be given explicitly. Usually this approach is used to determine the feedforward part of the control supplementary to the feedback part which takes care of the imperfect kinematic model, disturbances, and the initial pose error [4].

2.2.2 Bicycle Drive

Bicycle drive is shown in Fig. 2.4. It has a steering wheel where α is the steering angle and ω_S is wheel angular velocity around its axis (front-wheel drive). The ICR point is defined by the intersection of both wheel axes. In each moment of time the bicycle circles around ICR with angular velocity ω, radius R, and distance between the wheels d:

$$R(t) = d \tan\left(\frac{\pi}{2} - \alpha(t)\right) = \frac{d}{\tan(\alpha(t))}$$

The steering wheel circles around ICR with ω so we can write

$$\omega(t) = \dot{\varphi} = \frac{v_s(t)}{\sqrt{d^2 + R^2}} = \frac{v_s(t)}{d} \sin(\alpha(t))$$

where $v_s(t) = \omega_s(t)r$ and r are the rim velocity and the radius of the front wheel, respectively.

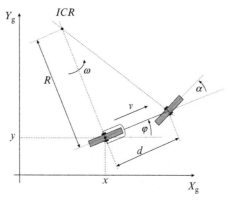

Fig. 2.4 Bicycle drive kinematics.

Internal robot kinematics (in robot frame) is

$$\dot{x}_m = v_s(t) \cos(\alpha(t))$$
$$\dot{y}_m = 0$$
$$\dot{\varphi} = \frac{v_s(t)}{d} \sin(\alpha(t))$$

(2.15)

and the external kinematic model is

$$\dot{x} = v_s(t) \cos(\alpha(t)) \cos(\varphi(t))$$
$$\dot{y} = v_s(t) \cos(\alpha(t)) \sin(\varphi(t))$$
$$\dot{\varphi} = \frac{v_s(t)}{d} \sin(\alpha(t))$$

(2.16)

or in compact form

$$\begin{bmatrix} \dot{x} \\ \dot{y} \\ \dot{\varphi} \end{bmatrix} = \begin{bmatrix} \cos(\varphi(t)) & 0 \\ \sin(\varphi(t)) & 0 \\ 0 & 1 \end{bmatrix} \begin{bmatrix} v(t) \\ \omega(t) \end{bmatrix}$$

(2.17)

where $v = v_s(t) \cos(\alpha(t))$ and $\omega(t) = \frac{v_s(t)}{d} \sin(\alpha(t))$.

Rear-Wheel Bicycle Drive

Usually vehicles are powered by the rear wheels and steered by the front wheel (e.g., bicycle, tricycles, and some cars). The control variables in this case are rear velocity $v_r(t)$ and steering angle of the front wheel $\alpha(t)$. The internal kinematic model can simply be derived from Eq. (2.15) where substituting $v_r(t) = v_s(t) \cos(\alpha(t))$ leads to the following:

$$\dot{x}_m(t) = v_r(t)$$
$$\dot{y}_m(t) = 0$$
$$\omega(t) = \dot{\varphi}(t) = \frac{v_r(t)}{d} \tan(\alpha(t))$$

(2.18)

and the external kinematic model becomes

$$\dot{x} = v_r(t) \cos(\varphi(t))$$
$$\dot{y} = v_r(t) \sin(\varphi(t))$$
$$\dot{\varphi} = \frac{v_r(t)}{d} \tan(\alpha(t))$$

(2.19)

or in equivalent matrix form

$$\begin{bmatrix} \dot{x} \\ \dot{y} \\ \dot{\varphi} \end{bmatrix} = \begin{bmatrix} \cos(\varphi(t)) & 0 \\ \sin(\varphi(t)) & 0 \\ 0 & 1 \end{bmatrix} \begin{bmatrix} v_r(t) \\ \omega(t) \end{bmatrix}$$

(2.20)

where $\omega(t) = \frac{v_r(t)}{d} \tan(\alpha(t))$.

Direct and Inverse Kinematics

Considering Eq. (2.17) bicycle direct kinematics (front-wheel drive case) is obtained by Eq. (2.4) similarly as in the case of differential drive.

Inverse kinematics is in general very difficult to solve unless a special case of motion strategy with two motion patterns is applied as follows. By the first motion pattern the robot can only move straight ahead $(\alpha(t) = 0)$ and by the second motion pattern the robot can rotate in the spot $(\alpha(t) = \pm\frac{\pi}{2})$. For the straight motion robot, velocities simplify to $v(t) = v_s(t)$ and $\omega(t) = 0$. By inserting these velocities to Eq. (2.17) and discretization, the motion equations reduce to

$$x(k+1) = x(k) + v_s(k) \cos(\varphi(k)) T_s$$
$$y(k+1) = y(k) + v_s(k) \sin(\varphi(k)) T_s$$
$$\varphi(k+1) = \varphi(k)$$

(2.21)

For rotation at the spot robot velocities simplify to $v(t) = 0$ and $\omega(t) = \frac{v_s(t)}{d}$. After inserting these velocities to Eq. (2.17) and discretization the motion model is given by

$$x(k+1) = x(k)$$
$$y(k+1) = y(k)$$
$$\varphi(k+1) = \varphi(k) + \frac{v_s(t)}{d} T_s \qquad (2.22)$$

Control variables can be expressed from Eqs. (2.21), (2.22) for required motion during each sample time.

2.2.3 Tricycle Drive

Tricycle drive (see Fig. 2.5) has the same kinematics as bicycle drive.

$$\dot{x} = v_s(t) \cos(\alpha(t)) \cos(\varphi(t))$$
$$\dot{y} = v_s(t) \cos(\alpha(t)) \sin(\varphi(t))$$
$$\dot{\varphi} = \frac{v_s(t)}{d} \sin(\alpha(t)) \qquad (2.23)$$

where $v = v_s(t) \cos(\alpha(t))$, $\omega(t) = \frac{v_s(t)}{d} \sin(\alpha(t))$, and v_s is the rim velocity of the steering wheel. Tricycle drive is more common in mobile robotics because of the three wheels that make the robot stable in a vertical direction by itself.

2.2.4 Tricycle With a Trailer

The tricycle part of kinematics is already described in Section 2.2.3. For a trailer ICR_2 is found by the intersection of the tricycle and the trailer

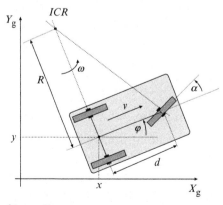

Fig. 2.5 Tricycle drive kinematics.

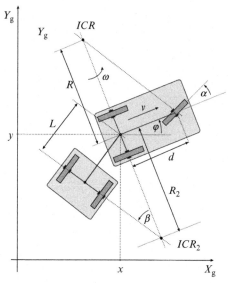

Fig. 2.6 Tricycle with a trailer drive kinematics.

wheel axes. The angular velocity at which the trailer wheels circle around the ICR_2 point is

$$\omega_2(t) = \frac{v}{R_2} = \frac{v_s \cos \alpha}{R_2} = \frac{v_s \cos \alpha \sin \beta}{L} = \dot{\beta}$$

The final kinematic model (for Fig. 2.6) is then obtained as follows:

$$\dot{x} = v_s(t) \cos(\alpha(t)) \cos(\varphi(t))$$
$$\dot{y} = v_s(t) \cos(\alpha(t)) \sin(\varphi(t))$$
$$\dot{\varphi} = \frac{v_s(t)}{d} \sin(\alpha(t)) \tag{2.24}$$
$$\dot{\beta} = \frac{v_s \cos \alpha \sin \beta}{L}$$

2.2.5 Car (Ackermann) Drive

Cars use the Ackermann steering principle. The idea behind the Ackermann steering is that the inner wheel (closer to ICR) should steer for a bigger angle than the outer wheel in order to allow the vehicle to rotate around the middle point between the rear wheel axis. Consequently the inner wheel travels with a slower speed than the outer wheel. The

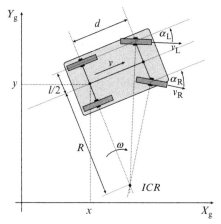

Fig. 2.7 Ackermann drive schematics.

Ackermann driving mechanism allows for the rear wheels to have no slip angle, which requires that the ICR point lies on a straight line defined by the rear wheels' axis. This driving mechanism therefore minimizes tire wear. In Fig. 2.7 the left wheel is on the outer side and the right wheel is on the inner side. The steering orientations of the front wheels are therefore determined from

$$\tan\left(\frac{\pi}{2} - \alpha_L\right) = \frac{R + \frac{l}{2}}{d}$$

$$\tan\left(\frac{\pi}{2} - \alpha_R\right) = \frac{R - \frac{l}{2}}{d}$$

and the final steering angles are expressed as

$$\alpha_L = \frac{\pi}{2} - \arctan\frac{R + \frac{l}{2}}{d}$$

$$\alpha_R = \frac{\pi}{2} - \arctan\frac{R - \frac{l}{2}}{d}$$

(2.25)

The inner and outer back wheels circle around the ICR with the same angular velocity ω; therefore their rim velocities are

$$lv_L = \omega\left(R + \frac{l}{2}\right)$$

$$v_R = \omega\left(R - \frac{l}{2}\right)$$

(2.26)

This type of kinematic drive is especially appropriate to model larger vehicles. A motion model can also be described using tricycle kinematics (Eq. 2.23) where the average the Ackermann angle $\alpha = \frac{\pi}{2} - \arctan\frac{R}{d}$ is used.

Inverse kinematics of Ackermann drive is complicated and exceeds the scope of this work.

2.2.6 Synchronous Drive

If a vehicle is able to steer each of its wheels around the vertical axis synchronously (wheels have the same orientation and are steered synchronously) then this is called synchronous drive (also synchro drive). Typical synchro drive has three wheels arranged symmetrically (equilateral triangle) around the vehicle center as in Fig. 2.8. All wheels are synchronously steered so their rotation axes are always parallel and the ICR point is at infinity. The vehicle can directly control the orientation of the wheels, which presents the third state in the state vector (2.27). The control variables are wheels' steering velocity ω, and forward speed v.

The kinematics has a similar form as in the case of differential drive:

$$\begin{bmatrix} \dot{x}(t) \\ \dot{y}(t) \\ \dot{\varphi}(t) \end{bmatrix} = \begin{bmatrix} \cos(\varphi(t)) & 0 \\ \sin(\varphi(t)) & 0 \\ 0 & 1 \end{bmatrix} \begin{bmatrix} v(t) \\ \omega(t) \end{bmatrix} \qquad (2.27)$$

where $v(t)$ and $\omega(t)$ (steering velocity of the wheels) are the control variables that can be controlled independently (which is not the case for differential drive).

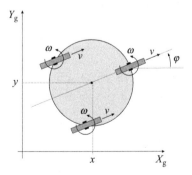

Fig. 2.8 Synchronous drive.

Forward and Inverse Kinematics

Forward kinematics is obtained by integration of the kinematic model (2.27):

$$lx(t) = \int_0^t v(t)\cos(\varphi(t))\,dt$$

$$y(t) = \int_0^t v(t)\sin(\varphi(t))\,dt \qquad (2.28)$$

$$\varphi(t) = \int_0^t \omega(t)\,dt$$

The general solution to the inverse kinematics is not possible because there exists many possible solutions to arrive at the desired pose. However, the inverse kinematics can be solved for a special case where the robot either rotates at the spot or moves straight in its current direction. When the robot rotates at the spot with the constant ω for some time Δt, its orientation changes for $\omega\Delta t$. In the case of straight motion for some time Δt, with a constant forward velocity v, the robot moves for $v\Delta t$ in the direction of its current orientation.

2.2.7 Omnidirectional Drive

In previous kinematic models simple wheels were used where the wheel can only move (roll) in the direction of its orientation. Such simple wheels have only one rolling direction. To allow omnidirectional rolling motion (in more directions) a complex wheel construction is required. An example of such a wheel is the Mecanum wheel (see Fig. 2.9) or Swedish wheel where the number of smaller rollers are arranged around the main wheel rim. The

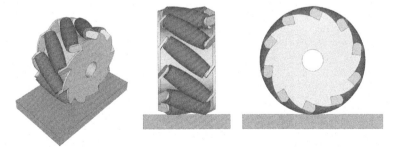

Fig. 2.9 Mecanum wheel with rollers mounted around the wheel rim. Each roller axis is at 45 degree from the main wheel plane and at 45 degree to the line parallel with the wheel's axis.

passive roller axes are not aligned with the main wheel axis (typically they are at a $\gamma = 45$ degree angle). This allows a number of different motion directions that result from an arbitrary combination of the main wheel's rotation direction and passive rollers' rotation.

Kinematics of Four-Wheel Omnidirectional Drive

Another example of a complex wheel is the omni wheel or poly wheel, which enables omnidirectional motion similarly to the Mecanum wheel. The omni wheel (Fig. 2.10) has passive rollers mounted around the wheel rim so that their axes are 90 degree to the wheel axis.

The popular four-wheel platform known as the Mecanum platform is shown in Fig. 2.11. The Mecanum platform has wheels with left- and right-handed rollers where diagonal wheels are of the same type. This enables the vehicle to move in any direction with arbitrary rotation by commanding a proper velocity to each main wheel. Forward or backward motion is obtained by setting all main wheels to the same velocity. If velocities of the main wheels on one side are opposite to the velocities of the wheels on the other side then the platform will rotate. Side motion of the platform is obtained if the wheels on one diagonal have opposite velocity to the wheels on the other diagonal. With combination of the motions described, motion of the platform in any direction and with any rotation can be achieved.

The inverse internal kinematics of the four-wheel Mecanum drive in Fig. 2.12 can be derived as follows. The first wheel velocity in robot coordinates is obtained from the main wheel velocity $v_1(t)$ and the passive roller's velocity $v_R(t)$. In the sequel the time dependency notation is omitted to have more compact and easy-to-follow equations (e.g., $v_1(t) = v_1$). The total wheel velocity in x_m and y_m direction of the robot coordinate

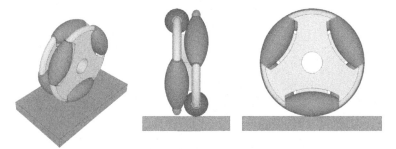

Fig. 2.10 Omni wheel with six free rotating rollers mounted around the wheel rim. The wheel can rotate and slide laterally.

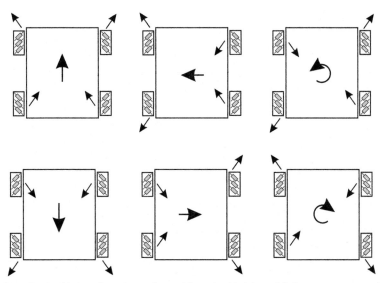

Fig. 2.11 Basic driving directions of omnidirectional drive with four Mecanum wheels where the main wheels are rotating forwards or backwards. Resulting passive rollers velocities are shown.

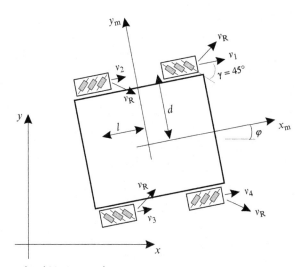

Fig. 2.12 Four-wheel Mecanum drive.

frame are $v_{m1x} = v_1 + v_R \cos(\frac{\pi}{4}) = v_1 + \frac{v_R}{\sqrt{2}}$ and $v_{m1y} = v_R \sin(\frac{\pi}{4}) = \frac{v_R}{\sqrt{2}}$, from where the main wheel velocity is obtained: $v_1 = v_{m1x} - v_{m1y}$. The first wheel velocity in the robot coordinate frame direction can also be expressed with the robot's translational velocity $v_m = \sqrt{\dot{x}_m^2 + \dot{y}_m^2}$ and its angular velocity $\dot{\varphi}$ as follows: $v_{m1x} = \dot{x}_m - \dot{\varphi}d$ and $v_{m1y} = \dot{y}_m + \dot{\varphi}l$ (the meaning of distances d and l can be deduced from Fig. 2.12). From the latter relations the main wheel velocity can be expressed by the robot body velocity as $v_1 = \dot{x}_m - \dot{y}_m - (l + d)\dot{\varphi}$. Similar equations can be obtained for v_2, v_3, and v_4. The inverse kinematics in local coordinates is

$$\begin{bmatrix} v_1 \\ v_2 \\ v_3 \\ v_4 \end{bmatrix} = \begin{bmatrix} 1 & -1 & -(l+d) \\ 1 & 1 & -(l+d) \\ 1 & -1 & (l+d) \\ 1 & 1 & (l+d) \end{bmatrix} \begin{bmatrix} \dot{x}_m \\ \dot{y}_m \\ \dot{\varphi} \end{bmatrix} \qquad (2.29)$$

The internal inverse kinematics (Eq. 2.29) in compact form reads $v = J\dot{q}_m$, where $v^T = [v_1, v_2, v_3, v_4]^T$ and $q_m^T = [x_m, y_m, \varphi]^T$.

To calculate the inverse kinematics in global coordinates the rotation matrix representing the orientation of the local coordinates with respect to the global coordinates ($q_m = R_G^L q$),

$$R_G^L = \begin{bmatrix} \cos\varphi & \sin\varphi & 0 \\ -\sin\varphi & \cos\varphi & 0 \\ 0 & 0 & 1 \end{bmatrix}$$

needs to be considered as follows: $v = JR_G^L \dot{q}$.

From the internal inverse kinematics $v = J\dot{q}_m$ (Eq. 2.29), the forward internal kinematics is obtained by $\dot{q}_m = J^+ v$, where $J^+ = (J^T J)^{-1} J^T$ is the pseudo-inverse of J. Forward internal kinematics for the four-wheel Mecanum platform reads as follows:

$$\begin{bmatrix} \dot{x}_m \\ \dot{y}_m \\ \dot{\varphi} \end{bmatrix} = \frac{1}{4} \begin{bmatrix} 1 & 1 & 1 & 1 \\ -1 & 1 & -1 & 1 \\ \frac{-1}{(l+d)} & \frac{-1}{(l+d)} & \frac{1}{(l+d)} & \frac{1}{(l+d)} \end{bmatrix} \begin{bmatrix} v_1 \\ v_2 \\ v_3 \\ v_4 \end{bmatrix} \qquad (2.30)$$

Forward kinematics in global coordinates is obtained by $\dot{q} = (R_G^L)^T J^+ v$.

Kinematics of Three-Wheel Omnidirectional Drive

A popular omnidirectional configuration for three-wheel drive is shown in Fig. 2.13. Its inverse kinematics (in global coordinates) is obtained by considering the robot translational velocity $v = \sqrt{\dot{x}^2 + \dot{y}^2}$ and its angular velocity $\dot{\varphi}$. Velocity of the first wheel $v_1 = v_{1t} + v_{1r}$ consists of a translational part $v_{1t} = -\dot{x}\sin(\varphi) + \dot{y}\cos(\varphi)$ and an angular part $v_{1r} = R\dot{\varphi}$. The first wheel common velocity therefore is $v_1 = -\dot{x}\sin(\varphi) + \dot{y}\cos(\varphi) + R\dot{\varphi}$.

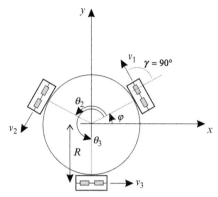

Fig. 2.13 Three-wheel omnidirectional drive ($\theta_2 = 120$ degree, $\theta_3 = 240$ degree).

Similarly, considering the global angle of the second wheel $\varphi + \theta_2$, its velocity is $v_2 = -\dot{x}\sin(\varphi + \theta_2) + \dot{y}\cos(\varphi + \theta_2) + R\dot{\varphi}$. And the velocity of the third wheel reads $v_3 = -\dot{x}\sin(\varphi + \theta_3) + \dot{y}\cos(\varphi + \theta_3) + R\dot{\varphi}$. The inverse kinematics in global coordinates of three-wheel drive is

$$
\begin{bmatrix} v_1 \\ v_2 \\ v_3 \end{bmatrix} = \begin{bmatrix} -\sin(\varphi) & \cos(\varphi) & R \\ -\sin(\varphi + \theta_2) & \cos(\varphi + \theta_2) & R \\ -\sin(\varphi + \theta_3) & \cos(\varphi + \theta_3) & R \end{bmatrix} \begin{bmatrix} \dot{x} \\ \dot{y} \\ \dot{\varphi} \end{bmatrix} \tag{2.31}
$$

The inverse kinematics (Eq. 2.31) in compact form reads $v = J\dot{q}$. Sometimes it is more convenient to steer the robot in its local coordinates, which can be obtained by considering rotation transformation as follows: $v = J(R_G^L)^T\dot{q}_m$.

Forward kinematics in global coordinates is obtained from the inverse kinematics (Eq. 2.31) by $\dot{q} = Sv$ where $S = J^{-1}$:

$$
\begin{bmatrix} \dot{x} \\ \dot{y} \\ \dot{\varphi} \end{bmatrix} = \frac{2}{3} \begin{bmatrix} -\sin(\theta_1) & -\sin(\theta_1 + \theta_2) & -\sin(\theta_1 + \theta_3) \\ \cos(\theta_1) & \cos(\theta_1 + \theta_2) & \cos(\theta_1 + \theta_3) \\ \frac{1}{2R} & \frac{1}{2R} & \frac{1}{2R} \end{bmatrix} \begin{bmatrix} v_1 \\ v_2 \\ v_3 \end{bmatrix} \tag{2.32}
$$

2.2.8 Tracked Drive

The kinematics of the tracked drive (Fig. 2.14) can be approximately described by the differential drive kinematics.

$$
\begin{bmatrix} \dot{x}(t) \\ \dot{y}(t) \\ \dot{\varphi}(t) \end{bmatrix} = \begin{bmatrix} \cos(\varphi(t)) & 0 \\ \sin(\varphi(t)) & 0 \\ 0 & 1 \end{bmatrix} \begin{bmatrix} v(t) \\ \omega(t) \end{bmatrix} \tag{2.33}
$$

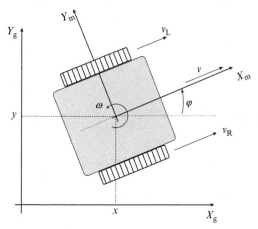

Fig. 2.14 Tracked drive with caterpillars.

However, the differential drive assumes perfect rolling contact between its wheels and the ground, which is not the case for tracked drive. The tracked drive (Caterpillar tractor) has a bigger contact surface between its wheels and the ground and requires wheels slipping to change its direction. The tracked drives can therefore move through more rough terrain where wheeled vehicles usually cannot. The amount of sleep between the Caterpillar tractor and the ground is not constant and depends on ground contact; therefore the odometry (direct kinematics) is even less reliable to estimate the robot pose compared to the differential drive.

2.3 MOTION CONSTRAINTS

Motion of WMR is constrained by dynamic and kinematic constraints. *Dynamic constraints* have origin in the system dynamic model where the system's response is limited due to its inertia or constraints of the actuators (e.g., limited torque of the motor drive due to its capabilities or to prevent wheel slipping). *Kinematic constraints* have origin in robot construction and its kinematic model. Kinematic constraints can be holonomic or nonholonomic. *Nonholonomic constraints* limit some driving directions of a mobile robot [5]. *Holonomic constraints* are related to the dimensionality of the system state description (generalized coordinates). If holonomic constraints are present, then some coordinates of the state description depend on the others and can therefore be used to eliminate those coordinates from the system's state description.

A system is holonomic if it has no kinematic constraints or if it only has holonomic constraints. Holonomic systems have no limitation in their velocity space so all directions of motion in the state space are possible. The system is nonholonomic if it has nonholonomic constraints and therefore it cannot move in arbitrary direction (e.g., a car can only move forwards or backwards and not in the lateral direction). For holonomic systems one can determine a subset of independent generalized coordinates which defines a space in which all directions of motion are possible. For noholonomic systems the motion of the system is not arbitrary; nonholonomic constraints define a subspace of possible velocity space in the current moment.

In holonomic systems their state directly depends on the configuration of the system's internal variables (wheel rotation, joint's angles). In nonholonomic systems the latter is not true because the returning of the system's internal variables to the initial values does not guarantee that the system will return to its initial state. The state of the nonholonomic system therefore depends on the traveled path (the sequence of the internal variables).

In the following, mechanical systems will be described whose configuration (system pose in the environment and relation of the systems parts) can be described by the vector of generalized coordinates q. At some given trajectory $q(t)$ a vector of generalized velocities is defined by $\dot{q}(t)$.

Holonomic constraints are expressed by equations that contain generalized coordinates. These equations can be used to eliminate some generalized coordinates in the effort to obtain a smaller set of required generalized coordinates for some system description. Noholonomic constraints do not reduce the generalized coordinate's dimension but only the dimension of generalized velocity space. Nonholonomic constraints therefore affect the path planning problem. Regarding nonholonomic constraints the following questions appear:

- How are nonholonomic constraints identified? If a constraint is integrable, then its velocity dependence can be integrated and expressed by a holonomic constraint.
- Does a nonholonomic constraint limit the space of reachable configurations (e.g., system pose)? No, with the use of control theory tools simple conditions are obtained under which arbitrary configuration can be obtained.
- How can a feasible path generator for a nonholonomic robot be constructed?

2.3.1 Holonomic Constraints

These constraints depend on generalized coordinates. For a system with n generalized coordinates $q = [q_1, \ldots, q_n]^T$ a holonomic constraint is expressed in the following form:

$$f(q) = f(q_1, \ldots, q_n) = 0 \tag{2.34}$$

where f and its derivatives are continuous functions. This constraint defines a subspace of all possible configurations in generalized coordinates for which Eq. (2.34) is true. Constraint (2.34) can be used to eliminate certain generalized coordinates (it can be expressed by $n - 1$ other coordinates).

In general we can have m holonomic constraints ($m < n$). If these constraints are linearly independent then they define ($n - m$)-dimensional subspace, which is true configuration space (work space, system has $n - m$ DOFs).

2.3.2 Nonholonomic Constraints

Nonholonomic constraints limit possible velocities of the system or possible directions of motion. The nonholonomic constraint can be formulated by

$$f(q, \dot{q}) = f(q_1, \ldots, q_n, \dot{q}_1, \ldots, \dot{q}_n) = 0 \tag{2.35}$$

where f is a smooth function with continuous derivatives and $\dot{q}$ is the vector of system velocities in the generalized coordinates. In case the system has no constraints (2.35), it has no limitation in motion directions.

A kinematic constraint (2.35) is holonomic if it is integrable, which means that velocities $\dot{q}_1, \ldots, \dot{q}_n$ can be eliminated from Eq. (2.35) and the constraint can be expressed in the form of Eq. (2.34). If the constraint (2.35) is not integrable, then it is nonholonomic.

If m linearly independent nonholonomic constraints exist in the form Eq. (2.35), the velocity space is ($n - m$)-dimensional. Nonholonomic constraints limit allowable velocities of the system. For example, a differential drive vehicle (e.g., wheelchair) can move in the direction of the current wheel orientation but not in the lateral direction.

Assuming linear constraints in $\dot{q} = [\dot{q}_1, \ldots, \dot{q}_n]^T$, Eq. (2.35) can be written as

$$f(q, \dot{q}) = a^T(q)\dot{q} = \begin{bmatrix} a_1(q) & \cdots & a_n(q) \end{bmatrix} \begin{bmatrix} \dot{q}_1 \\ \vdots \\ \dot{q}_n \end{bmatrix} = 0 \tag{2.36}$$

where $a(q)$ is a parameter vector of the constraint. For m nonholonomic constraints a constraint matrix is obtained as follows:

$$A(q) = \begin{bmatrix} a_1^T(q) \\ \vdots \\ a_m^T(q) \end{bmatrix} \tag{2.37}$$

and all nonholonomic constraints are given in matrix form

$$A(q)\dot{q} = 0 \tag{2.38}$$

In each time instant a matrix of reachable motion directions is $S(q) = [s_1(q), \dots, s_{n-m}(q)]$ (m constraints define $(n-m)$ reachable directions). This matrix defines the kinematic model as follows:

$$\dot{q}(t) = S(q)v(t) \tag{2.39}$$

where $v(t)$ is the control vector (see kinematic model (2.2)). The product of the constraint matrix A and the kinematic matrix S is a zero matrix:

$$AS = 0 \tag{2.40}$$

2.3.3 Integrability of Constraints

To determine if some constraint is nonholonomic, an integrability test needs to be done. If it cannot be integrated to obtain the form of holonomic constraint, the constraint is nonholonomic.

2.3.4 Vector Fields, Distribution, Lie Bracket

In current time t and current state q all possible motion directions can be obtained by a linear combination of vector fields in matrix S. *Distribution* is defined as a reachable subspace from q using motions defined by a linear combination of vector fields (columns in S).

Vector fields are time derivatives of generalized coordinates and therefore represent velocities or motion directions. A vector field is a mapping that is continuously differentiable and assigns a specific vector to each point in state space. A demonstration of reachable vector field determination for differential drive is given in Example 2.1.

EXAMPLE 2.1

For a differential drive robot with kinematic model (2.2), define reachable velocities (motion directions) and motion constraints.

Solution

The vector fields of reachable velocities (motion directions) are defined as

$$
s_1(q) = \begin{bmatrix} \cos\varphi \\ \sin\varphi \\ 0 \end{bmatrix} \qquad s_2(q) = \begin{bmatrix} 0 \\ 0 \\ 1 \end{bmatrix} \tag{2.41}
$$

This means that all possible motion directions in current time at the current pose q are obtained by the linear combination

$$
\dot{q} = u_1 s_1(q) + u_2 s_2(q) \tag{2.42}
$$

where u_1 in u_2 are arbitrary real numbers that represent the control inputs. Eq. (2.42) can be seen as an alternative form of the kinematic model (2.2).

If vector fields s_i are not given, they can be obtained from known constraints a_j considering the fact that constraint directions are orthogonal to the motion directions, so $s_i \perp a_j$. In Fig. 2.3 the constraint can be determined by identifying the directions where the robot cannot move; this is in the direction lateral to the wheels. The only constraint is

$$
a(q) = \begin{bmatrix} -\sin\varphi \\ \cos\varphi \\ 0 \end{bmatrix}
$$

which is perpendicular to the motion direction vector $s_1(q)$ (longitudinal motion in the direction of the wheels rotation) and to the vector $s_2(q)$ (rotation around axis normal to the surface).

If distribution of some vector fields span the whole space then it is *involutive*. For noninvolutive distribution of some vector fields a set of new vector fields (linearly independent from vector fields of the basic distribution) can be defined. These new vector fields (motion directions) can be obtained by a finite switching of the basic motion directions with movement length (in each basic direction) limited to zero. New motion directions can be calculated using *Lie brackets*, which operates over two vector fields as shown later in Eq. (2.44). Parallel parking of a car or differential drive is a practical example of obtaining a new motion direction by switching between the basic motion directions. The car cannot directly move in the lateral direction to the parking slot. However, it can still go into a parking slot by lateral direction movement from a series of forward, backward, and rotation motions as illustrated in Example 2.2.

EXAMPLE 2.2

Illustration of parallel parking maneuver of differential drive vehicle.

Solution

Using the basic motion directions defined by the vector fields $s_1(q)$ and $s_2(q)$ (see Eq. 2.41) and the initial vehicle pose $q_0 = q(0)$ new vehicle poses are obtained as follows. Starting from $q_0 = q(0)$ and driving in the direction s_1 for a short period of time ε, then driving in the direction s_2 for time ε followed by driving in the direction $-s_1$ for time ε, and finally for time ε in the direction $-s_2$, the final pose is obtained. Mathematically this is formulated as follows:

$$q(4\varepsilon) = \phi_\varepsilon^{-s_2}\left(\phi_\varepsilon^{-s_1}\left(\phi_\varepsilon^{s_2}\left(\phi_\varepsilon^{s_1}\left(q_0\right)\right)\right)\right)$$

which represents a nonlinear differential equation whose solution can be approximated by Taylor series expansion (see details in [1]) by

$$q(4\varepsilon) = q_0 + \varepsilon^2\left(\frac{\partial s_2}{\partial q}s_1(q_0) - \frac{\partial s_1}{\partial q}s_2(q_0)\right) + O(\varepsilon^3) \qquad (2.43)$$

where partial derivatives are estimated for q_0 and $O(\varepsilon^3)$ is the remaining part belonging to a higher order of derivatives that can be ignored for a short time ε. The obtained final motion distance of the parallel parking maneuver is ε^2 in the direction of the vector obtained by Lie brackets (defined in Eq. 2.44).

Illustration of this maneuver can be done by the following experiment. Suppose the starting vehicle pose is $q_0 = [0, 0, 0]^T$. By performing the first step of the parking maneuver the resulting pose is

$$q_1 = q_0 + \varepsilon s_1 = \begin{bmatrix} 0 \\ 0 \\ 0 \end{bmatrix} + \varepsilon \begin{bmatrix} \cos 0 \\ \sin 0 \\ 0 \end{bmatrix} = \begin{bmatrix} \varepsilon \\ 0 \\ 0 \end{bmatrix}$$

In the second step a rotation is performed

$$q_2 = q_1 + \varepsilon s_2 = \begin{bmatrix} \varepsilon \\ 0 \\ 0 \end{bmatrix} + \varepsilon \begin{bmatrix} 0 \\ 0 \\ 1 \end{bmatrix} = \begin{bmatrix} \varepsilon \\ 0 \\ \varepsilon \end{bmatrix}$$

The third step contains translation in the negative direction:

$$q_3 = q_2 - \varepsilon s_1 = \begin{bmatrix} \varepsilon \\ 0 \\ \varepsilon \end{bmatrix} - \varepsilon \begin{bmatrix} \cos \varepsilon \\ \sin \varepsilon \\ 0 \end{bmatrix} = \begin{bmatrix} \varepsilon - \varepsilon \cos \varepsilon \\ -\varepsilon \sin \varepsilon \\ \varepsilon \end{bmatrix}$$

(Continued)

EXAMPLE 2.2—cont'd

and in the last step there is a rotation in the negative direction:

$$q_4 = q_3 - \varepsilon s_2 = \begin{bmatrix} \varepsilon - \varepsilon \cos \varepsilon \\ -\varepsilon \sin \varepsilon \\ \varepsilon \end{bmatrix} - \varepsilon \begin{bmatrix} 0 \\ 0 \\ 1 \end{bmatrix} = \begin{bmatrix} \varepsilon - \varepsilon \cos \varepsilon \\ -\varepsilon \sin \varepsilon \\ 0 \end{bmatrix}$$

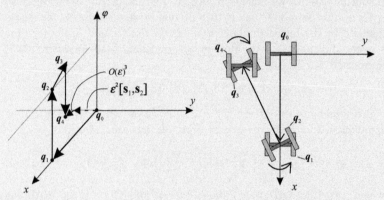

Fig. 2.15 Parallel parking maneuver from top view (*right*) and parking maneuver in configuration space (*left*), where z-axis represents orientation.

Computed motions and intermediate points are shown in Fig. 2.15. The final pose of this maneuver is in the lateral direction to the starting pose. The final pose, however, is not directly accessible with basic motion vector fields s_1 and s_2, but it can be achieved by a combination of the basic vector fields. The obtained final motion is not strictly in the lateral direction due to the final time ε and part $O(\varepsilon^3)$ in Eq. (2.43). In limit case when each motion time duration $\varepsilon \to 0$ the obtained final pose is exactly in the lateral direction.

$$q_4 = \begin{bmatrix} 0 \\ -\varepsilon^2 \\ 0 \end{bmatrix}$$

As seen from Example 2.2 the Lie brackets can be used to identify new motion directions that are not allowed by the basic distribution. This new motion directions can be achieved by a finite number of infinitely small motions in directions from the basic vector field's distribution. A Lie bracket is operating over two vector fields $i(q)$ and $j(q)$, and the result is a new vector field $[i, j]$ obtained as follows:

$$[i, j] = \frac{\partial j}{\partial q} i - \frac{\partial i}{\partial q} j \tag{2.44}$$

where

$$\frac{\partial \boldsymbol{i}}{\partial \boldsymbol{q}} = \begin{bmatrix} \frac{\partial i_1}{\partial q_1} & \frac{\partial i_1}{\partial q_2} & \cdots & \frac{\partial i_1}{\partial q_n} \\ \frac{\partial i_2}{\partial q_1} & \frac{\partial i_2}{\partial q_2} & \cdots & \frac{\partial i_2}{\partial q_n} \\ \vdots & \vdots & & \vdots \\ \frac{\partial i_n}{\partial q_1} & \frac{\partial i_n}{\partial q_2} & \cdots & \frac{\partial i_n}{\partial q_n} \end{bmatrix}$$

$$\frac{\partial \boldsymbol{j}}{\partial \boldsymbol{q}} = \begin{bmatrix} \frac{\partial j_1}{\partial q_1} & \frac{\partial j_1}{\partial q_2} & \cdots & \frac{\partial j_1}{\partial q_n} \\ \frac{\partial j_2}{\partial q_1} & \frac{\partial j_2}{\partial q_2} & \cdots & \frac{\partial j_2}{\partial q_n} \\ \vdots & \vdots & & \vdots \\ \frac{\partial j_n}{\partial q_1} & \frac{\partial j_n}{\partial q_2} & \cdots & \frac{\partial j_n}{\partial q_n} \end{bmatrix}$$

Distribution is *involutive* if new linearly independent vector fields cannot be obtained using Lie brackets over the existing vector fields in the distribution. If distribution is involutive then it is closed under Lie brackets [6]. Then the system is completely holonomic with no motion constraints or only with holonomic constraints. This statement is explained by the Frobenious theorem [7]: *If basic distribution is involutive then the system is holonomic and all existing constraints are integrable.* Let us suppose that from m constraints, k constraints are holonomic and $(m - k)$ nonholonomic. According to the Frobenious theorem there are three possibilities for k [8]:

- $k = m$, which means that the dimension of involutive distribution is $(n - m)$, which also equals to the dimension of the basic distribution. Distribution dimension equals the number of linearly independent vector fields.
- $0 < k < m$, k constraints are integrable, so k generalized coordinates can be eliminated from the system description. The dimension of involutive distribution is $n - k$.
- $k = 0$, the dimension of involutive distribution is n, and all constraints are nonholonomic.

EXAMPLE 2.3

For a wheel rolling on a plane, determine motion constraints and vector fields of possible motion directions. Determine the system's DOF and the number and type of its constraints.

Solution

The kinematic model of the wheel is the same as for the differential drive given in Fig. 2.3 and in kinematic model (2.2). Constrained motion

(*Continued*)

EXAMPLE 2.3—cont'd

directions and possible motion directions have already been determined in Example 2.1. We have one velocity constraint, which does not limit the accessibility of states q, and therefore the system has three DOFs. To prove that there is a velocity constraint (nonholonomic) Lie brackets and the Frobenious theorem could be used.

Using reachable basic vector fields s_1 and s_2 and determining the new vector field by Lie brackets (Eq. 2.41), the following is found:

$$
\begin{aligned}
s_3 = [s_1, s_2] &= \frac{\partial s_2}{\partial q} s_1 - \frac{\partial s_1}{\partial q} s_2 \\
&= \begin{bmatrix} 0 & 0 & 0 \\ 0 & 0 & 0 \\ 0 & 0 & 0 \end{bmatrix} \begin{bmatrix} \cos\varphi \\ \sin\varphi \\ 0 \end{bmatrix} - \begin{bmatrix} 0 & 0 & -\sin\varphi \\ 0 & 0 & \cos\varphi \\ 0 & 0 & 0 \end{bmatrix} \begin{bmatrix} 0 \\ 0 \\ 1 \end{bmatrix} \\
&= \begin{bmatrix} \sin\varphi \\ -\cos\varphi \\ 0 \end{bmatrix}
\end{aligned}
$$

A new possible direction of motion is obtained that is linearly independent from the vector fields s_1 and s_2 and therefore not present in the basic distribution. Therefore we can conclude that the constraint is nonholonomic; the dimension of involutive distribution is 3 (system has 3 DOFs), which is the same as the dimension of the system. It is also not possible to generate new linearly independent vector fields using additional Lie brackets ($[s_1, s_3]$, $[s_2, s_3]$).

EXAMPLE 2.4

Determine motion constraints and directions of possible motion for a car-like vehicle without a (or with a locked) steering mechanism (Fig. 2.16). Find out the number and the type of constraints and the system's DOF.

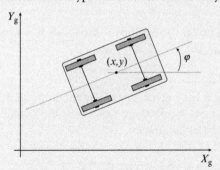

Fig. 2.16 Car-like vehicle without steering mechanism.

EXAMPLE 2.4—cont'd
Solution
The vehicle cannot move in the lateral direction and it cannot rotate (its orientation φ is fixed); therefore constraint motion directions are

$$a_1(q) = \begin{bmatrix} \sin\varphi \\ -\cos\varphi \\ 0 \end{bmatrix} \quad a_2(q) = \begin{bmatrix} 0 \\ 0 \\ 1 \end{bmatrix}$$

and the motion constraints are

$$a_1(q)^T \dot{q} = \dot{x}\sin\varphi - \dot{y}\cos\varphi = 0$$
$$a_2(q)^T \dot{q} = \dot{\varphi} = 0$$

Both constraints are integrable, which can simply be seen by calculating their integral:

$$\varphi = \varphi_0$$
$$(x - x_0)\sin\varphi - (y - y_0)\cos\varphi = 0$$

The vehicle can only move in the direction of the vector field:

$$s_1 = \begin{bmatrix} \cos\varphi \\ \sin\varphi \\ 0 \end{bmatrix}$$

Because both constraints are holonomic the distribution of the vector field s_1 is involutive, meaning that the system is completely holonomic and the system has 1 DOF. Using Lie brackets it is not possible to generate new vector fields as we have only one vector field that defines possible motion.

EXAMPLE 2.5
Using Matlab perform simulation of the differential drive platform. The vehicle parameters are as follows: sampling period $T_s = 0.033$ s, time of simulation 10 s, wheel radius $r = 0.04$ m, distance between the wheels $L = 0.08$ m.

Tasks:
1. Calculate analytically and by means of simulation the shape of the path done by the robot if the initial state is $q(0) = [x, y, \varphi]^T = [0, -0.5, 0]^T$ and robot inputs in three cases are the following:
   - $v(t) = 0.5$ m/s, $\omega(t) = 0$ rad/s;
   - $v(t) = 1$ m/s, $\omega(t) = 2$ rad/s;
   - the wheels angular velocities are $\omega_L(t) = 24$ rad/s, $\omega_R(t) = 16$ rad/s.

(Continued)

EXAMPLE 2.5—cont'd

What are the wheels' angular speeds? Is the calculated and simulated path the same. Why or why not?

2. Perform localization of the vehicle at constant input speeds ($v(t) = 0.5$ m/s, $\omega(t) = 1$ rad/s) using odometry. Compare the estimated path with the simulated path of the true robot. Compare the obtained results also in the ideal situation without noise and modeling error.

3. Perform the localization from question 2 also with initial state error (suppose the true initial state is unknown; therefore choose the initial state in localization to be different from the one in simulation).

4. Imagine a path composed of two or three straight line segments where the initial point coincides with the initial robot pose. Calculate the required sequence of inputs to drive the robot on that path (time of simulation is 10 s).

5. Choose some trajectory defined as parametric functions of time ($x(t) = f(t)$ and $y(t) = g(t)$), where f and g are smooth functions. Calculate required inputs to follow this path. What happens if the robot initially is not on the trajectory?

6. Simulate the parallel parking maneuver described in Example 2.2. Choose proper ε and constant velocities v and ω.

7. For a given initial state $q(0) = [0, -0.5, 0]^T$ calculate the kinematic matrix S and constraint matrix A.

Solution

A basic Matlab script for simulation of differential drive is given in Listing 2.1, which can be modified to obtain the desired solutions.

Listing 2.1 Differential drive simulation

```
1 r = 0.04; % Wheel radius
2 L = 0.08; % Axle length
3 Ts = 0.03; % Sampling time
4 t = 0:Ts:10; % Simulation time
5 q = [4; 0.5; pi/6]; % Initial pose
6
7 for k = 1:length(t)
8     wL = 12; % Left wheel velocity
9     wR = 12.5; % Right wheel velocity
10     v = r/2*(wR+wL); % Robot velocity
11     w = r/L*(wR-wL); % Robot angular velocity
12     dq = [v*cos(q(3)+Ts*w/2); v*sin(q(3)+Ts*w/2); w];
13     q = q + Ts*dq; % Integration
14     q(3) = wrapToPi(q(3)); % Map orientation angle to [-pi, pi]
15 end
```

EXAMPLE 2.6

Using Matlab perform a simulation of a tricycle drive (rear-powered) platform. The vehicle parameters are the following: sampling period $T_s = 0.033$ s, time of simulation 10 s, wheel radius $R = 0.2$ m, distance between the wheels $L = 0.08$ m, and distance between the front and rear wheels $D = 0.07$ m.

Tasks:

1. Calculate analytically and by means of simulation the shape of the path made by the robot if the initial state is $q(0) = [x, y, \varphi]^T = [0, -0.5, 0]^T$ and robot inputs in two cases are the following:
   - $v(t) = 0.5$ m/s, $\alpha(t) = 0$
   - $v(t) = 1$ m/s, $\alpha(t) = \frac{\pi}{6}$

   What are the wheels' angular speeds? Is the calculated and simulated path the same. Why or why not?

2. Implement localization of the vehicle at constant input speeds $(v(t) = 0.5$ m/s, $\alpha(t) = \frac{\pi}{6})$ using odometry. Compare the estimated path with the simulated path of the true robot. Compare the obtained results also in the ideal situation without noise and modeling error.

3. Repeat the localization from question 2 also with initial state error (suppose the true initial state is unknown; therefore choose initial state in localization different than in simulation).

4. Imagine a path composed of two or three straight line segments where the initial point coincides with the initial robot pose. Calculate the required sequence of inputs to drive the robot on that path (time of simulation is 10 s).

5. Choose some trajectory defined as parametric functions of time $(x(t) = f(t)$ and $y(t) = g(t))$, where f and g are smooth functions. Calculate required inputs to follow this path. What happens if the robot initially is not on the trajectory?

Solution

A basic Matlab script for simulation of the tricycle drive is given in Listing 2.2, which can be modified to obtain the desired solutions.

Listing 2.2 Tricycle drive simulation

```
1 D = 0.07; % Distance between the front wheel and rear axle
2 Ts = 0.03; % Sampling time
3 t = 0:Ts:10; % Simulation time
4 q = [4; 0.5; pi/6]; % Initial pose
5
6 for k = 1:length(t)
7     v = 0.5; % Robot velocity
```

(Continued)

EXAMPLE 2.6—cont'd

```
8     alpha = 0.04*(1+sin(k*Ts*pi/2)); % Front wheel orientation
9     w = v/D*tan(alpha); % Robot angular velocity
10    dq = [v*cos(q(3)+Ts*w/2); v*sin(q(3)+Ts*w/2); w];
11    q = q + Ts*dq; % Integration
12    q(3) = wrapToPi(q(3)); % Map orientation angle to [-pi, pi]
13 end
```

2.3.5 Controllability of Wheeled Mobile Robots

Before designing motion control we need to find the answer to the following question: Can the robot reach any point q in the configuration space by performing its available maneuvers? The answer to this question is related to the *controllability* of the system. If the system is controllable, it can reach any configuration q by combining available motion maneuvers.

If the robot has kinematic constraints then we need to analyze them and figure out if they affect the system controllability. The robot with non-holonomic constraints can only move in its basic motion directions (basic maneuvers like driving straight and rotating) but not in the constrained direction. However, such robot may still achieve any desired configuration q by combining basic maneuvers. New direction of motion is different than any basic direction or any linear combination of the basic directions as given in Eq. (2.39). Controllability of the robot is determined by analyzing the involutive (accessibility) distribution obtained by successive Lie bracket operations over the basic motion directions s_1, s_2, s_3, ... as follows [7, 9–12]:

$$\{s_1, s_2, s_3, [s_1, s_2], [s_1, s_3], [s_2, s_3], \ldots, [s_1, [s_1, s_2]], [s_1, [s_1, s_3]], \ldots\}$$
(2.45)

If the rank of the involutive distribution is n (n is the dimension of q) then the robot can reach any point q in the configuration space and is therefore controllable. The latter statement is also known as Chow's theorem: *The system is controllable if its involutive distribution regarding Lie brackets has rank equal to the dimension of the configuration space.* Some robots are controllable if all their constraints are nonholonomic and if their involutive distribution has rank n. This controllability test is for nonlinear systems but can also be applied to linear systems $\dot{q} = Aq + Bu$, where the basic vector fields are $s_1 = Aq$ and $s_2 = B$. Calculating the involutive distribution results in the Kalman test for controllability of the linear systems:

$$\text{rank}[B, [AB], [A^2 B], \ldots] = n$$

To obtain involutive distribution successive levels of Lie brackets are calculated. The number of those successive levels defines the difficulty index of maneuvering the WMR platform [13]. The higher the difficulty index, the higher the number of the basic maneuvers required to achieve any desired motion direction.

EXAMPLE 2.7
Show that the differential drive is controllable.

Solution
Differential drive has three states $(n = 3)$ $q = [x, y, \varphi]$ and two basic motion directions s_1 and s_2:

$$s_1(q) = \begin{bmatrix} \cos \varphi \\ \sin \varphi \\ 0 \end{bmatrix} \quad s_2(q) = \begin{bmatrix} 0 \\ 0 \\ 1 \end{bmatrix}$$

The new motion direction obtained by the Lie bracket is

$$s_3 = [s_1, s_2] = \frac{\partial s_2}{\partial q} s_1 - \frac{\partial s_1}{\partial q} s_2$$

$$= \begin{bmatrix} \sin \varphi \\ -\cos \varphi \\ 0 \end{bmatrix}$$

Calculating Lie brackets from any combination of s_1, s_2, and s_3 does not produce a new linearly independent motion direction; therefore the distribution $[s_1, s_2, s_3]$ is involutive (closed by Lie brackets) and has rank:

$$\text{rank}[s_1, s_2, s_3] = n = 3$$

so the system is controllable.

EXAMPLE 2.8
Determine motion constraints and possible motion directions for a rear-powered car. The car can be controlled by the rim velocity of the rear wheels v and by the angular speed of the steering mechanism γ (Fig. 2.17).

Find out the number and the type of motion constraints, the kinematic model, and the number of DOFs. Is the system controllable?

(Continued)

EXAMPLE 2.8—cont'd

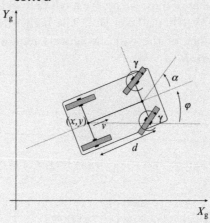

Fig. 2.17 Rear-powered car controlled by the rim velocity of the rear wheels v and by the angular speed of the steering mechanism γ.

Solution

The vehicle configuration can be described by four states $q = [x, y, \varphi, \alpha]^T$, because the control inputs are velocity v and the angular velocity of the steering mechanism γ. The fourth state ($\alpha = \int_0^t \gamma \, dt$) therefore describes the current state of the steering.

Note that bicycle kinematics with rear wheel drive has the same kinematics (Eq. 2.19) as the car in this example, but it has only three states because the angle of the steering mechanism is already defined by one of the inputs in Eq. (2.19).

The vehicle cannot move laterally to the rear and front wheels; therefore the motion constraints are

$$a_1(q) = \begin{bmatrix} \sin\varphi \\ -\cos\varphi \\ 0 \\ 0 \end{bmatrix} \qquad a_2(q) = \begin{bmatrix} \sin(\alpha + \varphi) \\ -\cos(\alpha + \varphi) \\ d\cos\alpha \\ 0 \end{bmatrix}$$

Vector fields that define the basic motion directions are defined by the rolling direction of the rear wheels (direction of motion when $\gamma = 0$, which means constant α) and by the rotation of the steering wheels (direction of motion when $v = 0$)

$$s_1 = \begin{bmatrix} \cos\varphi \\ \sin\varphi \\ \frac{1}{d}\tan\alpha \\ 0 \end{bmatrix} \qquad s_2 = \begin{bmatrix} 0 \\ 0 \\ 0 \\ 1 \end{bmatrix}$$

EXAMPLE 2.8—cont'd

The vehicle kinematic model is defined by

$$\dot{q} = [s_1, s_2]\begin{bmatrix} v \\ \gamma \end{bmatrix} = \begin{bmatrix} \cos\varphi & 0 \\ \sin\varphi & 0 \\ \frac{1}{d}\tan\alpha & 0 \\ 0 & 1 \end{bmatrix}\begin{bmatrix} v \\ \gamma \end{bmatrix}$$

which is similar to kinematics (Eq. 2.19) but with an additional state α and input $\gamma = \dot{\alpha}$.

So the system is described by four states; it has two constraints and two inputs. We will try to obtain new linearly independent motion directions s_i. Two are defined by the kinematic model, and the other two are obtained by Lie brackets:

$$s_3 = [s_1, s_2] = \frac{\partial s_2}{\partial q}s_1 - \frac{\partial s_1}{\partial q}s_2$$

$$= \begin{bmatrix} 0 & 0 & 0 & 0 \\ 0 & 0 & 0 & 0 \\ 0 & 0 & 0 & 0 \\ 0 & 0 & 0 & 0 \end{bmatrix}\begin{bmatrix} \cos\varphi \\ \sin\varphi \\ \frac{1}{d}\tan\alpha \\ 0 \end{bmatrix} - \begin{bmatrix} 0 & 0 & -\sin\varphi & 0 \\ 0 & 0 & \cos\varphi & 0 \\ 0 & 0 & 0 & \frac{1}{d\cos^2\alpha} \\ 0 & 0 & 0 & 0 \end{bmatrix}\begin{bmatrix} 0 \\ 0 \\ 0 \\ 1 \end{bmatrix}$$

$$= \begin{bmatrix} 0 \\ 0 \\ \frac{-1}{d\cos^2\alpha} \\ 0 \end{bmatrix}$$

and

$$s_4 = [s_1, s_3] = \frac{\partial s_3}{\partial q}s_1 - \frac{\partial s_1}{\partial q}s_3$$

$$= \begin{bmatrix} 0 & 0 & 0 & 0 \\ 0 & 0 & 0 & 0 \\ 0 & 0 & 0 & \frac{-2\sin\alpha}{d\cos^3\alpha} \\ 0 & 0 & 0 & 0 \end{bmatrix}\begin{bmatrix} \cos\varphi \\ \sin\varphi \\ \frac{1}{d}\tan\alpha \\ 0 \end{bmatrix} - \begin{bmatrix} 0 & 0 & -\sin\varphi & 0 \\ 0 & 0 & \cos\varphi & 0 \\ 0 & 0 & 0 & \frac{1}{d\cos^2\alpha} \\ 0 & 0 & 0 & 0 \end{bmatrix}\begin{bmatrix} 0 \\ 0 \\ \frac{-1}{d\cos^2\alpha} \\ 0 \end{bmatrix}$$

$$= \begin{bmatrix} \frac{-\sin\varphi}{d\cos^2\alpha} \\ \frac{\cos\varphi}{d\cos^2\alpha} \\ 0 \\ 0 \end{bmatrix}$$

Because all four vector fields s_i, $i = 1, \ldots, 4$ are linearly independent the system has four DOFs and both constraints are nonholonomic.

The involutive closure has a rank of 4, which is the number of the system states. The vehicle is therefore controllable.

2.4 DYNAMIC MODEL OF A MOBILE SYSTEM WITH CONSTRAINTS

The kinematic model only describes static transformation of some robot velocities (pseudo velocities) to the velocities expressed in global coordinates. However, the dynamic motion model of the mechanical system includes dynamic properties such as system motion caused by external forces and system inertia. Using Lagrange formulation, which is especially suitable to describe mechanical systems [14], the dynamic model reads

$$\frac{d}{dt}\left(\frac{\partial \mathcal{L}}{\partial \dot{q}_k}\right) - \frac{\partial \mathcal{L}}{\partial q_k} + \frac{\partial P}{\partial \dot{q}_k} + g_k + \tau_{d_k} = f_k \qquad (2.46)$$

where index k describes the general coordinates q_k ($k = 1, \ldots, n$), $\mathcal{L}$ defines the Lagrangian (difference between kinetic and potential energy of the system), P is the power dissipation function due to friction and damping in the system, g_k are the forces due to gravitation, τ_{d_k} are the system disturbances, and f_k are the general forces (external influences to the system) related to the general coordinate q_k. Eq. (2.46) is valid only for a nonconstrained system, that is, for the system without constraints that has n DOFs and no velocity constraints.

For systems with motion constraints the dynamic motion equations are given using Lagrange multipliers [15] as follows:

$$\frac{d}{dt}\left(\frac{\partial \mathcal{L}}{\partial \dot{q}_k}\right) - \frac{\partial \mathcal{L}}{\partial q_k} + \frac{\partial P}{\partial \dot{q}_k} + g_k + \tau_{d_k} = f_k - \sum_{j=1}^{m} \lambda_j a_{jk} \qquad (2.47)$$

where m is the number of linearly independent motion constraints, λ_j is the Lagrange multiplier associated with the jth constraint relation, and a_{jk} ($j = 1, \ldots, m$, $k = 1, \ldots, n$) are coefficients of the constraints.

The final set of equations consists of $n + m$ differential and algebraic equations (n Lagrange equations and m constraints equations) with $n + m$ unknowns (n general coordinates q_k and m Lagrange multipliers λ_j). Equations are differential in generalized coordinates and algebraic regarding Lagrange multipliers.

A dynamic model (2.47) of some mechanical system with constraints can be expressed in matrix form as follows:

$$\boldsymbol{M}(\boldsymbol{q})\ddot{\boldsymbol{q}} + \boldsymbol{V}(\boldsymbol{q}, \dot{\boldsymbol{q}}) + \boldsymbol{F}(\dot{\boldsymbol{q}}) + \boldsymbol{G}(\boldsymbol{q}) + \boldsymbol{\tau}_d = \boldsymbol{E}(\boldsymbol{q})\boldsymbol{u} - \boldsymbol{A}^T(\boldsymbol{q})\boldsymbol{\lambda} \qquad (2.48)$$

where the meaning of matrices is described in Table 2.1.

Table 2.1 Meaning of matrices in the dynamic model (2.48)

q	Vector of generalized coordinates (dimension $n \times 1$)
$M(q)$	Positive-definite matrix of masses and inertia (dimension $n \times n$)
$V(q, \dot{q})$	Vector of Coriolis and centrifugal forces (dimension $n \times 1$)
$F(\dot{q})$	Vector of friction and dumping forces (dimension $n \times 1$)
$G(q)$	Vector of forces and torques due to gravity (dimension $n \times 1$)
τ_d	Vector of unknown disturbances including unmodeled dynamics (dimension $n \times 1$)
$E(q)$	Transformation matrix from actuator space to generalized coordinate space (dimension $n \times r$)
u	Input vector (dimension $r \times 1$)
$A^T(q)$	Matrix of kinematic constraint coefficients (dimension $m \times n$)
λ	Vector of constraint forces (Lagrange multipliers) (dimension $m \times 1$)

2.4.1 State-Space Representation of the Dynamic Model of a Mobile System With Constraints

In the sequel a state-space dynamic model for a system with m kinematic constraints is derived. Then partial linearization of the system state-space description is performed using a nonlinear feedback relation [6] that enables the second-order kinematic model to be obtained.

A dynamic system with m kinematic constraints is described by

$$M(q)\ddot{q} + V(q, \dot{q}) + F(\dot{q}) + G(q) = E(q)u - A^T(q)\lambda \qquad (2.49)$$

where the influence of unknown disturbances and unmodeled dynamics from Eq. (2.48) is left out. The kinematic motion model is given by

$$\dot{q} = S(q)v(t) \qquad (2.50)$$

The dynamic model (2.49) and kinematic model (2.50) can be joined to a single state-space description. A unified description of nonholonomic and holonomic constraints can be found in [16], where holonomic constraints are expressed in differential (with velocities) form like nonholonomic constraints.

Due to shorter notation the dependence on q is omitted in the following equations (e.g. $M(q)$ or $S(q)$ is written as M or S, respectively). A time derivative of Eq. (2.50) is

$$\ddot{q} = \dot{S}v + S\dot{v} \qquad (2.51)$$

Inserting into Eq. (2.49) and expressing generalized coordinates q with pseudo velocities v results in

$$M\dot{S}v + MS\dot{v} + V + F + G = Eu - A^T\lambda \qquad (2.52)$$

The presence of Lagrangian multipliers $\boldsymbol{\lambda}$ due to motion constraints can be eliminated by considering relation $\boldsymbol{AS} = \boldsymbol{0}$ and its transform $\boldsymbol{S}^T \boldsymbol{A}^T = \boldsymbol{0}$. Multiplying Eq. (2.52) by $\boldsymbol{S}^T$ therefore gives a reduced dynamical model:

$$\boldsymbol{S}^T \boldsymbol{M} \dot{\boldsymbol{S}} \boldsymbol{v} + \boldsymbol{S}^T \boldsymbol{M} \boldsymbol{S} \dot{\boldsymbol{v}} + \boldsymbol{S}^T \boldsymbol{V} + \boldsymbol{S}^T \boldsymbol{F} + \boldsymbol{S}^T \boldsymbol{G} = \boldsymbol{S}^T \boldsymbol{E} \boldsymbol{u} \tag{2.53}$$

where Lagrange multipliers $\boldsymbol{\lambda}$ are eliminated. Introducing $\tilde{\boldsymbol{M}} = \boldsymbol{S}^T \boldsymbol{M} \boldsymbol{S}$, $\tilde{\boldsymbol{V}} = \boldsymbol{S}^T \boldsymbol{M} \dot{\boldsymbol{S}} \boldsymbol{v} + \boldsymbol{S}^T (\boldsymbol{V} + \boldsymbol{F} + \boldsymbol{G})$, and $\tilde{\boldsymbol{E}} = \boldsymbol{S}^T \boldsymbol{E}$ gives a more transparent notation,

$$\tilde{\boldsymbol{M}} \dot{\boldsymbol{v}} + \tilde{\boldsymbol{V}} = \tilde{\boldsymbol{E}} \boldsymbol{u} \tag{2.54}$$

from where a vector of pseudo accelerations $\dot{\boldsymbol{v}}$ is expressed:

$$\dot{\boldsymbol{v}} = \tilde{\boldsymbol{M}}^{-1} \left(\tilde{\boldsymbol{E}} \boldsymbol{u} - \tilde{\boldsymbol{V}} \right) \tag{2.55}$$

If condition $\det \boldsymbol{S}^T \boldsymbol{E} \neq 0$ is true (which in most of realistic examples is) then from Eq. (2.54) the system input can be expressed as

$$\boldsymbol{u} = \tilde{\boldsymbol{E}}^{-1} \left(\tilde{\boldsymbol{M}} \dot{\boldsymbol{v}} + \tilde{\boldsymbol{V}} \right) \tag{2.56}$$

By extending the state vector with pseudo velocities $\boldsymbol{x} = [\boldsymbol{q}^T \boldsymbol{v}^T]^T$ and presenting the system in nonlinear form $\dot{\boldsymbol{x}} = \boldsymbol{f}(\boldsymbol{x}) + \boldsymbol{g}(\boldsymbol{x}) \boldsymbol{u}$ (expression $\boldsymbol{f}(\boldsymbol{x})$ contains nonlinear dependence from states) the state-space representation of the system reads

$$\dot{\boldsymbol{x}} = \begin{bmatrix} \boldsymbol{S} \boldsymbol{v} \\ -\tilde{\boldsymbol{M}}^{-1} \tilde{\boldsymbol{V}} \end{bmatrix} + \begin{bmatrix} \boldsymbol{0}_{n \times r} \\ \tilde{\boldsymbol{M}}^{-1} \tilde{\boldsymbol{E}} \end{bmatrix} \boldsymbol{u} \tag{2.57}$$

where r is the number of inputs in vector $\boldsymbol{u}$ and state vector $\boldsymbol{x}$ has the dimension $(2n - m) \times 1$.

Using the inverse model (2.56) the required system input can be calculated for the desired pseudo accelerations of the system. Applying these inputs to Eq. (2.57) the resulting model becomes

$$\dot{\boldsymbol{x}} = \begin{bmatrix} \boldsymbol{S} \boldsymbol{v} \\ \boldsymbol{0}_{(n-m) \times 1} \end{bmatrix} + \begin{bmatrix} \boldsymbol{0}_{n \times (n-m)} \\ \boldsymbol{I}_{(n-m) \times (n-m)} \end{bmatrix} \boldsymbol{u}_z \tag{2.58}$$

where $\boldsymbol{u}_z$ are pseudo accelerations of the system. Relation (2.56) can therefore be used for calculating the required predicted system inputs. These inputs can be used to control the system without feedback (open-loop control) or they are better used as feedforward control in combination with some feedback control.

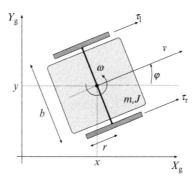

Fig. 2.18 Vehicle with differential drive.

2.4.2 Kinematic and Dynamic Model of Differential Drive Robot

The differential drive vehicle in Fig. 2.3 has two wheels usually powered by electric motors. Let us assume that the vehicle mass center coincides with its geometric center. The mass of the vehicle without the wheels is m_v, and the mass of the wheels is m_w. The vehicle moves on the plane and its moment of inertia around the z-axis is J_v (z is in the normal direction to the plane), and the moment of inertia for the wheels is J_w.

In practice usually the mass of the wheels is much smaller than the mass of the casing; therefore a common mass m and inertia J can be used. The vehicle is described by three generalized coordinates $q = [x, y, \varphi]$ and its input is the torque on the left and the right wheel, τ_l and τ_r, respectively (Fig. 2.18).

Kinematic Model and Constraints
The kinematic model (2.2) of the vehicle is

$$\begin{bmatrix} \dot{x} \\ \dot{y} \\ \dot{\varphi} \end{bmatrix} = \begin{bmatrix} \cos(\varphi) & 0 \\ \sin(\varphi) & 0 \\ 0 & 1 \end{bmatrix} \begin{bmatrix} v \\ \omega \end{bmatrix}$$

The nonholonomic motion constraint is

$$-\dot{x}\sin\varphi + \dot{y}\cos\varphi = 0$$

which means that the reachable motion directions are columns of the kinematic matrix

$$S = \begin{bmatrix} \cos(\varphi(t)) & 0 \\ \sin(\varphi(t)) & 0 \\ 0 & 1 \end{bmatrix} \tag{2.59}$$

and the constraint coefficient matrix is

$$A = \begin{bmatrix} -\sin\varphi & \cos\varphi & 0 \end{bmatrix} \qquad (2.60)$$

Dynamic Model

The dynamic model is derived by Lagrange formulation:

$$\frac{d}{dt}\left(\frac{\partial \mathcal{L}}{\partial \dot{q}_k}\right) - \frac{\partial \mathcal{L}}{\partial q_k} + \frac{\partial P}{\partial \dot{q}_k} = f_k - \sum_{j=1}^{m} \lambda_j a_{jk} \qquad (2.61)$$

where unknown disturbance τ_{d_k} from relation (2.47) is omitted. Similarly the forces and the torques due to the gravitation g_k are not present because the vehicle drives on the plane where the potential energy is constant (without loss of generality $W_P = 0$ can be assumed).

Overall, kinetic energy of the system reads

$$W_K = \frac{m}{2}\left(\dot{x}^2 + \dot{y}^2\right) + \frac{J}{2}\dot{\varphi}^2 \qquad (2.62)$$

The potential energy is $W_P = 0$, so the Lagrangian is

$$\mathcal{L} = W_K - W_P = \frac{m}{2}\left(\dot{x}^2 + \dot{y}^2\right) + \frac{J}{2}\dot{\varphi}^2 \qquad (2.63)$$

Additionally damping and friction at the wheels rotation can be neglected ($P = 0$). Forces and torques in Eq. (2.61) are

$$\frac{d}{dt}\left(\frac{\partial \mathcal{L}}{\partial \dot{x}}\right) = m\ddot{x}$$

$$\frac{d}{dt}\left(\frac{\partial \mathcal{L}}{\partial \dot{y}}\right) = m\ddot{y} \qquad (2.64)$$

$$\frac{d}{dt}\left(\frac{\partial \mathcal{L}}{\partial \dot{\varphi}}\right) = J\ddot{\varphi}$$

and

$$\frac{\partial \mathcal{L}}{\partial x} = 0$$

$$\frac{\partial \mathcal{L}}{\partial y} = 0 \qquad (2.65)$$

$$\frac{\partial \mathcal{L}}{\partial \varphi} = 0$$

According to Lagrange formulation (2.61) the following differential equations are obtained:

$$m\ddot{x} - \lambda_1 \sin\varphi = F_x$$
$$m\ddot{y} + \lambda_1 \cos\varphi = F_y \qquad (2.66)$$
$$J\ddot{\varphi} = M$$

where the resultant force on the left and the right wheel is $F = \frac{1}{r}(\tau_r + \tau_l)$ and r is the wheel radius. The force in the x-axis direction is $F_x = \frac{1}{r}(\tau_r + \tau_l)\cos\varphi$ and in the y-axis it is $F_y = \frac{1}{r}(\tau_r + \tau_l)\sin\varphi$. Torque on the vehicle is obtained by $M = \frac{L}{2r}(\tau_r - \tau_l)$, where L is the distance among the wheels. Therefore, the final model is

$$m\ddot{x} - \lambda_1 \sin\varphi - \frac{1}{r}(\tau_r + \tau_l)\cos\varphi = 0$$
$$m\ddot{y} + \lambda_1 \cos\varphi - \frac{1}{r}(\tau_r + \tau_l)\sin\varphi = 0 \qquad (2.67)$$
$$J\ddot{\varphi} - \frac{L}{2r}(\tau_r - \tau_l) = 0$$

The obtained dynamic model written in matrix form is

$$M(q)\ddot{q} + V(q,\dot{q}) + F(\dot{q}) = E(q)u - A^T(q)\lambda$$

where matrices are

$$M = \begin{bmatrix} m & 0 & 0 \\ 0 & m & 0 \\ 0 & 0 & J \end{bmatrix}$$

$$E| = \frac{1}{r}\begin{bmatrix} \cos\varphi & \cos\varphi \\ \sin\varphi & \sin\varphi \\ \frac{L}{2} & -\frac{L}{2} \end{bmatrix}$$

$$A = \begin{bmatrix} -\sin\varphi & \cos\varphi & 0 \end{bmatrix}$$

$$u = \begin{bmatrix} \tau_r \\ \tau_l \end{bmatrix}$$

and the remaining matrices are zero.

The common state-space model (2.57) that includes the kinematic and dynamic model is determined by matrices

$$\tilde{M} = \begin{bmatrix} m & 0 \\ 0 & J \end{bmatrix}$$

$$\tilde{V} = \begin{bmatrix} 0 \\ 0 \end{bmatrix}$$

$$\tilde{E} = \frac{1}{r} \begin{bmatrix} 1 & 1 \\ \frac{L}{2} & -\frac{L}{2} \end{bmatrix}$$

and according to Eq. (2.57) the system can be written in the state-space form $\dot{x} = f(x) + g(x)u$, where the state vector is $x = [q^T, v^T]^T$. The resulting model is

$$\begin{bmatrix} \dot{x} \\ \dot{y} \\ \dot{\varphi} \\ \dot{v} \\ \dot{\omega} \end{bmatrix} = \begin{bmatrix} v \cos \varphi \\ v \sin \varphi \\ \omega \\ 0 \\ 0 \end{bmatrix} + \begin{bmatrix} 0 & 0 \\ 0 & 0 \\ 0 & 0 \\ \frac{1}{mr} & \frac{1}{mr} \\ \frac{L}{2Jr} & \frac{-L}{2Jr} \end{bmatrix} \begin{bmatrix} \tau_r \\ \tau_l \end{bmatrix} \qquad (2.68)$$

The inverse system model is obtained by taking into account Eq. (2.56) as

$$\begin{bmatrix} \tau_r \\ \tau_l \end{bmatrix} = \begin{bmatrix} \frac{\dot{v} m r}{2} + \frac{\dot{\omega} J r}{L} \\ \frac{\dot{v} m r}{2} - \frac{\dot{\omega} J r}{L} \end{bmatrix} \qquad (2.69)$$

Using the inverse model the required input torques to the wheels can be obtained from the known robot velocities and accelerations. The obtained input torques can be applied in open-loop control or better as a feedforward part in a closed-loop control.

EXAMPLE 2.9

Derive the kinematic and dynamic model for a vehicle with differential drive in Fig. 2.19. Express the models with the coordinates of the mass center (x_c, y_c), which are distance d away from the geometric center (x, y).

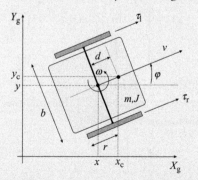

Fig. 2.19 Vehicle with differential drive with a different mass center (x_c, y_c) and geometric center (x, y).

EXAMPLE 2.9—cont'd
Solution

Considering the transformation between the geometric and mass center $x = x_c - d\cos\varphi$ and $y = y_c - d\sin\varphi$ and their time derivatives $\dot{x} = \dot{x}_c + d\dot\varphi\sin\varphi$ and $\dot{y} = \dot{y}_c - d\dot\varphi\cos\varphi$ and inserting the derivatives to kinematic model (Fig. 2.3), the final kinematic model and kinematic constraint for the mass center are obtained:

$$\begin{bmatrix} \dot{x}_c \\ \dot{y}_c \\ \dot\varphi \end{bmatrix} = \begin{bmatrix} \cos(\varphi) & -d\sin(\varphi) \\ \sin(\varphi) & d\cos(\varphi) \\ 0 & 1 \end{bmatrix} \begin{bmatrix} v \\ \omega \end{bmatrix}$$

$$- \dot{x}_c \sin\varphi + \dot{y}_c \cos\varphi - \dot\varphi d = 0$$

The Lagrangian for the mass center is $\mathcal{L} = \frac{m}{2}\left(\dot{x}_c^2 + \dot{y}_c^2\right) + \frac{J}{2}\dot\varphi^2$. Considering relation (2.61) the dynamic model is

$$m\ddot{x}_c - \lambda_1 \sin\varphi = \frac{1}{r}(\tau_r + \tau_l)\cos\varphi$$

$$m\ddot{y}_c + \lambda_1 \cos\varphi = F_y = \frac{1}{r}(\tau_r + \tau_l)\sin\varphi$$

$$J\ddot\varphi - \lambda_1 d = \frac{L}{2r}$$

and matrices for the model in matrix form are the following:

$$M = \begin{bmatrix} m & 0 & 0 \\ 0 & m & 0 \\ 0 & 0 & J \end{bmatrix}$$

$$E = \frac{1}{r}\begin{bmatrix} \cos\varphi & \cos\varphi \\ \sin\varphi & \sin\varphi \\ \frac{L}{2} & -\frac{L}{2} \end{bmatrix}$$

$$A = \begin{bmatrix} -\sin\varphi & \cos\varphi & -d \end{bmatrix}$$

$$S = \begin{bmatrix} \cos(\varphi(t)) & -d\sin(\varphi) \\ \sin(\varphi(t)) & d\cos(\varphi) \\ 0 & 1 \end{bmatrix}$$

Considering Eq. (2.57) the overall state-space representation of the system reads

$$\begin{bmatrix} \dot{x}_c \\ \dot{y}_c \\ \dot\varphi \\ \dot{v} \\ \dot\omega \end{bmatrix} = \begin{bmatrix} v\cos\varphi - \omega d\sin\varphi \\ v\sin\varphi + \omega d\cos\varphi \\ \omega \\ d\omega^2 \\ \frac{-dv\omega m}{md^2 + J} \end{bmatrix} + \begin{bmatrix} 0 & 0 \\ 0 & 0 \\ 0 & 0 \\ \frac{1}{mr} & \frac{1}{mr} \\ \frac{L}{2r(d^2 m + J)} & \frac{-L}{2r(d^2 m + J)} \end{bmatrix} \begin{bmatrix} \tau_r \\ \tau_l \end{bmatrix}$$

$$(2.70)$$

EXAMPLE 2.10

Derive the kinematic and dynamic model for a vehicle in Example 2.9 where the mass center and geometric center are not the same. Express the models using coordinates of the geometric center (x, y), which could be more appropriate in case that the mass center of the vehicle is changing due to a varying load.

Solution

The kinematic model and kinematic constraint for the geometric center are obtained:

$$\begin{bmatrix} \dot{x} \\ \dot{y} \\ \dot{\varphi} \end{bmatrix} = \begin{bmatrix} \cos(\varphi) & 0 \\ \sin(\varphi) & 0 \\ 0 & 1 \end{bmatrix} \begin{bmatrix} v \\ \omega \end{bmatrix}$$

$$-\dot{x}_c \sin\varphi + \dot{y}_c \cos\varphi = 0$$

The Lagrangian is $\mathcal{L} = \frac{m}{2}\left(\dot{x}_c^2 + \dot{y}_c^2\right) + \frac{J}{2}\dot{\varphi}^2$, where the mass center is expressed by the geometric center. The final state–space model is

$$\begin{bmatrix} \dot{x} \\ \dot{y} \\ \dot{\varphi} \\ \dot{v} \\ \dot{\omega} \end{bmatrix} = \begin{bmatrix} v\cos\varphi \\ v\sin\varphi \\ \omega \\ d\omega^2 \\ \frac{-d\omega m(\dot{x}\cos\varphi + \dot{y}\sin\varphi - v\cos(2\varphi) - d\omega\sin(2\varphi))}{md^2 + J} \end{bmatrix}$$

$$+ \begin{bmatrix} 0 & 0 \\ 0 & 0 \\ 0 & 0 \\ \frac{1}{mr} & \frac{1}{mr} \\ \frac{L - 2d\sin(2\varphi)}{2r(d^2m + J)} & \frac{-L - 2d\sin(2\varphi)}{2r(d^2m + J)} \end{bmatrix} \begin{bmatrix} \tau_r \\ \tau_l \end{bmatrix} \qquad (2.71)$$

EXAMPLE 2.11

Using Matlab calculate the required torques so that the mobile robot from Example 2.9 will drive along the reference trajectory $x_r = 1.1 + 0.7\sin(2\pi/30)$, $y_r = 0.9 + 0.7\sin(4\pi/30)$ without using sensors (open-loop control). In the simulation calculate robot torques using the inverse dynamic model, and depict system trajectory. Robot parameters are $m = 0.75$ kg, $J = 0.001$ kg m^2, $L = 0.075$ m, $r = 0.024$ m, and $d = 0.01$ m.

Solution

The inverse robot model considering relation (2.56) is

$$\begin{bmatrix} \tau_r \\ \tau_l \end{bmatrix} = \begin{bmatrix} \frac{r(\dot{v}m - d\omega^2 m)}{2} + \frac{r(\dot{\omega}(md^2 + J) + d\omega^2 m)}{L} \\ \frac{r(\dot{v}m - d\omega^2 m)}{2} - \frac{r(\dot{\omega}(md^2 + J) + d\omega^2 m)}{L} \end{bmatrix}$$

EXAMPLE 2.11—cont'd

From the reference, trajectory reference velocity v_r and reference angular velocity ω_r and their time derivatives $\dot{v}_r$ and $\dot{\omega}_r$ are calculated.

Obtained responses using Matlab simulation in Listing 2.3 are shown in Figs. 2.20 and 2.21.

Listing 2.3 Inverse robot dynamic model

```
1  Ts = 0.033; % Sampling time
2  t = 0:Ts:30; % Simulation time
3
4  % Reference
5  freq = 2*pi/30;
6    xRef = 1.1 + 0.7*sin(freq*t);      yRef = 0.9 + 0.7*sin(2*freq*t);
7   dxRef = freq*0.7*cos(freq*t);      dyRef = 2*freq*0.7*cos(2*freq*t);
8  ddxRef =-freq^2*0.7*sin(freq*t);  ddyRef =-4*freq^2*0.7*sin(2*freq*t);
9  dddxRef =-freq^3*0.7*cos(freq*t); dddyRef =-8*freq^3*0.7*cos(2*freq*t);
10 qRef = [xRef; yRef; atan2(dyRef, dxRef)]; % Reference trajectory
11 vRef = sqrt(dxRef.^2+dyRef.^2);
12 wRef = (dxRef.*ddyRef-dyRef.*ddxRef)./(dxRef.^2+dyRef.^2);
13 dvRef = (dxRef.*ddxRef+dyRef.*ddyRef)./vRef;
14 dwRef = (dxRef.*dddyRef-dyRef.*dddxRef)./vRef.^2 - 2.*wRef.*dvRef./vRef;
15
16 q = [qRef(:,1); vRef(1); wRef(2)]; % Inital robot state
17 m = 0.75; J = 0.001; L = 0.075; r = 0.024; d = 0.01; % Robot parameters
18
19 for k = 1:length(t)
20    % Calculate torques from the trajectory and inverse model
21    v = vRef(k); w = wRef(k); dv = dvRef(k); dw = dwRef(k);
22    tau = [(r*(dv*m-d*w*m*w))/2 + (r*(dw*(m*d^2+J) + d*w*m*v))/L; ...
23           (r*(dv*m-d*w*m*w))/2 - (r*(dw*(m*d^2+J) + d*w*m*v))/L];
24
25    % Robot motion simulation using kinematic and dynamic model
26    phi = q(3); v = q(4); w = q(5);
27    F = [v*cos(phi) - d*w*sin(phi); ...
28         v*sin(phi) + d*w*cos(phi); ...
29         w; ...
30         d*w^2; ...
31        -(d*w*v*m)/(m*d^2 + J)];
32    G = [0,                        0; ...
33         0,                        0; ...
34         0,                        0; ...
35         1/(m*r),                  1/(m*r); ...
36         L/(2*r*(m*d^2 + J)), -L/(2*r*(m*d^2 + J))];
37    dq = F + G*tau; % State space model
38    q = q + dq*Ts; % Euler integration
39    q(3) = wrapToPi(q(3)); % Map orientation angle to [-pi, pi]
40 end
```

(Continued)

EXAMPLE 2.11—cont'd

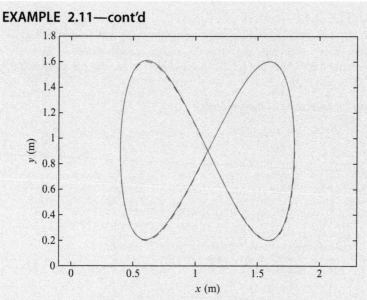

Fig. 2.20 Obtained robot path (*solid line*) and the reference path (*dashed line*).

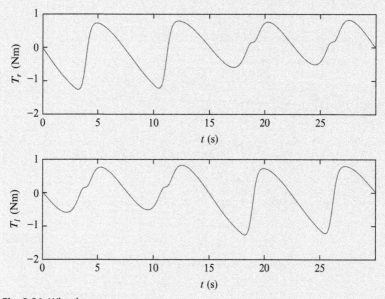

Fig. 2.21 Wheel torques.

REFERENCES

[1] H. Choset, K. Lynch, S. Hutchinson, G. Kantor, W. Burgard, L. Kavraki, S. Thrun, Principles of Robot Motion: Theory, Algorithms, and Implementations, MIT Press, Boston, MA, 2005.

[2] B. Siciliano, O. Khatib, Springer Handbook of Robotics, Springer, Berlin, 2008.

[3] G. Dudek, M. Jenkin, Computational Principles of Mobile Robotics, Cambridge University Press, New York, NY, 2010.

[4] G. Klančar, I. Škrjanc, Tracking-error model-based predictive control for mobile robots in real time, J. Intell. Robot. Syst. 55 (2007) 460–469.

[5] I. Kolmanovsky, N.H. McClamroch, Developments in nonholonomic control problems, J. Intell. Robot. Syst. 15 (6) (1995) 20–36.

[6] A. De Luca, G. Oriolo, Kinematics and dynamics of multi-body systems, in: Modelling and Control of Nonholonomic Mechanical Systems, CISM International Centre for Mechanical Sciences, Springer, Vienna, 1995, pp. 277–342.

[7] J.C. Latombe, Robot Motion Planning, Kluwer Academic Publishers, New York, 1991.

[8] N. Sarkar, X. Yun, V. Kumar, Control of mechanical systems with rolling constraints application to dynamic control of mobile robots, Int. J. Robot. Res. 13 (1) (1994) 55–69.

[9] R. Hermann, A.J. Krener, Nonlinear controllability and observability, IEEE Trans. Autom. Control 22 (5) (1977) 728–740.

[10] R.W. Brockett, Asymptotic stability and feedback stabilization, in: Differential Geometric Control Theory, Birkhuser, Boston, MA, 1983, pp. 181–191.

[11] J.P. Laumond, Robot Motion Planning and Control, in: Lecture Notes in Control and Information Sciences, vol. 229, Springer, 1998.

[12] W.S. Levine, The Control Handbook, in: Electrical Engineering Handbook, second ed., CRC Press, Boca Raton, FL, 2010.

[13] A. De Luca, G. Oriolo, M. Vendittelli, Control of wheeled mobile robots: an experimental overview, in: S. Nicosia, B. Siciliano, A. Bicchi, P. Valigi (Eds.), Ramsete, Lecture Notes in Control and Information Sciences, vol. 270, Springer, Berlin, 2001, pp. 181–226.

[14] O. Egeland, J.T. Gravdahl, Modeling and Simulation for Automatic Control, Marine Cybernetics, Trondheim, Norway, 2002.

[15] Z. Li, J.F. Canny, Nonholonomic Motion Planning, Kluwer Academic Publishers, Boston, Dordrecht, London, 1993.

[16] X. Yun, State space representation of holonomic and nonholonomic constraints resulting from rolling contacts, in: IEEE International Conference on Robotics and Automation, IEEE, Nagoya, 1995, pp. 2690–2694.

CHAPTER 3

Control of Wheeled Mobile Systems

3.1 INTRODUCTION

Motion control of wheeled mobile robots in the environment without obstacles can be performed by controlling motion from some start pose to some goal pose (classic control, where intermediate state trajectory is not prescribed) or by reference trajectory tracking. For wheeled mobile robots with nonholonomic constrains the control to the reference pose is harder than following the reference trajectory that connects the stat pose and the goal pose. For a successful reference pose or trajectory tracking control the use of a nonsmooth or time varying controller needs to be applied [1] because the controlled system is nonlinear and time varying. When moving the robot, nonholonomic constraints need to be considered so its path cannot be arbitrary. An additional reason favoring the trajectory tracking control approaches is also the fact that robots are usually driving in the environments with various limitations, obstacles, and some demands that all somehow define the desired path that leads the robot to the goal pose.

When controlling nonholonomic systems the control action should be decomposed to the *feedforward* control part and the *feedback* control part. This is also known as *two-degree-of-freedom control*. The feedforward control part is calculated from the reference trajectory and those calculated inputs can then be fed to the system to drive on the reference trajectory in the open loop (without the feedback sensor). However, only feedforward control is not practical as it is not robust to disturbances and initial state errors; therefore, a feedback part needs to be included. Two-degree-of-freedom control is natural and usually used when controlling nonholonomic mechanic systems. Therefore, it will also be applied to the majority of examples that follow.

Wheeled mobile robots are *dynamic* systems where an appropriate torque needs to be applied to the wheels to obtain desired motion of the platform. Motion control algorithms therefore need to consider the system's dynamic properties. Usually this problem is tackled using cascade control schemas with the outer controller for velocity control and the inner torque (force, motor current, etc.) control. The outer controller determines the required system velocities for the system to navigate to the reference pose or to follow the reference trajectory. While the inner faster controller calculates the required torques (force, motor current, etc.) to achieve the system velocities determined from the outer controller. The inner controller needs to be fast enough so that additional phase lag introduced to the system is not problematic. In the majority of available mobile robot platforms the inner torque controller is already implemented in the platform and the user only commands the desired system's velocities by implementing control considering the system kinematics only.

The rest of these chapters are divided based on the control approaches achieving the reference pose and the control approaches for the reference trajectory tracking. The former will include the basic idea and some simple examples applied to different platforms. The latter will deal with some more details because these approaches are more natural to wheeled mobile robots driving in environments with known obstacles, as already mentioned.

3.2 CONTROL TO REFERENCE POSE

In this section some basic approaches for controlling a wheeled mobile robot to the reference pose are explained, where the reference pose is defined by position and orientation. In this case the path or trajectory to the reference pose is not prescribed, and the robot can drive on any feasible path to arrive to the goal. This path can be either defined explicitly and adapted during motion or it is defined implicitly by the realization of applied control algorithm to achieve the reference pose.

Because only the initial (or current) and the final (or reference) poses are given, and the path between these two positions is arbitrary, this opens new possibilities, such as choosing an "optimal" path. It needs not to be stressed that a feasible path should be chosen where all the constraints such as kinematic, dynamic, and environmental are taken into account. This usually still leaves an infinite selection of possible paths where a particular one is chosen respecting some additional criteria such as time, length, curvature, energy consumption, and the like. In general path planning is a challenging task, and these aspects will not be considered in this section.

In the following, the control to the reference pose will be split into two separate tasks, namely orientation control and forward-motion control. These two building blocks cannot be used individually, but by combining them, several control schemes for achieving the reference pose are obtained. These approaches are general and can be implemented to different mobile robot kinematics. In this section they will be illustrated by examples on a differential and Ackermann drive.

3.2.1 Orientation Control

In any planar movement, orientation control is necessary to perform the motion in a desired orientation. The fact that makes orientation control even more important is that a wheeled robot can only move in certain directions due to nonholonomic constraints. Although orientation control cannot be performed independently from the forward-motion control, the problem itself can also be viewed from the classical control point of view to gain some additional insight. This will show how control gains influence classical control-performance criteria of the orientation control feedback loop.

Let us assume that the orientation of the wheeled robot at some time t is $\varphi(t)$, and the reference or desired orientation is $\varphi_{\text{ref}}(t)$. The control error can be defined as

$$e_\varphi(t) = \varphi_{\text{ref}}(t) - \varphi(t)$$

As in any control system, a control variable is needed that can influence or change the controlled variable, which is the orientation in this case. The control goal is to drive the control error to 0. Usually the convergence to 0 should be fast but still respecting some additional requirements such as energy consumption, actuator load, robustness of the system in the presence of disturbances, noise, parasitic dynamics, etc. The usual approach in the control design is to start with the model of the system to be controlled. In this case we will build upon the kinematic model, more specifically on its orientation equation.

Orientation Control for Differential Drive
Differential drive kinematics is given by Eq. (2.2). The third equation in Eq. (2.2) is the orientation equation of the kinematic model and is given by

$$\dot{\varphi}(t) = \omega(t) \tag{3.1}$$

From the control point of view, Eq. (3.1) describes the system with control variable $\omega(t)$ and integral nature (its pole lies in the origin of the complex plane s). It is very well known that a simple proportional controller is able to drive the control error of an integral process to 0. The control law is then given by

$$\omega(t) = K(\varphi_{\text{ref}}(t) - \varphi(t)) \tag{3.2}$$

where the control gain K is an arbitrary positive constant. The interpretation of the control law (3.3) is that the angular velocity $\omega(t)$ of the platform is set proportionally to the robot orientation error. Combining Eqs. (3.1), (3.2), the dynamics of the orientation control loop can be rewritten as:

$$\dot{\varphi}(t) = K(\varphi_{\text{ref}}(t) - \varphi(t))$$

from where closed-loop transfer function of the controlled system can be obtained:

$$G_{\text{cl}}(s) = \frac{\phi(s)}{\phi_{\text{ref}}(s)} = \frac{1}{\frac{1}{K}s + 1}$$

with $\phi(s)$ and $\phi_{\text{ref}}(s)$ being the Laplace transforms of $\varphi(t)$ and $\varphi_{\text{ref}}(t)$, respectively. The transfer function $G_{\text{cl}}(s)$ is of the first order, which means that in the case of a constant reference the orientation exponentially approaches the reference (with time constant $T = \frac{1}{K}$). The closed-loop transfer function also has unity gain, so there is no orientation error in steady state.

It has been shown that the controller (3.3) made the closed-loop transfer function behave as the first-order system. Sometimes the second–order transfer function $G_{\text{cl}}(s) = \frac{\phi(s)}{\phi_{\text{ref}}(s)}$ is desired because it gives more degrees of freedom to the designer in terms of the transients shape. We begin the construction by setting the angular acceleration $\dot{\omega}(t)$ of the platform proportional to the robot orientation error:

$$\dot{\omega}(t) = K(\varphi_{\text{ref}}(t) - \varphi(t)) \tag{3.3}$$

The obtained controlled system

$$\dot{\omega} = \ddot{\varphi}(t) = K(\varphi_{\text{ref}}(t) - \varphi(t))$$

with the transfer function

$$G_{\text{cl}}(s) = \frac{\phi(s)}{\phi_{\text{ref}}(s)} = \frac{K}{s^2 + K}$$

is the second-order system with natural frequency $\omega_n = \sqrt{K}$ and damping coefficient $\zeta = 0$. Such a system is marginally stable and its oscillating responses are unacceptable. Damping is achieved by the inclusion of an additional term to the controller (3.3):

$$\dot{\omega}(t) = K_1(\varphi_{\text{ref}}(t) - \varphi(t)) - K_2\dot{\varphi}(t) \tag{3.4}$$

where K_1 and K_2 are arbitrary positive control gains. Combining Eqs. (3.1), (3.4) the closed-loop transfer function becomes,

$$G_{\text{cl}}(s) = \frac{\phi(s)}{\phi_{\text{ref}}(s)} = \frac{K_1}{s^2 + K_2 s + K_1}$$

where natural frequency is $\omega_n = \sqrt{K_1}$, damping coefficient $\zeta = \frac{K_2}{2\sqrt{K_1}}$, and closed-loop poles $s_{1,2} = -\zeta\omega_n \pm j\omega_n\sqrt{1 - \zeta^2}$.

Orientation Control for the Ackermann Drive

Controller design in the case of the Ackermann drive is very similar to the differential drive controller design; the only difference is due to a different kinematic model (2.19) for orientation that reads

$$\dot{\varphi} = \frac{v_r(t)}{d}\tan(\alpha(t)) \tag{3.5}$$

The control variable is α, which can again be chosen proportionally to the orientation error:

$$\alpha(t) = K(\varphi_{\text{ref}}(t) - \varphi(t)) \tag{3.6}$$

The dynamics of the orientation error can be described by the following differential equation obtained by combining Eqs. (3.5), (3.6):

$$\dot{\varphi}(t) = \frac{v_r(t)}{d}\tan(K(\varphi_{\text{ref}}(t) - \varphi(t)))$$

The system is therefore nonlinear. For small angles $\alpha(t)$ and a constant velocity of the rear wheels $v_r(t) = V$, a linear model approximation is obtained,

$$\dot{\varphi}(t) = \frac{V}{d}K(\varphi_{\text{ref}}(t) - \varphi(t))$$

which can be converted to the transfer function form:

$$G_{\text{cl}}(s) = \frac{\phi(s)}{\phi_{\text{ref}}(s)} = \frac{1}{\frac{d}{VK}s + 1}$$

Similarly as in the case of differential drive, the orientation error converges to 0 exponentially (for constant reference orientation) and the decay time constant is $T = \frac{d}{VK}$.

If the desired transfer function of the orientation feedback loop is of the second order, we follow a similar approach as in the case of the differential drive. Control law given by Eq. (3.4) can be adapted to the Ackermann drive as well:

$$\dot{\alpha}(t) = K_1(\varphi_{\text{ref}}(t) - \varphi(t)) - K_2\dot{\varphi}(t) \tag{3.7}$$

Assuming again a constant velocity $v_r(t) = V$ and small angles α, a linear approximation of Eq. (3.5) is obtained:

$$\dot{\varphi}(t) = \frac{V}{d}\alpha(t) \tag{3.8}$$

Deriving $\alpha(t)$ from Eq. (3.8) and introducing it into Eq. (3.7) yields

$$\ddot{\varphi}(t) = K_1\frac{V}{d}(\varphi_{\text{ref}}(t) - \varphi(t)) - K_2\frac{V}{d}\dot{\varphi}(t)$$

The closed-loop transfer function therefore becomes

$$G_c(s) = \frac{\phi(s)}{\phi_{\text{ref}}(s)} = \frac{K_1\frac{V}{d}}{s^2 + K_2\frac{V}{d}s + K_1\frac{V}{d}}$$

The obtained response of the robot orientation is damped with the damping coefficient $\zeta = \frac{K_2}{2}\sqrt{\frac{V}{dK_1}}$ and has a natural frequency $\omega_n = \sqrt{K_1\frac{V}{d}}$.

3.2.2 Forward-Motion Control

By forward-motion control we refer to control algorithms that define the translational velocity of the mobile robot $v(t)$ to achieve some control goal. Forward-motion control per se cannot be used as a mobile robot control strategy. For example, if the desired orientation can be achieved by orientation control taking the angular velocity $\omega(t)$ as a control in the case of the differential drive, forward-motion control alone cannot drive the robot to a desired position unless the robot is directed to its goal initially. This means that forward-motion control is inevitably interconnected with orientation control.

Nevertheless, the translational velocity has to be controlled to achieve the control goal. In the case of trajectory tracking, the velocity is more or less governed by the trajectory while in the reference pose control the velocity should decrease when we approach the final goal. A reasonable idea is to apply the control that is proportional to the distance to the reference point $(x_{\text{ref}}(t), y_{\text{ref}}(t))$:

$$v(t) = K\sqrt{(x_{\text{ref}}(t) - x(t))^2 + (y_{\text{ref}}(t) - y(t))^2} \qquad (3.9)$$

Note that the reference position can be constant or it can change according to some reference trajectory. The control (3.9) certainly has some limitations and special treatment is needed in case of very large or very small distances to the reference:

- If the distance to the reference point is large, the control command given by Eq. (3.9) also becomes large. It is advisable to introduce some limitations on the maximum velocity command. In practice the limitations are dictated by actuator limitations, driving surface conditions, path curvature, etc.
- If the distance to the reference point is very small, the robot can in practice "overtake" the reference point (due to noise or imperfect model of the vehicle). As the robot is moving away from the reference point, the distance is increasing, and the robot is accelerating according to Eq. (3.9), which is a problem that will be dealt with when the forward-motion controllers are combined with orientation controllers.

Velocity is inseparably related to acceleration. The latter is also limited in any practical implementation due to the limited forces and torques exerted by the actuators. This aspect is also very important when designing a forward-motion control. One possibility is to limit the acceleration. Usually it is sufficient to just low-pass filter the output of the controller $v(t)$ before the command is sent to the robot in the form of the signal $v^*(t)$. The simplest filter of the first order with the DC gain of 1 can be used for this purpose. It is given by a differential equation:

$$\tau_f \dot{v}^*(t) + v^*(t) = v(t) \qquad (3.10)$$

or equivalently with a transfer function

$$G_f(s) = \frac{V^*(s)}{V(s)} = \frac{1}{\tau_f s + 1}$$

where τ_f is a time constant of the filter.

In Example 3.1 a simple control algorithm is implemented to drive the robot with Ackermann drive to the reference point. The algorithm contains orientation control and forward-motion control presented previously.

EXAMPLE 3.1

Write a control algorithm for a tricycle robot with a rear-powered pair of wheels to achieve the reference position $x_{ref} = 4$ m and $y_{ref} = 4$ m. The vehicle is controlled with the front-wheel steering angle α and by the rear wheels' velocity v_r. The distance between the front wheel axis and the rear wheels axis is $d = 0.1$ m. The initial pose of the vehicle is $[x(0), y(0), \varphi(0)] = [1 \text{ m}, 0, -\pi]$. The control algorithm should consider the vehicle constraints $v_{max} = 0.8$ m/s and $\alpha_{max} = \frac{\pi}{4}$.

Write a control algorithm and test it on a simulation of the vehicle kinematics using the Euler integration method.

Solution

The orientation control and forward-motion control presented in this section can be used simultaneously. An implementation of solution in Matlab is given in Listing 3.1. Simulation results are shown in Figs. 3.1 and 3.2. It is seen that the vehicle reaches the reference point and stops there. In Fig. 3.2 the control variables are limited according to the vehicle's physical constraints.

Listing 3.1

```
1  Ts = 0.03; % Sampling time
2  t = 0:Ts:30; % Simulation time
3  d = 0.1; % Distance between axes
4  xyRef = [4; 4]; % Reference position
5  q = [1; 0; -pi]; % Initial state
6
7  for k = 1:length(t)
8      phi_ref = atan2(xyRef(2)-q(2), xyRef(1)-q(1)); %Reference orientation
9      qRef = [xyRef; phi_ref];
10
11     e = qRef - q; %Position end orientation error
12
13     % Controller
14     v     = 0.3*sqrt(e(1)^2+e(2)^2);
15     alpha = 0.2*e(3);
16
17     % Physical constraints
18     if abs(alpha)>pi/4, alpha = pi/4*sign(alpha); end
19     if abs(v)>0.8, v = 0.8*sign(v); end
20
21     % Robot motion simulation
22     dq = [v*cos(q(3)); v*sin(q(3)); v/d*tan(alpha)];
23     noise = 0.00; % Set to experiment with noise (e.g. 0.001)
24     q = q + Ts*dq + randn(3,1)*noise; % Euler integration
25 end
```

EXAMPLE 3.1—cont'd

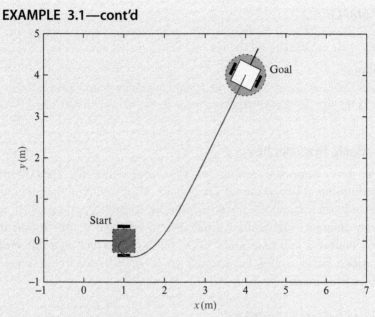

Fig. 3.1 Robot path to the goal position from Example 3.1.

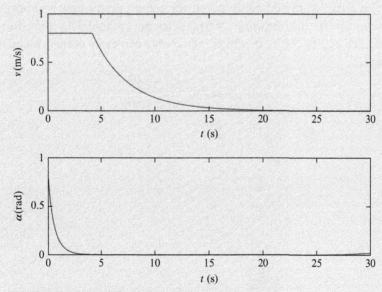

Fig. 3.2 Control signals from Example 3.1.

EXAMPLE 3.2

The same problem as in Example 3.1 has to be solved for the differential drive with a maximum vehicle velocity of $v_{max} = 0.8$ m/s.

Solution

Make necessary adjustments to the control algorithm from Example 3.1 in the controller part and correct kinematics in the simulation part.

3.2.3 Basic Approaches

This section introduces several practical approaches for controlling a wheeled mobile robot to the reference pose. They combine the previously introduced orientation and forward-motion control (Sections 3.2.1 and 3.2.2) in different ways and are applicable to wheeled mobile robots that need to achieve a reference pose. They will be illustrated on a differential drive robot but can also be adapted to other wheeled robot types as illustrated previously.

Control to Reference Position

In this case the robot is required to arrive at a reference (final) position where the final orientation is not prescribed, so it can be arbitrary. In order to arrive to the reference point, the robot's orientation is controlled continuously in the direction of the reference point. This direction is denoted by φ_r (see Fig. 3.3), which can be obtained easily using geometrical relations:

$$\varphi_r(t) = \arctan \frac{y_{ref} - y(t)}{x_{ref} - x(t)}$$

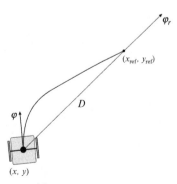

Fig. 3.3 Control to reference position.

Angular velocity control $\omega(t)$ is therefore commanded as

$$\omega(t) = K_1(\varphi_r(t) - \varphi(t)) \tag{3.11}$$

where K_1 is a positive controller gain. The basic control principle is similar as in Example 3.1 and is illustrated in Fig. 3.3.

Translational robot velocity is first commanded as proposed by Eq. (3.9):

$$v(t) = K_2\sqrt{(x_{\text{ref}}(t) - x(t))^2 + (y_{\text{ref}}(t) - y(t))^2} \tag{3.12}$$

We have already mentioned that the maximum velocity should be limited in the starting phase due to many practical constraints. Specifically, velocity and acceleration limits should be taken into account.

The control law (3.12) also hides a potential danger when the robot approaches the target pose. The velocity command (3.12) is always positive, and when decelerating toward the final position the robot can accidentally drive over it. The problem is that after crossing the reference pose the velocity command will start increasing because the distance between the robot and the reference starts increasing. The other problem is that crossing the reference also makes the reference orientation opposite, which leads to fast rotation of the robot. Some simple solutions to this problem exist:

- When the robot drives over the reference point, the orientation error abruptly changes (±180 degrees). The algorithm will therefore check if the absolute value of orientation error exceeds 90 degree. The orientation error will then be either increased or decreased by 180 degree (so that it is in the interval $[-180$ degree$, 180$ degree$]$) before it enters the controller. Moreover, the control output (3.12) changes its sign in this case. The mentioned problems are therefore circumvented by an upgraded versions of control laws (3.11), (3.12):

$$e_\varphi(t) = \varphi_r(t) - \varphi(t)$$
$$\omega(t) = K_1 \arctan(\tan(e_\varphi(t)))$$
$$v(t) = K_2\sqrt{(x_{\text{ref}}(t) - x(t))^2 + (y_{\text{ref}}(t) - y(t))^2} \cdot \text{sgn}(\cos(e_\varphi(t)))$$
$$\tag{3.13}$$

- When the robot reaches a certain neighborhood of the reference point, the approach phase is finished and zero velocity commands are sent. This mechanism to fully stop the vehicle needs to be implemented even in the case of the modified control law (3.13), especially in case of noisy measurements.

Control to Reference Pose Using an Intermediate Point

This control algorithm is easy to implement because we use a simple controller given by Eqs. (3.11), (3.12), which drives the robot to the desired reference point. But in this case not only the reference point $(x_{ref}(t), y_{ref}(t))$ but also the reference orientation φ_{ref} is required in the reference point. The idea of this approach is to add an intermediate point that will shape the trajectory in a way that the correct final orientation is obtained. The intermediate point (x_t, y_t) is placed on the distance r from the reference point such that the direction from the intermediate point toward the reference point coincides with the reference orientation (as shown in Fig. 3.4). The intermediate point is determined by

$$x_t = x_{ref} - r \cos \varphi_{ref}$$
$$y_t = y_{ref} - r \sin \varphi_{ref}$$

The control algorithm consists of two phases. In the first phase the robot is driven toward the intermediate point. When the distance to the intermediate point becomes sufficiently low (checked by condition $\sqrt{(x - x_t)^2 + (y - y_t)^2} < d_{tol}$), the algorithm switches to the second phase where the robot is controlled to the reference point. This procedure ensures that the robot arrives to the reference position with the required orientation (a very small orientation error is possible in the reference pose). It is possible to make many variations of this algorithm and also make more intermediate points to improve the operation.

The described algorithm is very simple and applicable in many areas of use. According to the application, appropriate selection of parameters r and d_{tol} should be made.

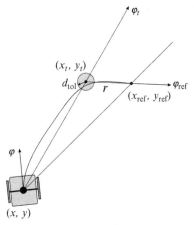

Fig. 3.4 Control to the reference pose using an intermediate point.

EXAMPLE 3.3

Write a control algorithm for a differential type robot to achieve the reference pose $[x_{ref}, y_{ref}, \varphi_{ref}] = [4\text{ m}, 4\text{ m}, 0\text{ degree}]$ using an intermediate point. Find appropriate values for parameters r and d_{tol}. The initial pose of the vehicle is $[x(0), y(0), \varphi(0)] = [1\text{ m}, 0\text{ m}, 100\text{ degree}]$.

Test the control algorithm by means of simulation of the differential drive vehicle kinematics.

Solution

M–script code of a possible solution is given in Listing 3.2. Simulation results are shown in Figs. 3.5 and 3.6.

Listing 3.2

```
1  Ts = 0.03; % Sampling time
2  t = 0:Ts:15; % Simulation time
3  r = 0.5; % Distance parameter for the intermediate point
4  dTol = 0.05; % Tolerance distance (to the intermediate point) for switch
5  qRef = [4; 4; 0]; % Reference pose
6  q = [1; 0; 100/180*pi]; % Initial pose
7
8  % Intermediate point
9  xT = qRef(1) - r*cos(qRef(3));
10 yT = qRef(2) - r*sin(qRef(3));
11
12 state = 0; % State: 0 - go to intermediate point, 1 - go to reference point
13 for k = 1:length(t)
14     D = sqrt((qRef(1)-q(1))^2 + (qRef(2)-q(2))^2);
15     if D<dTol % Stop when close to the goal
16         v = 0;
17         w = 0;
18     else
19         if state==0
20             d = sqrt((xT-q(1))^2+(yT-q(2))^2);
21             if d<dTol, state = 1; end
22
23             phiT = atan2(yT-q(2), xT-q(1));
24             ePhi = phiT - q(3);
25         else
26             ePhi = qRef(3) - q(3);
27         end
28         ePhi = wrapToPi(ePhi);
29
30         % Controller
31         v = D*0.8;
32         w = ePhi*5;
33     end
34
35     % Robot motion simulation
36     dq = [v*cos(q(3)); v*sin(q(3)); w];
37     noise = 0.00; % Set to experiment with noise (e.g. 0.001)
38     q = q + Ts*dq + randn(3,1)*noise; % Euler integration
39     q(3) = wrapToPi(q(3)); % Map orientation angle to [-pi, pi]
40 end
```

(Continued)

EXAMPLE 3.3—cont'd

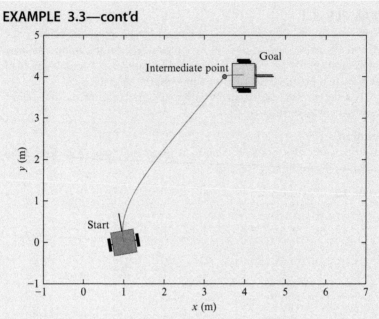

Fig. 3.5 Robot path to the goal pose based on intermediate point from Example 3.3.

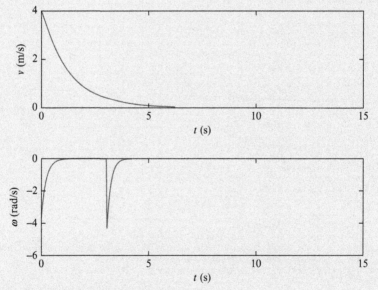

Fig. 3.6 Control signals from Example 3.3.

Control to Reference Pose Using an Intermediate Direction

A robot needs to arrive from its initial pose to the reference pose where the position (x_{ref}, y_{ref}) and the orientation φ_{ref} are given. The idea of the algorithm with an intermediate direction is illustrated in Fig. 3.7 [2]. We begin with the construction of a right-angled triangle. The reference point is put to the vertex with the right angle. One cathetus connects the current robot position with the reference one and has the length $D = +\sqrt{(x_{ref}(t) - x(t))^2 + (y_{ref}(t) - y(t))^2}$. This cathetus defines the orientation φ_r that is facing from the robot toward the target. The other cathetus has a fixed length of $r > 0$, which is a designer parameter of this method. The angle between cathetus D and the hypotenuse is denoted by $\beta(t)$. Two angles play a crucial role in this approach, and they are defined as follows:

$$\alpha(t) = \varphi_r(t) - \varphi_{ref}$$

$$\beta(t) = \begin{cases} \arctan \frac{+r}{D} & \alpha(t) > 0 \\ -\arctan \frac{r}{D} & \text{otherwise} \end{cases} \tag{3.14}$$

Note that the angles $\alpha(t)$ and $\beta(t)$ are always of the same sign (in the case depicted in Fig. 3.7 they are both positive). If α becomes 0 at a certain time, β becomes irrelevant, so its sign can be arbitrary. Large values (in the absolute sense) of α mean that driving straight to the reference is not a good idea because the orientation error will be large in the reference pose. The angle α should therefore be reduced (in the absolute sense). This is achieved by defining an intermediate direction that is shifted from φ_r, and the shift is always away from the reference orientation φ_{ref} (if the reference orientation is seen pointing to the right from the robot's perspective, the robot should approach the reference position from the left side, and vice versa). While

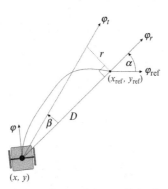

Fig. 3.7 Control to reference pose using an intermediate direction.

$\alpha(t)$ normally decreases (in the absolute sense) when approaching the target, $\beta(t)$ increases (in the absolute sense) as the distance to the reference decreases. We will make use of this when constructing the control algorithm.

The algorithm again has two phases. In the first phase (when $|\alpha(t)|$ is large), the robot's orientation is controlled toward the intermediate direction $\varphi_t(t) = \varphi_r(t) + \beta(t)$ (this case is depicted in Fig. 3.7). When the angles α and β become the same, the current reference orientation switches to $\varphi_t(t) = \varphi_r(t) + \alpha(t)$. Note that this switch does not cause any bump because both directions are the same at the time of the switch. The control law for orientation is therefore defined as follows:

$$e_\varphi(t) = \varphi_r(t) - \varphi(t) + \begin{cases} \alpha(t) & |\alpha(t)| < |\beta(t)| \\ \beta(t) & \text{otherwise} \end{cases}$$ (3.15)

$$\omega(t) = Ke_\varphi(t)$$

Note that the current reference heading is never toward the reference point but is always slightly shifted. The shift is chosen so that the angle $\alpha(t)$ is driven toward 0. This in turn implies that the reference heading is forced toward the reference point and the robot will arrive there with the correct reference orientation. Note that this algorithm is also valid for negative values of angles α and β. We only need to ensure that all the angles are in the $(-\pi, \pi]$ interval.

Robot translational velocity can be defined similarly as in the previous sections.

EXAMPLE 3.4

Write a control algorithm for a differential type robot to achieve the reference pose $[x_{ref}, y_{ref}, \varphi_{ref}] = [4\ m, 4\ m, 0\ degree]$ using the presented control to the reference pose using an intermediate direction. Find an appropriate value for parameter r. The initial pose of the vehicle is $[x(0), y(0), \varphi(0)] = [1\ m, 0\ m, 100\ degree]$. Test the algorithm on a simulated kinematic model.

Solution

The M-script code of a possible solution is in Listing 3.3. Simulation results are shown in Figs. 3.8 and 3.9.

Listing 3.3

```
1 Ts = 0.03; % Sampling time
2 t = 0:Ts:15; % Simulation time
3 r = 0.2; % Distance parameter
4 qRef = [4; 4; 0]; % Reference pose
5 q = [1; 0; 100/180*pi]; % Initial pose
```

EXAMPLE 3.4—cont'd

```
6
7  for k = 1:length(t)
8      % Compute intermidieate direction shift
9      phiR = atan2(qRef(2)-q(2), qRef(1)-q(1));
10     D = sqrt((qRef(1)-q(1))^2 + (qRef(2)-q(2))^2);
11
12     alpha = wrapToPi(phiR - qRef(3));
13     beta = atan(r/D);
14     if alpha<0, beta = -beta; end
15
16     % Controller
17     if abs(alpha) < abs(beta)
18         ePhi = wrapToPi(phiR - q(3) + alpha); % The second part
19     else
20         ePhi = wrapToPi(phiR - q(3) + beta); % The first part
21         int.setPose(q);
22     end
23     v = D*0.8;
24     w = ePhi*5;
25
26     % Robot motion simulation
27     dq = [v*cos(q(3)); v*sin(q(3)); w];
28     noise = 0.00; % Set to experiment with noise (e.g. 0.001)
29     q = q + Ts*dq + randn(3,1)*noise; % Euler integration
30     q(3) = wrapToPi(q(3)); % Map orientation angle to [-pi, pi]
31 end
```

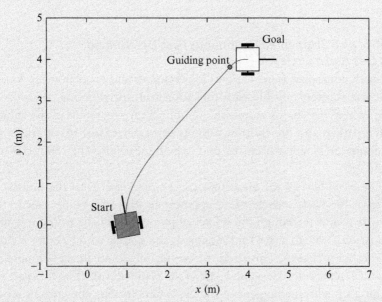

Fig. 3.8 Robot path to the goal pose based on intermediate direction from Example 3.4.

(*Continued*)

EXAMPLE 3.4—cont'd

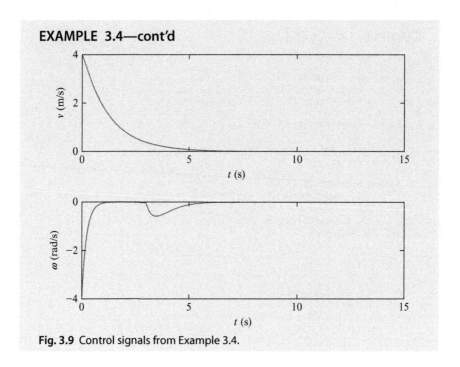

Fig. 3.9 Control signals from Example 3.4.

Control on a Segmented Continuous Path Determined by a Line and a Circle Arc

The path composed from straight line segments and circular arcs is known to be the shortest possible path for Ackermann drive robots as shown in [3–5], where the circle's radius is the minimal turning radius of the vehicle. Such a path is also the shortest possible for a differential robot where the minimum circle radius limits to zero, which means that the robot can turn on the spot.

The basic idea of the algorithm can be described with the illustration in Fig. 3.10. First, a circle of an appropriate (this will be discussed later) radius R is drawn through the reference point tangentially to the reference orientation. Note that two solutions exist, and a circle whose center is closer to the robot is chosen. This circle, more precisely a certain arc belonging to this circle, represents the second part of the planned path. The first part will run on a straight line tangent to the circle and crossing the current robot's position. Again, two tangents exist, and we select the one that gives proper direction of driving along the circular arc. The direction of driving along the arc is defined with the reference orientation (in the case illustrated in

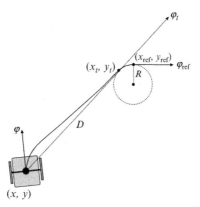

Fig. 3.10 Control on a segmented continuous path determined by a line and a circle arc.

Fig. 3.10 the robot will drive along the circle in the clockwise direction). The solution can be chosen easily by checking the sign of the cross products between the radius vectors and the tangential vectors.

In the first part of the algorithm the goal is to drive toward the point (x_t, y_t) where the straight segment meets the arc segment. The orientation controller is a very simple one:

$$\omega(t) = K(\varphi_t(t) - \varphi(t))$$

where $\varphi_t(t) = \arctan \frac{y_t - y(t)}{x_t - x(t)}$. When the distance to the intermediate point is small enough, the second phase begins. This includes driving along the trajectory, and the controller changes to

$$\omega(t) = \frac{v(t)}{R} + K(\varphi_{\text{tang}}(t) - \varphi(t))$$

where R is the circle radius, $v(t)$ is the desired translational velocity, and $\varphi_{\text{tang}}(t)$ is the direction of the tangent to the arc in the current robot position. Note that the first term in the control is a feedforward part that ensures the robot will drive along the circular arc with radius R, while the second term is a feedback control that corrects the control errors.

To achieve higher robustness the reference path is calculated in each iteration, which ensures that the robot is always on the reference straight line or on the reference circle. The obtained path slightly differs from the ideal one composed of a straight line segment and a circular arc. The mentioned difference is due to inial robot orientation that does not coincide with the constructed tangent. Other reasons for this difference are noise and disturbances (wheels slipping and the like).

This reference path is relatively easy to determine in real time. The path itself is continuous, but the required inputs are not. Due to transition from the straight line to the circle the robot angular velocity instantly changes from zero to $\frac{v(t)}{R}$. In reality this is not achievable (limited robot acceleration), and therefore some tracking error appears in this transition.

Such a control is appropriate for situations where a robot needs to arrive to the reference pose quickly and the shape of the path is not prescribed but the length should be as short as possible (e.g., robot soccer). Especially if we deal with the robots with a limited turning radius (e.g., Ackermann drive), such a path is the shortest possible to the goal pose if the minimal robot turning radius is used for parameter R. But of course, this leads to high values of radial accelerations, so perhaps a larger value of R is desired.

EXAMPLE 3.5

Write a control algorithm for a differential type robot to achieve the reference pose $[x_{ref}, y_{ref}, \varphi_{ref}] = [0 \text{ m}, 0 \text{ m}, 0 \text{ degree}]$ using the control algorithm proposed in this section. The radius of the circle should be $R = 0.4$ m. The initial pose of the vehicle is $[x(0), y(0), \varphi(0)] = [-3 \text{ m}, -3 \text{ m}, 100 \text{ degree}]$. Test the algorithm on a simulated kinematic model.

Solution

Although the basic idea of the control algorithm is quite simple, the implementation becomes a bit more complicated as it needs to calculate appropriate circle centers, tangent points on the circle, and some other parameters. A possible solution is given in Listing 3.4, and the simulation results are shown in Figs. 3.11 and 3.12.

Listing 3.4

```
 1 Ts = 0.03; % Sampling time
 2 t = 0:Ts:15; % Simulation time
 3 r = 0.2; % Distance parameter
 4 qRef = [4; 4; 0]; % Reference pose
 5 q = [1; 0; 100/180*pi]; % Initial pose
 6
 7 aMax = 5; % Maximum acceleration
 8 vMax = 0.4; % Maximum speed
 9 accuracy = vMax*Ts; % Accuracy
10 curveZone = 0.6; % Radius
11 Rr = 0.99*curveZone/2; % Radius
12 slowDown = false; v = 0; vDir = 1; w = 0; % Initial states
13 X = [0, 1; -1, 0]; % Support matrix: a.'*X*b = a(1)*b(2) - a(2)*b(1)
14
```

EXAMPLE 3.5—cont'd

```
15  for k = 1:length(t)
16      fin = [cos(qRef(3)); sin(qRef(3))];
17      D = qRef(1:2); % Destination point
18      S = q(1:2); % Robot position
19      M = (D + S)/2;
20      Ov = [cos(q(3)); sin(q(3))]; % Orientation vector
21      SDv = D - S; % SD vector
22      l2 = norm(SDv); % Distance
23
24      if slowDown
25          v = v - aMax*Ts; if v < 0, v = 0; end
26          w = 0;
27      else
28          if fin.'*X*SDv > SDv.'*X*fin
29              Ps = D - Rr*X.'*fin; % Circle centre
30          else
31              Ps = D - Rr*X*fin; % Circle centre
32          end
33
34          l = norm(Ps-S);
35          if l < curveZone/2
36              Dv = fin;
37          else
38              d = sqrt(sum((S-Ps).^2) - Rr^2);
39              alpha = atan(Rr/d);
40              phi = wrapTo2Pi(atan2(Ps(2)-S(2), Ps(1)-S(1)));
41              U1 = S + d*[cos(phi+alpha); sin(phi+alpha)];
42              U2 = S + d*[cos(phi-alpha); sin(phi-alpha)];
43              if ((U1 - S).'*X*(Ps - U1)) * (fin.'*X*(Ps - D)) >= 0
44                  D = U1;
45              else
46                  D = U2;
47              end
48              M = (D + S)/2;
49              SDv = D - S;
50              Dv = SDv/(norm(SDv)+eps);
51          end
52
53          if l2 > accuracy % If the position is not reached
54              v = v + aMax*Ts; if v > vMax, v = vMax; end
55
56              Ev = X*(D-S);
57              DTv = X*Dv;
58              if abs(DTv.'*X*Ev) < 0.000001 % Go on a straight line
59                  gamma = 0;
60                  Sv = SDv/(norm(SDv)+eps);
61              else % Go on a circle
62                  C = DTv * Ev.'*X*(D - M)/(DTv.'*X*Ev) + D; %
63                  Circle centre
64                  if SDv.'*X*Dv > 0, a = 1; else a = -1; end
65                  Sv = a*X*(C-S);
66                  Sv = Sv/(norm(Sv)+eps);
67                  gamma = a*acos(Dv.'*Sv);
68                  if a*Sv.'*X*Dv < 0, gamma = a*2*pi - gamma; end
69                  l = abs(gamma*norm(S-C)); % Curve length
70              end
71
```

(Continued)

EXAMPLE 3.5—cont'd

```
72              if v > eps
73                  if Ov.'*Sv < 0, vDir = -1; else vDir = 1; end %
74                   Direction
75                  ePhi = acos(vDir*Sv.'*Ov); % Angular error
76                  if vDir*Ov.'*X*Sv < 0, ePhi = -ePhi; end
77                  dt = 1/v; if dt < 0.00001, dt = 0.00001; end
78                  w = gamma/dt + ePhi/dt*10*(1-exp(-12/0.1)); %
79                   Angular speed
80              else
81                  w = 0;
82              end
83          else
84              slowDown = true;
85          end
86      end
87      u = [vDir*v; w]; % Tangential and angular velocity
88
89      % Robot motion simulation
90      dq = [u(1)*cos(q(3)); u(1)*sin(q(3)); u(2)];
91      noise = 0.00; % Set to experiment with noise (e.g. 0.001)
92      q = q + Ts*dq + randn(3,1)*noise; % Euler integration
93      q(3) = wrapToPi(q(3)); % Map orientation angle to [-pi, pi]
94  end
```

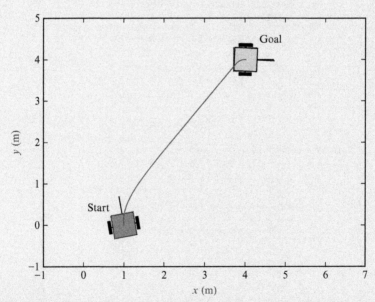

Fig. 3.11 Robot path to the goal pose based on lines and circles from Example 3.5.

EXAMPLE 3.5—cont'd

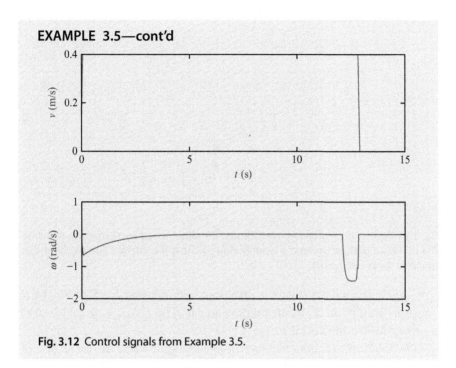

Fig. 3.12 Control signals from Example 3.5.

From Reference Pose Control to Reference Path Control

Often the control goal is given with a sequence of reference points that a robot should drive through. In these cases we no longer speak about reference pose control but rather a reference path is constructed through these points. Often straight line segments are used between the individual points, and the control goal is to get to each reference point with the proper orientation, and then automatically switch to the next reference point. This approach is easy to implement and usually sufficient for use in practice. Its drawback is the nonsmooth transition between neighboring line segments, which causes jumps of control error in these discontinuities.

The path is defined by a sequence of points $T_i = [x_i, y_i]^T$, where $i \in 1, 2, \ldots, n$, and n is the number of points. In the beginning the robot should follow the first line segment (between points T_1 and T_2), and it should arrive to T_2 with orientation defined by the vector $\overrightarrow{T_2 T_2}$. When it reaches the end of this segment it starts to follow the next line segment (between the points T_2 and T_3), and so on. Fig. 3.13 shows actual line segment between points T_i and T_{i+1} with variables marked. Vector $v = T_{i+1} - T_i =$

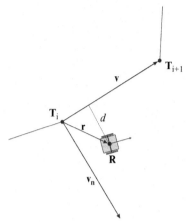

Fig. 3.13 Control on segmented continuous path determined by a sequence of points. The reference path between the neighboring points is the straight line segment that connects those two points.

$[\Delta x, \Delta y]^T$ defines the segment direction, and vector $\boldsymbol{r} = \boldsymbol{R} - \boldsymbol{T}_i$ is the direction from point $\boldsymbol{T}_i$ to the robot center $\boldsymbol{R}$. The vector $\boldsymbol{v}_n = [\Delta y, -\Delta x]$ is orthogonal to the vector $\boldsymbol{v}$.

The robot must follow the current line segment while the projection of vector $\boldsymbol{r}$ to vector $\boldsymbol{v}$ is inside the interval defined by points $\boldsymbol{T}_i$ in $\boldsymbol{T}_{i+1}$. This condition can be expressed as follows:

$$
\begin{cases}
\text{Follow the current segment } (\boldsymbol{T}_i, \boldsymbol{T}_{i+1}) & \text{if } 0 < u < 1 \\
\text{Follow the next segment } (\boldsymbol{T}_{i+1}, \boldsymbol{T}_{i+2}) & \text{if } u > 1
\end{cases}
$$

where u is the dot product

$$
u = \frac{\boldsymbol{v}^T \boldsymbol{r}}{\boldsymbol{v}^T \boldsymbol{v}}
$$

Variable u is used to check if the current line segment is still valid or the switch to the next line segment is necessary.

Orthogonal distance of the robot from the line segment is defined using the normal vector $\boldsymbol{v}_n$:

$$
d = \frac{\boldsymbol{v}_n^T \boldsymbol{r}}{\sqrt{\boldsymbol{v}_n^T \boldsymbol{v}_n}}
$$

If we normalize distance d by the distance length we obtain normalized orthogonal distance d_n between the robot and the straight line:

$$
d_n = \frac{\boldsymbol{v}_n^T \boldsymbol{r}}{\boldsymbol{v}_n^T \boldsymbol{v}_n}
$$

Normal distance d_n is zero if the robot is on the line segment and is positive if the robot is on the right side of the segment (according to vector $\mathbf{v}$) and vice versa. Normal distance d_n is used to define desired direction or robot motion. If the robot is on the line segment or very close to it then it needs to follow the line segment. If, however, the robot is far away from the line segment then it needs to drive perpendicularly to line segment in order to reach the segment faster. The reference orientation of driving in a certain moment can be defined as follows:

$$\varphi_{ref} = \varphi_{lin} + \varphi_{rot}$$

where $\varphi_{lin} = \text{arctan2}(\Delta y, \Delta x)$ (note that the function arctan2 is the four-quadrant inverse tangent function) is the orientation of the line segment and $\varphi_{rot} = \arctan(K_1 d_n)$ is an additional reference rotation correction that enables the robot to reach the line segment. Gain K_1 changes the sensitivity of φ_{rot} according to d_n. Because φ_{ref} is obtained by addition of two angles it needs to be checked to be in the valid range $[-\pi, \pi]$.

So far we define a reference orientation that the robot needs to follow using some control law. The control error is defined as

$$e_\varphi = \varphi_{ref} - \varphi$$

where φ is the robot orientation. From orientation error the robot's angular velocity is calculated using a proportional controller:

$$\omega = K_2 e_\varphi$$

where K_2 is a proportional gain. Similarly, also a PID controller can be implemented, where the integral part speeds up the decay of e_φ while the differential part damps oscillations due to the integral part. Translational robot velocity can be controlled by basic approaches discussed in the preceding sections.

EXAMPLE 3.6

Write a control algorithm for a differentially driven robot to drive on the sequence of line segments defined by points $T_1 = [3, 0]$, $T_2 = [6, 4]$, $T_3 = [3, 4]$, $T_4 = [3, 1]$, and $T_5 = [0, 3]$. Find appropriate values for parameters K_1 and K_2. The initial pose of the vehicle is $[x(0), y(0), \varphi(0)] = [5\ m, 1\ m, 108\ degree]$. Test the algorithm on a simulated kinematic model.

(Continued)

EXAMPLE 3.6—cont'd
Solution

An M-script code of a possible solution is given in Listing 3.5. The reference path and the actual robot motion curve are shown in Fig. 3.14, and the control signals are shown in Fig. 3.15.

Listing 3.5

```
1  Ts = 0.03; % Sampling time
2  t = 0:Ts:30; % Simulation time
3  T = [3, 0; 6, 4; 3, 4; 3, 1; 0, 3].'; % Reference points of line segments
4  q = [5; 1; 0.6*pi]; % Initial pose
5
6  i = 1; % Index of the first point
7  for k = 1:length(t)
8      % Reference segment determination
9      dx = T(1,i+1) - T(1,i);
10     dy = T(2,i+1) - T(2,i);
11
12     v = [dx; dy]; % Direction vector of the current segment
13     vN = [dy; -dx]; % Orthogonal direction vector of the current segment
14     r = q(1:2) - T(:,i);
15     u = v.'*r/(v.'*v);
16
17     if u>1 && i<size(T,2)-1 % Switch to the next line segment
18         i = i + 1;
19         dx = T(1,i+1) - T(1,i);
20         dy = T(2,i+1) - T(2,i);
21         v = [dx; dy];
22         vN = [dy; -dx];
23         r = q(1:2) - T(:,i);
24     end
25
26     dn = vN.'*r/(vN.'*vN); % Normalized orthogonal distance
27
28     phiLin = atan2(v(2), v(1)); % Orientation of the line segment
29     phiRot = atan(5*dn); % If we are far from the line then we need
30     % additional rotation to face towards the line. If we are on the left
31     % side of the line we turn clockwise, otherwise counterclock wise.
32     % Gain 5 increases the sensitivity ...
33
34     phiRef = wrapToPi(phiLin + phiRot);
35
36     % Orientation error for control
37     ePhi = wrapToPi(phiRef - q(3));
38
39     % Controller
40     v = 0.4*cos(ePhi);
41     w = 3*ePhi;
42
43     % Robot motion simulation
44     dq = [v*cos(q(3)); v*sin(q(3)); w];
45     noise = 0.00; % Set to experiment with noise (e.g. 0.001)
46     q = q + Ts*dq + randn(3,1)*noise; % Euler integration
47     q(3) = wrapToPi(q(3)); % Map orientation angle to [-pi, pi]
48 end
```

EXAMPLE 3.6—cont'd

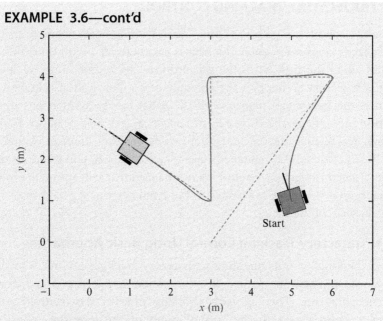

Fig. 3.14 Reference path and obtained robot motion from Example 3.6.

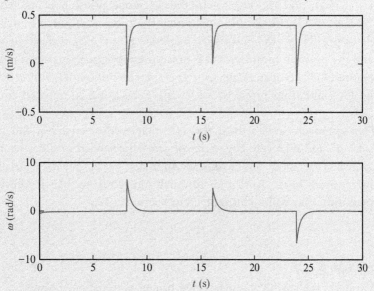

Fig. 3.15 Control signals from Example 3.6.

Try to extend the code given in the solution so that the robot could also drive in reverse if $|e_\varphi| > \frac{\pi}{2}$.

3.3 TRAJECTORY TRACKING CONTROL

In mobile robotics, a path is a "line" that a robot needs to drive in the generalized coordinates space. If a path is parameterized with time, that is, motion along the path has to be synchronized with time, then we speak about a trajectory. Whenever the schedule of a robot motion is known in advance, the (reference) trajectory of the robot can be written as a time-function in the generalized coordinate's space: $q_{ref}(t) = [x_{ref}(t), y_{ref}, \varphi_{ref}]^T$. For practical reasons, the trajectory is always defined on a finite time interval $t \in [0, T]$, that is, the reference trajectory has a start point and an end point. Trajectory tracking control is a mechanism that will ensure the robot trajectory $q(t)$ is as close as possible to the reference one $q_{ref}(t)$ despite any difficulties encountered.

3.3.1 Trajectory Tracking Control Using Basic Approaches

When thinking of implementing a trajectory tracking control, the first idea would probably be to imagine the reference trajectory as a moving reference position. Then in each sampling time of the controller, the reference point moves to the current point of the reference trajectory $(x_{ref}(t), y_{ref}(t))$, and the control to the reference position is applied via control laws (3.11), (3.12). Note that here the robot is required to drive as closely as possible to this imaginary reference point. When the velocity is low and the position measurement is noisy, the robot position measurement can be found ahead of the trajectory. It is therefore extremely important to handle such situations properly, for example, by using an updated control law (3.13).

This approach suffers from the fact that feedback carries the main burden here and relatively large control gains are needed in order to make the control errors small. This makes the approach susceptible to disturbances in the control loop. It is therefore advantageous to use feedforward compensation that will be introduced in Section 3.3.2.

EXAMPLE 3.7

A tricycle robot with a rear-powered pair of wheels from Example 3.1 has to be controlled to follow the reference trajectory $x_{ref} = 1.1 + 0.7 \sin(\frac{2\pi t}{30})$ and $y_{ref} = 0.9 + 0.7 \sin(\frac{4\pi t}{30})$. The initial pose of the vehicle

EXAMPLE 3.7—cont'd

is $[x(0), y(0), \varphi(0)] = [1.1, 0.8, 0]$. Write two control algorithms and test them on a simulated kinematic model:

- The first approach uses the basic control laws (3.11), (3.12).
- The second approach uses the upgraded control law (3.13).

Solution

By changing the value of the variable UpgradedLaw the control can switch between the so-called basic one, given by Eqs. (3.11), (3.12), and the upgraded one given by Eq. (3.13). The results of Example 3.7 are shown in Figs. 3.16 and 3.17 (in the tested case the basic and the upgraded control law behave the same).

The Matlab code is given in Listing 3.6.

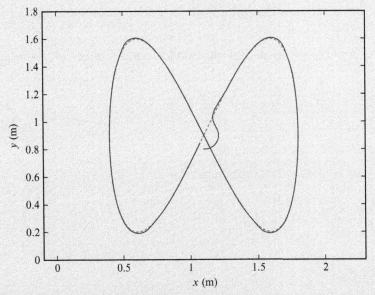

Fig. 3.16 Simple trajectory tracking control of Ackermann drive: reference path (*dashed line*) and actual path (*solid line*).

(*Continued*)

EXAMPLE 3.7—cont'd

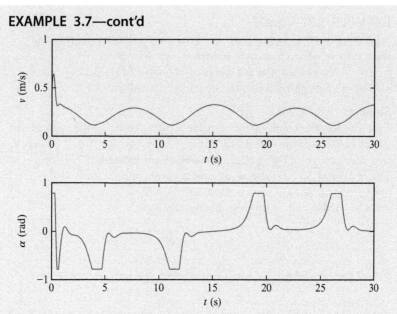

Fig. 3.17 Control signals of simple trajectory tracking control of Ackermann drive.

Listing 3.6

```
1  Ts = 0.03; % Sampling time
2  t = 0:Ts:30; % Simulation time
3  d = 0.1; % Distance between the front and the rear axis
4  q = [1.1; 0.8; 0]; % Initial robot pose
5
6  % Reference trajectory
7  freq = 2*pi/30;
8  xRef = 1.1 + 0.7*sin(freq*t);
9  yRef = 0.9 + 0.7*sin(2*freq*t);
10
11 % Control gains
12 Kphi = 2;
13 Kv = 5;
14
15 upgradedControl = true; % This setting can be changed to false
16 for k = 1:length(t)
17     % Reference
18     phiRef = atan2(yRef(k)-q(2), xRef(k)-q(1));
19     qRef = [xRef(k); yRef(k); phiRef];
20
21     % Error with respect to the (moving) reference
22     e = qRef - q;          % Error on x, y and orientation
23     e(3) = wrapToPi(e(3)); % Mapped to the [-pi, pi] interval
24
```

EXAMPLE 3.7—cont'd

```
25     % Control
26     alpha = e(3)*Kphi;          % Orientation control (basic)
27     v = sqrt(e(1)^2+e(2)^2)*Kv; % Forward-motion control (basic)
28     if upgradedControl
29          % If e(3) is not on the [-pi/2, pi/2], +/- pi should be added
30          % to e(3), and negative velocity should be commanded
31          v = v*sign(cos(e(3)));  % Changing sign of v if necessary
32          e(3) = atan(tan(e(3))); % Mapped to the [-pi/2, pi/2] interval
33          alpha = e(3)*Kphi;      % Orientation control (upgraded)
34     end
35
36     % Physical limitations of the vehicle
37     if abs(alpha)>pi/4, alpha = pi/4*sign(alpha); end
38     if abs(v)>0.8, v = 0.8*sign(v); end
39
40     % Robot motion simulation
41     dq = [v*cos(q(3)); v*sin(q(3)); v/d*tan(alpha)];
42     noise = 0.00; % Set to experiment with noise (e.g. 0.001)
43     q = q + Ts*dq + randn(3,1)*noise; % Euler integration
44     q(3) = wrapToPi(q(3)); % Map orientation angle to [-pi, pi]
45 end
```

3.3.2 Decomposing Control to Feedforward and Feedback Action

Trajectory tracking control is significant not only from a practical point of view, but also from a theoretical point of view. Its importance stems from the famous result that according to Brockett's condition nonholonomic systems cannot be asymptotically stabilized around the equilibrium using smooth time-invariant feedback [1]. This result can be checked easily for the case of the differential drive, and it can be extended to other kinematics including Ackermann kinematics. First, we start by checking whether the system is driftless. Since the derivative of the vector field $\dot{q}(t)$ is zero in the case of no control ($v = 0$, $\omega = 0$), the system is indeed driftless. Brockett [1] showed that one of the conditions for a driftless system to be stabilized by a continuous time-invariant feedback is that the number of inputs should be at least the same as the number of states. In the case of the differential drive this condition is clearly violated and other types of feedback should be sought. Nevertheless, completely nonholonomic, driftless systems are still controllable in a nonlinear sense. Therefore, asymptotic stabilization can be obtained using time-varying, discontinuous, or hybrid control laws. One of the possibilities to circumvent the limitation imposed by Brockett's condition is to introduce a different control structure. In the case of trajectory tracking control very often a two-degree-of-freedom control is used where one path corresponds to the feedforward path and the other corresponds to the feedback path from the control system's terminology.

Before introducing feedforward and feedback control, we need to define another important system property. A system is called differentially flat if there exist a set of flat outputs, and all the system states and system inputs can be rewritten as functions of these flat outputs and a finite number of their time derivatives. This means that some nonlinear functions f_x and f_u have to exist satisfying

$$x = f_x \left(z_f, \dot{z}_f, \ddot{z}_f, \ldots, \frac{d^p}{dt^p} z_f \right)$$

$$u = f_u \left(z_f, \dot{z}_f, \ddot{z}_f, \ldots, \frac{d^p}{dt^p} z_f \right)$$

where vectors x, u, and z_f represent the system states, its inputs, and flat outputs, respectively, while p is a finite integer. We should also mention that flat outputs should be functions of system states, its inputs, and a finite number of input derivatives. This means that in general flat outputs are fictitious; they do not correspond to the actual outputs.

In case of the differential drive kinematic model given by Eq. (2.2) the flat outputs are actual system outputs x and y. It can easily be shown that all the inputs (both velocities) and the third state (robot orientation) can be represented as functions of x and y, and their derivatives. We know that $\dot{x}$ and $\dot{y}$ can be interpreted as Cartesian coordinates of the robot's translational velocity. Therefore, the translational velocity is obtained by the Pythagorean sum of its component:

$$v(t) = \sqrt{\dot{x}^2(t) + \dot{y}^2(t)} \tag{3.16}$$

Due to nonholonomic constraints, a differentially driven wheeled robot always moves in the direction of its orientation, which means that the tangent of the orientation equals the quotient of Cartesian components of the translational velocity:

$$\varphi(t) = \arctan \left(\frac{\dot{y}(t)}{\dot{x}(t)} \right) \tag{3.17}$$

Finally, angular velocity $\omega(t)$ is determined as the time derivative of the orientation $\varphi(t)$ given by Eq. (3.17)

$$\omega(t) = \frac{d}{dt}\left[\arctan\left(\frac{\dot{y}(t)}{\dot{x}(t)}\right)\right] = \frac{\dot{x}(t)\ddot{y}(t) - \dot{y}(t)\ddot{x}(t)}{\dot{x}^2(t) + \dot{y}^2(t)} \qquad (3.18)$$

and is not defined only if the translational velocity becomes 0.

Cartesian coordinates of the position x and y are also the flat outputs of the rear-wheel-powered bicycle kinematic model given by Eq. (2.19). The first two equations of this kinematic model are the same as in the case of the differential drive equation (2.2), and therefore the velocity of the rear wheel and the orientation can be obtained by expressions in Eqs. (3.16), (3.17). The third equation in Eq. (2.19) is

$$\dot{\varphi} = \frac{v_r(t)}{d}\tan(\alpha(t)) \qquad (3.19)$$

from where it follows

$$\alpha(t) = \arctan\frac{d\dot{\varphi}(t)}{v_r(t)} \qquad (3.20)$$

The input $\alpha(t)$ can be rewritten by flat outputs and their derivatives by introducing the expressions in Eqs. (3.16), (3.18) into Eq. (3.20) instead of $v_r(t)$ and $\dot{\varphi}(t)$:

$$\alpha(t) = \arctan\frac{d\left(\dot{x}(t)\ddot{y}(t) - \dot{y}(t)\ddot{x}(t)\right)}{\left(\dot{x}^2(t) + \dot{y}^2(t)\right)^{(3/2)}} \qquad (3.21)$$

A very important consequence of the above analysis is that the Ackermann drive is structurally equivalent to the differential drive. To illustrate this, if a certain command $\{v(t), \omega(t)\}$ is applied to a differentially driven robot, the obtained trajectory is the same as if a command $\{v_r(t), \omega(t)\} = \{v(t), \arctan\frac{d\omega(t)}{v(t)}\}$ is applied to the robot with Ackermann drive. A large majority of the examples that follow are performed on a differentially driven robot. Those results can therefore be easily extended to robots with Ackermann drive and to some other kinematic structures as well.

If the system is flat, then all the system variables can be deduced from flat outputs without integration. A useful consequence of this fact is that based on the reference trajectory, the required control inputs can be calculated analytically. In the case of a differentially driven wheeled robot, Eqs. (3.16),

(3.18) provide formulae for obtaining reference velocities $v_{ref}(t)$ and $\omega_{ref}(t)$ from the reference trajectory given by $x_{ref}(t)$ and $y_{ref}(t)$:

$$v_{ref}(t) = \sqrt{\dot{x}_{ref}^2(t) + \dot{y}_{ref}^2(t)^2} \qquad (3.22)$$

$$\omega_{ref}(t) = \frac{\dot{x}_{ref}(t)\ddot{y}_{ref}(t) - \dot{y}_{ref}(t)\ddot{x}_{ref}(t)}{\dot{x}_{ref}^2(t) + \dot{y}_{ref}^2(t)} \qquad (3.23)$$

Note that similar formulae can also be obtained for other kinematic structures associated with flat systems.

Eqs. (3.22), (3.23) provide open-loop controls that ensure that the robot drives along the reference trajectory in the ideal case where the kinematic model of the robot exactly describes the motion, and there are no disturbances, measurement errors, and initial pose error. These assumptions are never met completely, and some feedback control is necessary. In these cases the reference velocities from Eqs. (3.22), (3.23) are used in the feedforward part of the control law while a wide spectrum of possible feedback control laws can be used. Some of them will also be discussed in the following sections.

3.3.3 Feedback Linearization

The idea of feedback linearization is to introduce some transformation (usually to the system input) that makes the system between new input and output linear. Thus any linear control design is made possible. First, we have to make sure that the system is differentially flat [6, 7]. We have shown in Section 3.3.2 that many kinematic structures are flat. Then, the procedure of designing a feedback linearization is as follows:

- The appropriate flat outputs should be chosen. Their number is the same as the number of the system inputs.
- Flat outputs should be differentiated and the obtained derivatives checked for the functional presence of inputs. This step is repeated until all the inputs (or their derivatives) appear in the flat output derivatives. If all the inputs (more precisely their highest derivatives) can be derived from this system of equations, we can proceed to the next step.
- The system of equations is solved for the highest derivatives of individual inputs. To obtain actual system inputs, a chain of integrators has to be used on the input derivatives. The output derivatives, on the other hand, serve as new inputs to the system.
- Since the newly obtained system is linear, a wide selection of possible control laws can be used on these new inputs.

In the case of a wheeled mobile robot with differential drive, the flat outputs are $x(t)$ and $y(t)$. Their first derivative according to the kinematic model (2.2) is

$$\dot{x} = v \cos \varphi$$
$$\dot{y} = v \sin \varphi$$

In the first derivatives, only the translational velocity v appears, and the differentiation continues:

$$\ddot{x} = \dot{v} \cos \varphi - v \dot{\varphi} \sin \varphi$$
$$\ddot{y} = \dot{v} \sin \varphi + v \dot{\varphi} \cos \varphi$$

In the second derivatives both velocities (v and $\omega = \dot{\varphi}$) are present. Now the system of equations is rewritten so that the second derivatives of the flat outputs are described as functions of the highest derivatives of individual inputs ($\dot{v}$ and ω in this case):

$$\begin{bmatrix} \ddot{x} \\ \ddot{y} \end{bmatrix} = \begin{bmatrix} \cos \varphi & -v \sin \varphi \\ \sin \varphi & v \cos \varphi \end{bmatrix} \begin{bmatrix} \dot{v} \\ \omega \end{bmatrix} = F \begin{bmatrix} \dot{v} \\ \omega \end{bmatrix} \qquad (3.24)$$

The matrix F has been introduced, which is nonsingular if $v \neq 0$. The system of equations can therefore be solved for $\dot{v}$ and ω:

$$\begin{bmatrix} \dot{v} \\ \omega \end{bmatrix} = F^{-1} \begin{bmatrix} \ddot{x} \\ \ddot{y} \end{bmatrix} = \begin{bmatrix} \cos \varphi & \sin \varphi \\ -\frac{\sin \varphi}{v} & \frac{\cos \varphi}{v} \end{bmatrix} \begin{bmatrix} \ddot{x} \\ \ddot{y} \end{bmatrix} \qquad (3.25)$$

The solution ω from Eq. (3.25) is the actual input to the robot, while the solution $\dot{v}$ should be integrated before it can be used as an input. The newly obtained linear system has inputs $[u_1, u_2]^T = [\ddot{x}, \ddot{y}]^T$, and the states $z = [x, y, \dot{x}, \dot{y}]^T$ (the kinematic model (2.2) has three states; the fourth one is due to an additional integrator). The dynamics of the new system can be described conveniently by the state-space representation:

$$\begin{bmatrix} \dot{x} \\ \ddot{x} \\ \dot{y} \\ \ddot{y} \end{bmatrix} = \begin{bmatrix} 0 & 1 & 0 & 0 \\ 0 & 0 & 0 & 0 \\ 0 & 0 & 0 & 1 \\ 0 & 0 & 0 & 0 \end{bmatrix} \begin{bmatrix} x \\ \dot{x} \\ y \\ \dot{y} \end{bmatrix} + \begin{bmatrix} 0 & 0 \\ 1 & 0 \\ 0 & 0 \\ 0 & 1 \end{bmatrix} \begin{bmatrix} u_1 \\ u_2 \end{bmatrix} \qquad (3.26)$$

or in a compact form as

$$\dot{z} = Az + Bu \qquad (3.27)$$

The system (3.27) is controllable because its controllability matrix,

$$Q_c = [B, AB] \qquad (3.28)$$

has full rank, and therefore the state controller exists for an arbitrarily chosen characteristic polynomial of the closed loop. An additional requirement is to design the control law so the robot will follow a reference trajectory. In the context of flat systems, a reference trajectory is given for the flat outputs, in this case $x_{ref}(t)$ and $y_{ref}(t)$. Then the reference can easily be obtained for the system state $z_{ref}(t) = [x_{ref}, \dot{x}_{ref}, y_{ref}, \dot{y}_{ref}]^T$ and the system input $u_{ref} = [\ddot{x}_{ref}, \ddot{x}_{ref}]^T$. Eq. (3.27) can also be written for the reference signals:

$$\dot{z}_{ref} = Az_{ref} + Bu_{ref} \tag{3.29}$$

The error between the actual states and the reference states is defined as $\tilde{z} = z - z_{ref}$. Subtracting Eq. (3.29) from Eq. (3.27) yields

$$\dot{\tilde{z}} = A\tilde{z} + B(u - u_{ref}) \tag{3.30}$$

Eq. (3.30) describes the dynamics of the state error. These dynamics should be stable and appropriately fast. One way to prescribe the closed-loop dynamics is to require specific closed-loop poles. We have already shown that the pair (A, B) is a controllable one, and therefore by properly selecting a constant control gain matrix K (of dimension 2×4), arbitrary locations of the closed-loop poles in the left half-plane of the complex plane s can be achieved. Eq. (3.30) can be rewritten:

$$\dot{\tilde{z}} = (A-BK)\tilde{z}+BK\tilde{z}+B(u-u_{ref}) = (A-BK)\tilde{z}+B(K\tilde{z}+u-u_{ref}) \tag{3.31}$$

If the last term in Eq. (3.31) is zero, the state errors converge to 0 with the prescribed dynamics, given by the closed-loop system matrix $(A - BK)$. Forcing this term to 0 defines the control law for this approach:

$$u(t) = K(z_{ref}(t) - z(t)) + u_{ref}(t) \tag{3.32}$$

Schematic representation of the complete control system is given in Fig. 3.18.

Due to a specific form of matrices A and B in Eq. (3.26), where u_1 only influences states z_1 and z_2 while u_2 only influences states z_3 and z_4, the control gain matrix takes a special form:

$$K = \begin{bmatrix} k_1 & k_2 & 0 & 0 \\ 0 & 0 & k_3 & k_4 \end{bmatrix} \tag{3.33}$$

The control law (3.32) can therefore be completely decomposed:

$$\begin{aligned} u_1(t) &= \ddot{x}(t) = k_1(x_{ref}(t) - x(t)) + k_2(\dot{x}_{ref}(t) - \dot{x}(t)) + \ddot{x}_{ref}(t) \\ u_2(t) &= \ddot{y}(t) = k_3(y_{ref}(t) - y(t)) + k_4(\dot{y}_{ref}(t) - \dot{y}(t)) + \ddot{y}_{ref}(t) \end{aligned} \tag{3.34}$$

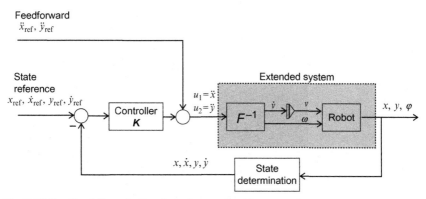

Fig. 3.18 Feedback linearization for reference tracking.

The proposed approach requires that all the states are known. While x and y are usually measured, their derivatives are not. Numerical differentiation amplifies noise and should be avoided in practice. Two possible solutions are as follows:

- The unmeasured states can be estimated by state observers.
- If the robot orientation φ is measured, the derivatives can be calculated as $\dot{x} = v\cos\varphi$, $\dot{y} = v\sin\varphi$.

A practical example of this approach is shown in Example 3.8.

EXAMPLE 3.8

A differentially driven vehicle is to be controlled to follow the reference trajectory $x_{\text{ref}} = 1.1 + 0.7\sin(\frac{2\pi t}{30})$ and $y_{\text{ref}} = 0.9 + 0.7\sin(\frac{4\pi t}{30})$. The sampling time is $T_s = 0.033$ s. The initial pose is $[x(0), y(0), \varphi(0)] = [1.1, 0.8, 0]$. Implement the algorithm presented in this section in a Matlab code and show the graphical results.

Solution

The code is shown in Listing 3.7. The results of Example 3.8 are shown in Figs. 3.19 and 3.20. In this approach the problems of periodic orientation do not appear (there is no need to map angles to the $(-\pi, \pi]$ interval). This stems from the fact that the orientation always appears within trigonometric functions that are periodic by themselves.

(Continued)

EXAMPLE 3.8—cont'd
Listing 3.7

```
1  Ts = 0.033; % Sampling time
2  t = 0:Ts:30; % Simulation time
3
4  % Reference
5  freq = 2*pi/30;
6    xRef = 1.1 + 0.7*sin(freq*t);     yRef = 0.9 + 0.7*sin(2*freq*t);
7   dxRef = freq*0.7*cos(freq*t);      dyRef = 2*freq*0.7*cos(2*freq*t);
8  ddxRef =-freq^2*0.7*sin(freq*t); ddyRef =-4*freq^2*0.7*sin(2*freq*t);
9  qRef = [xRef; yRef; atan2(dyRef, dxRef)];
10 uRef = [ddxRef; ddyRef];
11
12 q = [xRef(1)+.05; yRef(1)-0.1; 0]; % Initial robot pose
13 z1 = [q(1); dxRef(1)]; % Initial state [x, x']
14 z2 = [q(2); dyRef(1)]; % Initial state [y, y']
15 v = sqrt(z1(2)^2+z2(2)^2); % Initial state of velocity integrator
16
17 % Matrices of linearized system
18 A = [0, 1; 0, 0]; B = [0; 1]; C = [1, 0];
19 % State feedback controller
20 desPoles = [-2-1i; -2+1i]; % Desired poles (of the controller)
21 K = acker(A, B, desPoles); % Control gains obtained by pole placement
22
23 for k = 1:length(t)
24     % Reference states
25     zRef1 = [xRef(k); dxRef(k)];
26     zRef2 = [yRef(k); dyRef(k)];
27
28     % Error and control
29     ez1 = zRef1 - z1;
30     ez2 = zRef2 - z2;
31     uu = [ddxRef(k); ddyRef(k)] + [K*ez1; K*ez2];
32
33     % Compute inputs to the robot
34     F = [cos(q(3)), -v*sin(q(3)); ...
35          sin(q(3)),  v*cos(q(3))];
36     vv = F\uu; % Translational acceleration and angular velocity
37     v = v + Ts*vv(1); % Integrate translational acceleration
38     u = [v; vv(2)]; % Robot input
39
40     % Robot motion simulation
41     dq = [u(1)*cos(q(3)); u(1)*sin(q(3)); u(2)];
42     noise = 0.00; % Set to experiment with noise (e.g. 0.001)
43     q = q + Ts*dq + randn(3,1)*noise; % Euler integration
44     q(3) = wrapToPi(q(3)); % Map orientation angle to [-pi, pi]
45
46     % Take known (measured) orientation and velocity to compute states
47     z1 = [q(1); u(1)*cos(q(3))];
48     z2 = [q(2); u(1)*sin(q(3))];
49 end
```

EXAMPLE 3.8—cont'd

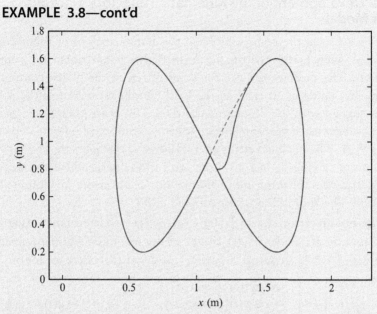

Fig. 3.19 Trajectory tracking control of differential drive based on feedback linearization from Example 3.8: reference path (*dashed line*) and actual path (*solid line*).

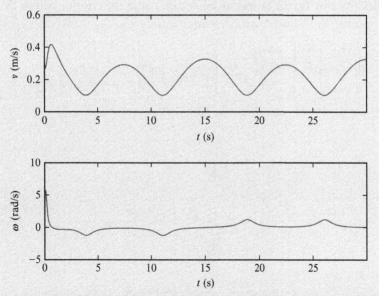

Fig. 3.20 Control signals of trajectory tracking control of differential drive based on feedback linearization from Example 3.8.

3.3.4 Development of the Kinematic Trajectory-Tracking Error Model

In order to solve the control problem the transformation of robot coordinates is often performed on the form that is better suited for control purposes. The posture of the error is not given in the global coordinate system, but rather as an error in the local coordinate system of the robot that is aligned with the driving mechanism. This error is expressed as the deviation between a virtual reference robot and the actual robot as depicted in Fig. 3.21. The obtained errors are as follows: e_x that gives the error in the direction of driving, e_y that gives the error in the perpendicular direction, and e_φ that gives the error in the orientation. These errors are illustrated in Fig. 3.21. The approach was first adopted in [8].

The posture error $e(t) = \left[e_x(t), e_y(t), e_\varphi(t)\right]^T$ is determined using the actual posture $q(t) = [x(t), y(t), \varphi(t)]^T$ of the real robot and the reference posture $q_{\text{ref}}(t) = [x_{\text{ref}}(t), y_{\text{ref}}(t), \varphi_{\text{ref}}(t)]^T$ of the virtual reference robot:

$$\begin{bmatrix} e_x(t) \\ e_y(t) \\ e_\varphi(t) \end{bmatrix} = \begin{bmatrix} \cos(\varphi(t)) & \sin(\varphi(t)) & 0 \\ -\sin(\varphi(t)) & \cos(\varphi(t)) & 0 \\ 0 & 0 & 1 \end{bmatrix} \left(q_{\text{ref}}(t) - q(t)\right) \quad (3.35)$$

Assuming that the actual and the reference robot have the same kinematic model given by Eq. (2.2) and taking into account the transformation (3.35), the posture error model can be written as follows:

$$\begin{bmatrix} \dot{e}_x \\ \dot{e}_y \\ \dot{e}_\varphi \end{bmatrix} = \begin{bmatrix} \cos e_\varphi & 0 \\ \sin e_\varphi & 0 \\ 0 & 1 \end{bmatrix} \begin{bmatrix} v_{\text{ref}} \\ \omega_{\text{ref}} \end{bmatrix} + \begin{bmatrix} -1 & e_y \\ 0 & -e_x \\ 0 & -1 \end{bmatrix} u \quad (3.36)$$

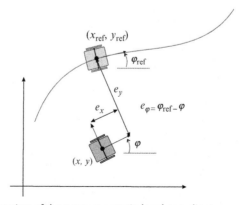

Fig. 3.21 The illustration of the posture error in local coordinates.

where v_{ref} and ω_{ref} are the linear and the angular reference velocities given by Eqs. (3.22), (3.23). The input $\boldsymbol{u} = [v, \omega]^T$ is to be commanded by the controller. Very often [9] the control $\boldsymbol{u}$ is decomposed as

$$\boldsymbol{u} = \begin{bmatrix} v \\ \omega \end{bmatrix} = \begin{bmatrix} v_{ref} \cos e_\varphi + v_{fb} \\ \omega_{ref} + \omega_{fb} \end{bmatrix} \tag{3.37}$$

where v_{fb} and ω_{fb} are the feedback signals to be defined later, while $v_{ref} \cos e_\varphi$ and ω_{ref} are the feedforward signals, although technically speaking, $v_{ref} \cos e_\varphi$ is modulated with the orientation error that originates from the output. On the other hand, when the orientation error is driven to 0, $v_{ref} \cos e_\varphi$ becomes a "true" feedforward. Inserting the control (3.37) into Eq. (3.36) results in the tracking-error model:

$$\begin{aligned}
\dot{e}_x &= \omega_{ref} e_y - v_{fb} + e_y \omega_{fb} \\
\dot{e}_y &= -\omega_{ref} e_x + v_{ref} \sin e_\varphi - e_x \omega_{fb} \\
\dot{e}_\varphi &= -\omega_{fb}
\end{aligned} \tag{3.38}$$

The control goal is to drive the errors of the error model (3.38) toward 0 by appropriately choosing the controls v_{fb} and ω_{fb}. This is the topic of the following sections.

3.3.5 Linear Controller

The error model (3.38) is nonlinear. In this section this model will be linearized to enable the use of a linear controller. The linearization has to take place around some equilibrium point. Here the obvious choice is the zero error ($e_x = e_y = 0, e_\varphi = 0$). This point is an equilibrium point of Eq. (3.38) if both feedback velocities are also 0 ($v_{fb} = 0, \omega_{fb} = 0$). Linearizing Eq. (3.38) around the zero-error yields the following:

$$\begin{bmatrix} \dot{e}_x \\ \dot{e}_y \\ \dot{e}_\varphi \end{bmatrix} = \begin{bmatrix} 0 & \omega_{ref} & 0 \\ -\omega_{ref} & 0 & v_{ref} \\ 0 & 0 & 0 \end{bmatrix} \begin{bmatrix} e_x \\ e_y \\ e_\varphi \end{bmatrix} + \begin{bmatrix} -1 & 0 \\ 0 & 0 \\ 0 & -1 \end{bmatrix} \begin{bmatrix} v_{fb} \\ \omega_{fb} \end{bmatrix} \tag{3.39}$$

which is a linear time-varying system due to $v_{ref}(t)$ and $\omega_{ref}(t)$ being time dependent.

The system given by Eq. (3.39) is a state-space representation of a dynamic error-system where all the states (errors in this case) are accessible. A full state feedback is therefore possible if the system is controllable. If it is assumed that v_{ref} and ω_{ref} are constant (the reference path consists of straight line segments and circular arcs), it is easy to confirm that the controllability matrix defined by Eq. (3.28) is of the full rank, and consequently all the errors can be forced to 0 by a static state feedback. If v_{ref} and ω_{ref} are not constant, controllability is retained if either of the reference signals is different from 0, although the analysis becomes more complicated.

Due to a special structure of the system (3.39), a static state feedback with a simple form of gain matrix is often used:

$$
\begin{bmatrix} v_{fb} \\ \omega_{fb} \end{bmatrix} = \begin{bmatrix} k_x & 0 & 0 \\ 0 & k_y & k_\varphi \end{bmatrix} \begin{bmatrix} e_x \\ e_y \\ e_\varphi \end{bmatrix} \tag{3.40}
$$

We see that the error in the direction of driving is corrected by v_{fb}, while the errors in the orientation and in the lateral directions are corrected by ω_{fb}.

Controller gains (k_x, k_y, k_φ) can be determined by trial and error, by optimizing them on a system model, by using the pole placement approach, etc. Here, the control gains are chosen such that the poles of the control system lie in appropriate locations in the complex plane s. The system has three poles of which one has to be real while the other two will be conjugate complex. First, the poles will be placed on fixed locations $-2\zeta\omega_n$, and $-\zeta\omega_n \pm \omega_n\sqrt{1 - \zeta^2}$ where the undamped natural frequency $\omega_n > 0$ and the damping coefficient $0 < \zeta < 1$ are designer parameters. If the characteristic polynomial of the closed-loop system

$$
\left| s\boldsymbol{I}_{3\times3} - \begin{bmatrix} 0 & \omega_{ref} & 0 \\ -\omega_{ref} & 0 & v_{ref} \\ 0 & 0 & 0 \end{bmatrix} - \begin{bmatrix} -1 & 0 \\ 0 & 0 \\ 0 & -1 \end{bmatrix} \begin{bmatrix} k_x & 0 & 0 \\ 0 & k_y & k_\varphi \end{bmatrix} \right|
$$

is compared to the desired one,

$$
(s + 2\zeta\omega_n)(s^2 + 2\zeta\omega_n s + \omega_n^2)
$$

the solution for control gains can be obtained [6]:

$$
k_x = k_\varphi = 2\zeta\omega_n
$$
$$
k_y(t) = \frac{\omega_n^2 - \omega_{ref}^2(t)}{v_{ref}(t)} \tag{3.41}
$$

Note that ω_n should be larger than the maximum of $|\omega_{ref}(t)|$. Control gains given by Eq. (3.41) are not practically applicable because $k_y(t)$ becomes extremely large when the reference velocity $v_{ref}(t)$ is low. To overcome this problem, the undamped natural frequency ω_n will become time-varying. Since it is natural to adapt the transient response settling times to the reference velocities, the following choice seems sensible: $\omega_n(t) = \sqrt{\omega_{ref}^2(t) + g v_{ref}^2(t)}$, $g > 0$. Repeating a similar procedure as above, the following control gains are obtained:

$$k_x(t) = k_\varphi(t) = 2\zeta \sqrt{\omega_{ref}^2(t) + g v_{ref}^2(t)}$$
$$k_y(t) = g v_{ref}(t) \tag{3.42}$$

Two remarks are important in the context of control algorithms shown in this section:

- The control laws are designed based on the linearized models. A linearized model is only valid in the vicinity of the operating point (zero error in this case), and the control performance might not be as expected in case of large control errors.
- Even if we deal with a linear but time-varying system, some of the results known from linear time invariant systems are no longer valid. Here we should mention that even if all the poles lie in the (fixed) locations of the left half-plane of the complex plane s, the system might be unstable.

In spite of the potential difficulties mentioned above, linear control laws are often used in practice due to their simplicity, relatively easy tuning, and acceptable performance and robustness. A simulated example of the use is given below.

EXAMPLE 3.9

A differentially driven vehicle is to be controlled to follow the reference trajectory $x_{ref} = 1.1 + 0.7 \sin(\frac{2\pi t}{30})$ and $y_{ref} = 0.9 + 0.7 \sin(\frac{4\pi t}{30})$. Sampling time is $T_s = 0.033$ s. The initial pose is $[x(0), y(0), \varphi(0)] = [1.1, 0.8, 0]$. Implement the algorithm presented in this section with the control gains given by Eq. (3.41) in a Matlab code and show the graphical results.

Solution

The code is given in Listing 3.8. Simulation results are shown in Figs. 3.22 and 3.23 where good tracking can be observed.

(*Continued*)

EXAMPLE 3.9—cont'd

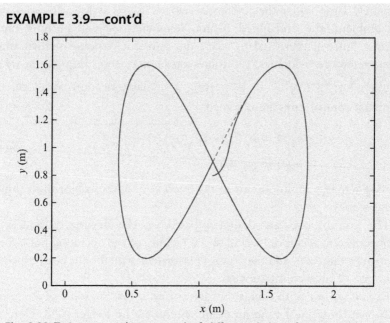

Fig. 3.22 Trajectory tracking control of differential drive from Example 3.9: reference path (*dashed line*) and actual path (*solid line*).

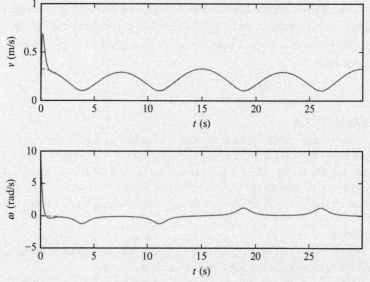

Fig. 3.23 Control signals of trajectory tracking control of differential drive from Example 3.9.

EXAMPLE 3.9—cont'd

Listing 3.8

```
1  Ts = 0.033; % Sampling time
2  t = 0:Ts:30; % Simulation time
3  q = [1.1; 0.8; 0]; % Initial robot pose
4
5  % Reference
6  freq = 2*pi/30;
7    xRef = 1.1 + 0.7*sin(freq*t);     yRef = 0.9 + 0.7*sin(2*freq*t);
8   dxRef = freq*0.7*cos(freq*t);      dyRef = 2*freq*0.7*cos (2*freq*t);
9  ddxRef =-freq^2*0.7*sin(freq*t); ddyRef =-4*freq^2*0.7* sin(2*freq*t);
10 qRef = [xRef; yRef; atan2(dyRef, dxRef)]; % Reference trajectory
11 vRef = sqrt(dxRef.^2+dyRef.^2);
12 wRef = (dxRef.*ddyRef-dyRef.*ddxRef)./(dxRef.^2+dyRef.^2);
13 uRef = [vRef; wRef]; % Reference inputs
14
15 for k = 1:length(t)
16     e = [cos(q(3)), sin(q(3)), 0; ...
17          -sin(q(3)), cos(q(3)), 0; ...
18          0,          0,         1]*(qRef(:,k) - q); % Error vector
19     e(3) = wrapToPi(e(3)); % Correct angle
20
21     % Current reference inputs
22     vRef = uRef(1,k);
23     wRef = uRef(2,k);
24
25     % Control
26     eX = e(1); eY=e(2); ePhi=e(3);
27     zeta = 0.9; % Experiment with this control design parameter
28     g = 85;     % Experiment with this control design parameter
29     Kx = 2*zeta*sqrt(wRef^2+g*vRef^2);
30     Kphi = Kx;
31     Ky = g*vRef;
32     % Gains can also be constant e.g.: Kx = Kphi = 3; Ky = 30;
33
34     % Control: feedforward and feedback
35     v = vRef*cos(e(3))+ Kx*e(1);
36     w = wRef + Ky*e(2) + Kphi*e(3);
37
38     % Robot motion simulation
39     dq = [v*cos(q(3)); v*sin(q(3)); w];
40     noise = 0.00; % Set to experiment with noise (e.g. 0.001)
41     q = q + Ts*dq + randn(3,1)*noise; % Euler integration
42     q(3) = wrapToPi(q(3)); % Map orientation angle to [-pi, pi]
43 end
```

3.3.6 Lyapunov-Based Control Design

We have already mentioned that the error model (3.38) is inherently nonlinear. Nonlinear systems are best controlled using a nonlinear controller that takes into account all the properties of the original system during control design. Theory based on Lyapunov functions is often used for nonlinear system stabilization problems. In our case, the (asymptotic) stability of the error model (3.38) will be analyzed in the presence of different control laws.

Lyapunov Stability

The second method of Lyapunov will be introduced briefly. The method gives sufficient conditions for (asymptotic) stability of equilibrium points of a nonlinear dynamical system given by $\dot{x} = f(x)$, $x \in \mathbb{R}$. First, it is assumed that the equilibrium lies at $x = 0$. The approach is based on positive definite scalar functions $V(x)$: $\mathbb{R}^n \to \mathbb{R}$ that fulfill the following: $V(x) = 0$ if $x = 0$, and $V(x) > 0$ if $x \neq 0$. The stability of the equilibrium point is analyzed by checking the derivative of the function V. It is very important that the derivative is obtained along the solutions of the differential equation of the system:

$$\dot{V} = \frac{\partial V}{\partial x}\dot{x} = \frac{\partial V}{\partial x}f(x) \qquad (3.43)$$

If $\dot{V} \leq 0$ ($\dot{V}$ is negative semidefinite), the equilibrium is (locally) stable. If $\dot{V} < 0$ except at $x = 0$ ($\dot{V}$ is negative definite), the equilibrium is (locally) asymptotically stable. If $\lim_{|x|\to\infty} V(x) = \infty$, the results are global. The approach is therefore based on searching for the functions with the above properties that are referred to as Lyapunov functions. Very often a quadratic Lyapunov function candidate is chosen, and if we are able to show its derivative is negative or at least zero, stability of the system is concluded.

A classical interpretation of Lyapunov functions is based on the system energy point of view. If system energy is used as a Lyapunov function, and the system is dissipative, the energy in the system cannot increase (the derivative of the function is nonpositive). Consequently, all the signals remain bounded, and the system stability can be confirmed. But it needs to be stressed that a Lyapunov function might not be related to system energy. Above all, "system stability" is an incorrect term in the nonlinear framework. Rather, stability of equilibrium points, or more generally of invariant sets, is to be analyzed. It is easy to find the systems where stable and unstable equilibria coexist.

Control Design in the Lyapunov Stability Framework

We shall show in the following how the Lyapunov stability theorem can be used for control design purposes. Our nonlinear system (3.38) has three states with equilibrium points at $e = 0$. We would like to design a control that would make this point stable and if possible asymptotically stable, which means that all the trajectories would eventually converge to the reference trajectory and stay there forever. The most obvious Lyapunov function candidate is a sum of three squared errors:

$$V(e) = \frac{k_y}{2}(e_x^2 + e_y^2) + \frac{1}{2}e_\varphi^2 \qquad (3.44)$$

which can be interpreted as a weighted sum of the squared error in distance and the squared error in orientation. A positive constant k_y has to be added because of different units, but it will later be shown that this constant plays an important role in the control law design. The time derivative of V is

$$\dot{V} = k_y e_x \dot{e}_x + k_y e_y \dot{e}_y + e_\varphi \dot{e}_\varphi \qquad (3.45)$$

but this derivative has to be evaluated along the solutions of Eq. (3.38), which means that the error derivatives from Eq. (3.38) should be introduced

$$\dot{V}(e) = k_y e_x \left(\omega_{\text{ref}} e_y - v_{\text{fb}} + e_y \omega_{\text{fb}}\right) + k_y e_y \left(-\omega_{\text{ref}} e_x + v_{\text{ref}} \sin e_\varphi - e_x \omega_{\text{fb}}\right) + e_\varphi \left(-\omega_{\text{fb}}\right)$$
$$= -k_y e_x v_{\text{fb}} + k_y v_{\text{ref}} e_y \sin e_\varphi - e_\varphi \omega_{\text{fb}}$$

$$(3.46)$$

The idea of Lyapunov-based control design is to make the derivative of the Lyapunov function negative by choosing the control laws properly. It is therefore obvious how the control could be constructed in this case. The linear velocity v_{fb} will make the first term on the right-hand side in Eq. (3.46) negative (by completing it to the negative square), while the angular velocity ω_{fb} will cancel the second term in Eq. (3.46) and will make the third one negative (by completing it to the negative square). The control law that achieves that is

$$v_{\text{fb}} = k_x e_x$$
$$\omega_{\text{fb}} = k_y v_{\text{ref}} \frac{\sin e_\varphi}{e_\varphi} e_y + k_\varphi e_\varphi \qquad (3.47)$$

This control law is a known and well-established control law from the literature [6, 10]. Now introducing the proposed control (3.47), $\dot{V}$ becomes

$$\dot{V} = -k_x k_y e_x^2 - k_\varphi e_\varphi^2 \qquad (3.48)$$

Control gains are positive, and it will later be shown that k_x and k_φ can be arbitrary, uniformly continuous positive functions, while k_y is required to be a positive constant. The derivative of the Lyapunov function is clearly nonpositive, but since it is evaluated to zero at $e_x = 0$, $e_\varphi = 0$, irrespective of e_y, the derivative is negative semidefinite, and the equilibrium is stable. This result means that the error will stay bounded, but its convergence to 0 has not been proven.

The analysis of error convergence is considerably more difficult. In order to cope with it, we first need to introduce some additional mathematical tools. Signal norms will play an important role. The $\mathcal{L}_p$ norm of a function $x(t)$ is defined as

$$\|x\|_p = \left(\int_0^\infty |x(\tau)|^p \, d\tau \right)^{1/p} \tag{3.49}$$

where $|\cdot|$ is the vector (scalar) length. If the above integral exists (is finite), the function $x(t)$ is said to belong to $\mathcal{L}_p$. Limiting p toward infinity provides a very important class of functions $\mathcal{L}_\infty$: bounded functions.

Two very well-known lemmas will be used to prove the stability of control laws. The first one is Barbălat's lemma and the other one is a derivation of Barbălat's lemma. Both lemmas are taken from [11] and are given below for the sake of completeness.

Lemma 3.1 (Barbălat's Lemma). *If* $\lim_{t \to \infty} \int_0^t f(\tau) d\tau$ *exists and is finite, and* $f(t)$ *is a uniformly continuous function, then* $\lim_{t \to \infty} f(t) = 0$.

Lemma 3.2. *If* $f, \dot{f} \in \mathcal{L}_\infty$ *and* $f \in \mathcal{L}_p$ *for some* $p \in [1, \infty)$, *then* $f(t) \to 0$ *as* $t \to \infty$.

Now, we are ready to address the problem of convergence of the errors in Eq. (3.38). Due to Eq. (3.48), $\dot{V} \leq 0$, and therefore the Lyapunov function is nonincreasing and thus has the limit $\lim_{t \to \infty} V(t)$. Consequently, the states of Eq. (3.38) are bounded:

$$e_x, e_y, e_\varphi \in \mathcal{L}_\infty \tag{3.50}$$

Additionally, it follows from Eq. (3.47) that the control signals are bounded, and from Eq. (3.38) that the derivatives of the errors are bounded:

$$v_{\text{fb}}, \omega_{\text{fb}}, \dot{e}_x, \dot{e}_y, \dot{e}_\varphi \in \mathcal{L}_\infty \tag{3.51}$$

where we also took into account that v_{ref}, ω_{ref}, k_x, and k_φ are bounded. The latter is true in the case of smooth reference trajectories $(x_{\text{ref}}, y_{\text{ref}}, \varphi_{\text{ref}})$.

In order to show asymptotic stability of Eq. (3.38), let us first calculate the following integral of $\dot{V}$ from Eq. (3.48):

$$\int_0^\infty \dot{V} dt = V(\infty) - V(0) = - \int_0^\infty k_x k_y e_x^2 dt - \int_0^\infty k_\varphi e_\varphi^2 dt \tag{3.52}$$

Since V is a positive definite function, the following inequality holds:

$$V(0) \geq \int_0^\infty k_x k_y e_x^2 \, dt + \int_0^\infty k_\varphi e_\varphi^2 \, dt \geq \underline{k}_x k_y \int_0^\infty e_x^2 \, dt + \underline{k}_\varphi \int_0^\infty e_\varphi^2 \, dt \quad (3.53)$$

where the lower bounds of functions $k_x(t)$ and $k_\varphi(t)$ have been introduced:

$$k_x(t) \geq \underline{k}_x > 0$$
$$k_\varphi(t) \geq \underline{k}_\varphi > 0 \qquad (3.54)$$

It follows from Eq. (3.53) that the errors $e_x(t)$ and $e_\varphi(t)$ belong to $\mathcal{L}_2$. Based on Lemma 3.2 it is easy to show that the errors $e_x(t)$ and $e_\varphi(t)$ converge to 0. Since the limit $\lim_{t \to \infty} V(t)$ exists, then also $\lim_{t \to \infty} e_y(t)$ exists.

We have seen that the convergence of the errors $e_x(t)$ and $e_\varphi(t)$ to 0 is relatively simple to show and the conditions for convergence are relatively mild—control gains and reference trajectories have to be bounded. Convergence of e_y to 0 is more difficult to show, and the requirements are much more difficult to achieve, as is shown next. Besides the control gains being uniformly continuous, the reference velocities have to be persistently exciting, which means that either v_{ref} or ω_{ref} should not limit to 0. Two cases will therefore be distinguished. In the first one, $v_{ref} \nrightarrow 0$ will be assumed, and in the second, $\omega_{ref} \nrightarrow 0$ will be assumed.

Now, it will be assumed that $\lim_{t \to \infty} v_{ref}(t) \neq 0$. Applying Lemma 3.1 on $\dot{e}_\varphi(t)$ from Eq. (3.38) ensures that $\lim_{t \to \infty} \dot{e}_\varphi(t) = 0$ since $\lim_{t \to \infty} e_\varphi(t)$ exists and is finite and $\dot{e}_\varphi(t)$ is uniformly continuous. The latter is true due to Eq. (3.38) if ω_{fb} is uniformly continuous. The easiest way to check the uniform continuity of $f(t)$ on $[0, \infty)$ is to see if $f, \dot{f} \in \mathcal{L}_\infty$. It has already been shown that e_y and e_φ are uniformly continuous, while the control gain k_φ and the reference velocity v_{ref} are uniformly continuous by the assumption, so it follows from Eq. (3.47) that $\dot{e}_\varphi(t)$ is also uniformly continuous. The statement $\lim_{t \to \infty} \dot{e}_y(t) = 0$ (which is identical to $\lim_{t \to \infty} \omega_{fb}(t) = 0$) has therefore been proven. The convergence of e_y to 0 follows from Eq. (3.47):

$$e_\varphi \to 0, \quad k_\varphi \in \mathcal{L}_\infty, \quad \omega_{fb} \to 0 \Rightarrow k_y v_{ref} \frac{\sin e_\varphi}{e_\varphi} e_y \to 0$$
$$k_y v_{ref} \frac{\sin e_\varphi}{e_\varphi} e_y \to 0, \quad \frac{\sin e_\varphi}{e_\varphi} \to 1, \quad k_y > 0, \quad v_{ref} \to 0 \Rightarrow e_y \to 0 \qquad (3.55)$$

Now, it will be assumed that $\lim_{t\to\infty} \omega_{ref}(t) \neq 0$. Again one has to guarantee that $\lim_{t\to\infty} \omega_{fb} = 0$. This is true if v_{ref} and k_φ are uniformly continuous, as shown before. Then Barbălat's lemma (Lemma 3.1) is applied on $\dot{e}_x$ in Eq. (3.38). It has already been shown that e_x, e_y, and ω_{fb} are uniformly continuous; v_{fb} is uniformly continuous since k_x is uniformly continuous by the assumption; and ω_{ref} is also continuous from the assumption of this paragraph. This proves the statement $\lim_{t\to\infty} \dot{e}_x(t) = 0$. Like in the last paragraph, it can be concluded that the last two terms in Eq. (3.38) for $\dot{e}_x$ go to 0 as t goes to infinity. Consequently, the product $\omega_{ref} e_y$ also goes to 0. Since ω_{ref} is persistently exciting and does not go to 0, e_y has to go to 0.

Once again it should be stressed that for convergence of e_x and e_φ, only boundedness of v_{ref} or ω_{ref} is required. A considerably more difficult task is to drive e_y to 0. This is achieved by persistent excitation from either v_{ref} or ω_{ref}. All the results are valid globally, which means that the convergence is guaranteed irrespective of the initial pose.

EXAMPLE 3.10
A differentially driven vehicle is to be controlled to follow the reference trajectory $x_{ref} = 1.1 + 0.7\sin(\frac{2\pi t}{30})$ and $y_{ref} = 0.9 + 0.7\sin(\frac{4\pi t}{30})$. Sampling time is $T_s = 0.033\,\text{s}$. The initial pose is $[x(0), y(0), \varphi(0)] = [1.1, 0.8, 0]$. Implement the control algorithm presented in this section in a Matlab code, experiment with control gains, and show the graphical results.

Solution
The code is given in Listing 3.9. Simulation results of Example 3.10 are shown in Figs. 3.24 and 3.25 where good tracking can be observed. Note that any positive functions can be used for $k_x(t)$ and $k_\varphi(t)$. Here the functions that are chosen make the linearized model of the system the same as in the case of the linear tracking controller (Example 3.9). Control laws are not the same, but they become the same in the limit case ($e_\varphi \to 0$). The result of this choice is that the shape of transients is the same near the reference trajectory irrespective of the reference velocities.

EXAMPLE 3.10—cont'd

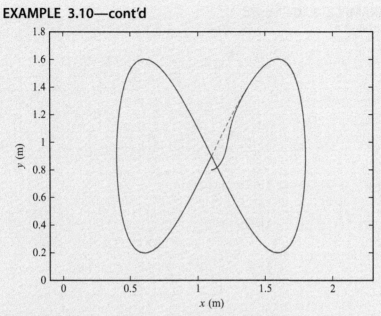

Fig. 3.24 Nonlinear trajectory tracking control of differential drive from Example 3.9: reference path (*dashed line*) and actual path (*solid line*).

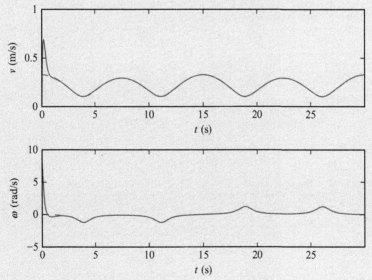

Fig. 3.25 Control signals of nonlinear trajectory tracking control of differential drive from Example 3.9.

(*Continued*)

EXAMPLE 3.10—cont'd

Listing 3.9

```
1  Ts = 0.033; % Sampling time
2  t = 0:Ts:30; % Simulation time
3  q = [1.1; 0.8; 0]; % Initial robot pose
4
5  % Reference
6  freq = 2*pi/30;
7    xRef = 1.1 + 0.7*sin(freq*t);    yRef = 0.9 + 0.7*sin(2*freq*t);
8   dxRef = freq*0.7*cos(freq*t);    dyRef = 2*freq*0.7* cos(2*freq*t);
9  ddxRef =-freq^2*0.7*sin(freq*t); ddyRef =-4*freq^2*0.7* sin(2*freq*t);
10 qRef = [xRef; yRef; atan2(dyRef, dxRef)]; % Reference trajectory
11 vRef = sqrt(dxRef.^2+dyRef.^2);
12 wRef = (dxRef.*ddyRef-dyRef.*ddxRef)./(dxRef.^2+dyRef.^2);
13 uRef = [vRef; wRef]; % Reference inputs
14
15 for k = 1:length(t)
16     e = [cos(q(3)), sin(q(3)), 0; ...
17          -sin(q(3)), cos(q(3)), 0; ...
18           0,          0,          1]*(qRef(:,k) - q); % Error vector
19     e(3) = wrapToPi(e(3)); % Correct angle
20
21     % Current reference inputs
22     vRef = uRef(1,k);
23     wRef = uRef(2,k);
24
25     % Control
26     zeta = 0.9; % Experiment with this control design parameter
27     g = 85;     % Experiment with this control design parameter
28     Kx = 2*zeta*sqrt(wRef^2 + g*vRef^2);
29     Kphi = Kx;
30     Ky = g;
31     % Gains Kx and Kphi could also be constant.
32     % This form is used to have the same damping of the transient
33     % irrespective of ref. velocities.
34
35     % Control: feedforward and feedback
36     v = vRef*cos(e(3)) + Kx*e(1);
37     w = wRef + Ky*vRef*sinc(e(3)/pi)*e(2) + Kphi*e(3);
38
39     % Robot motion simulation
40     dq = [v*cos(q(3)); v*sin(q(3)); w];
41     noise = 0.00; % Set to experiment with noise (e.g. 0.001)
42     q = q + Ts*dq + randn(3,1)*noise; % Euler integration
43     q(3) = wrapToPi(q(3)); % Map orientation angle to [-pi, pi]
44 end
```

Periodic Control Law Design

The problem of tracking is clearly periodic with respect to the orientation. This can be observed from the kinematic model by using an arbitrary control input and an arbitrary initial condition, resulting in a certain robot trajectory. If the same control input is applied to the robot and the initial condition only differs from the previous one by a multiple of 2π, the same response is obtained for $x(t)$ and $y(t)$, while $\varphi(t)$ differs from the previous solution for the same multiple of 2π. The periodic nature should also be reflected in the control law used for the tracking. This should mean that one searches for a control law that is periodic with respect to the error in the orientation e_φ (the period is 2π) and ensures the convergence of the posture error e to one of the points $\begin{bmatrix} 0 & 0 & 2k\pi \end{bmatrix}^T$ ($k \in \mathbb{Z}$). Thus, all the usual problems with orientation mapping to the $(-\pi, \pi]$ are alleviated. These problems can become critical around ± 180 degree in certain applications, for example, when using an observer to estimate the robot pose from the delayed measurements. Note that certain control laws are periodic in the sense discussed above, for example, feedback linearization control law given by Eqs. (3.25), (3.34) is periodic.

Obviously, the functions that are used in the section for the convergence analysis should also be periodic in e_φ. This means that these functions have multiple local minima and therefore do not satisfy the properties of the classic Lyapunov functions. Although the stability analysis resembles Lyapunov's direct method (the second method of Lyapunov), the convergence is not proven by this stability theory because the convergence of e to zero is not needed in our approach. Nevertheless, the functions used in this section for the convergence analysis will still be referred to as "Lyapunov functions."

Our goal is to bring the position error to zero, while the orientation error should converge to any multiple of 2π. In order to do this, a Lyapunov function that is periodic with respect to e_φ (with a natural period of 2π) will be used. First, the concept will be shown on one Lyapunov function, and later this will be extended to a more general case. The first Lyapunov-function candidate is chosen as

$$V = \frac{k_y}{2}\left(e_x^2 + e_y^2\right) + \frac{1}{2}\left(\frac{\tan\frac{e_\varphi}{2}}{\frac{1}{2}}\right)^2 \tag{3.56}$$

where k_y is a positive constant. Its derivative along the solutions of Eq. (3.38) is the following:

$$\dot{V} = k_y e_x \left(\omega_{\text{ref}} e_y - v_{\text{fb}} + e_y \omega_{\text{fb}} \right)$$

$$+ k_y e_y \left(-\omega_{\text{ref}} e_x + v_{\text{ref}} \sin e_\varphi - e_x \omega_{\text{fb}} \right) - 2 \frac{\tan \frac{e_\varphi}{2}}{\cos^2 \frac{e_\varphi}{2}} \omega_{\text{fb}}$$

$$= -k_y e_x v_{\text{fb}} + k_y v_{\text{ref}} e_y \sin e_\varphi - 2 \frac{\tan \frac{e_\varphi}{2}}{\cos^2 \frac{e_\varphi}{2}} \omega_{\text{fb}} \tag{3.57}$$

If the following control law is applied

$$v_{\text{fb}} = k_x e_x$$

$$\omega_{\text{fb}} = k_y v_{\text{ref}} e_y \cos^4 \frac{e_\varphi}{2} + k_\varphi \sin e_\varphi \tag{3.58}$$

where k_x and k_φ are positive bounded functions, the derivative $\dot{V}$ from Eq. (3.57) becomes

$$\dot{V} = -k_x k_y e_x^2 - k_\varphi \left(\frac{\tan \frac{e_\varphi}{2}}{\frac{1}{2}} \right)^2 \tag{3.59}$$

Following the same lines as in the analysis of the control law (3.47) it can easily be concluded that e_x and $\tan \frac{e_\varphi}{2}$ converge to 0 (this means that $e_\varphi \to 2k\pi$, $k \in \mathbb{Z}$) in the case of bounded control gains and a bounded trajectory. The convergence of e_y to 0 can also be concluded after a lengthy analysis if the same conditions are met as in the case of the control law (3.47).

EXAMPLE 3.11

A differentially driven vehicle is to be controlled to follow the reference trajectory $x_{\text{ref}} = 1.1 + 0.7 \sin(\frac{2\pi t}{30})$ and $y_{\text{ref}} = 0.9 + 0.7 \sin(\frac{4\pi t}{30})$. Sampling time is $T_s = 0.033$ s. The initial pose is $[x(0), y(0), \varphi(0)] = [1.1, 0.8, 0]$. Implement the control algorithm presented in this section in a Matlab code, experiment with control gains, and show the graphical results.

Solution

The code is given in Listing 3.10. Simulation results are shown in Figs. 3.26 and 3.27, where good tracking can be observed.

EXAMPLE 3.11—cont'd

Listing 3.10

```
1  Ts = 0.033; % Sampling time
2  t = 0:Ts:30; % Simulation time
3  q = [1.1; 0.8; 0]; % Initial robot pose
4
5  % Reference
6  freq = 2*pi/30;
7   xRef = 1.1 + 0.7*sin(freq*t);      yRef = 0.9 + 0.7*sin(2*freq*t);
8   dxRef = freq*0.7*cos(freq*t);      dyRef = 2*freq*0.7*cos(2*freq*t);
9  ddxRef =-freq^2*0.7*sin(freq*t); ddyRef =-4*freq^2*0.7*sin(2*freq*t);
10 qRef = [xRef; yRef; atan2(dyRef, dxRef)]; % Reference trajectory
11 vRef = sqrt(dxRef.^2+dyRef.^2);
12 wRef = (dxRef.*ddyRef-dyRef.*ddxRef)./(dxRef.^2+dyRef.^2);
13 uRef = [vRef; wRef]; % Reference inputs
14
15 for k = 1:length(t)
16     e = [cos(q(3)), sin(q(3)), 0; ...
17         -sin(q(3)), cos(q(3)), 0; ...
18          0,         0,         1]*(qRef(:,k) - q); % Error vector
19     e(3) = wrapToPi(e(3)); % Correct angle
20
21     % Current reference inputs
22     vRef = uRef(1,k);
23     wRef = uRef(2,k);
24
25     % Control
26     eX = e(1); eY = e(2); ePhi = e(3);
27     zeta = 0.9; % Experiment with this control design parameter
28     g = 85;     % Experiment with this control design parameter
29     Kx = 2*zeta*sqrt(wRef^2+g*vRef^2);
30     Kphi = Kx;
31     Ky = g;
32     % Gains Kx and Kphi could also be constant.
33     % This form is used to have the same damping of the transient
34     % irrespective of ref. velocities.
35
36     % Control: feedforward and feedback
37     v = vRef*cos(e(3)) + Kx*eX;
38     w = wRef + Ky*vRef*(cos(ePhi/2))^4*eY + Kphi*sin(ePhi);
39
40     % Robot motion simulation
41     dq = [v*cos(q(3)); v*sin(q(3)); w];
42     noise = 0.00; % Set to experiment with noise (e.g. 0.001)
43     q = q + Ts*dq + randn(3,1)*noise; % Euler integration
44     q(3) = wrapToPi(q(3)); % Map orientation angle to [-pi, pi]
45 end
```

(Continued)

EXAMPLE 3.11—cont'd

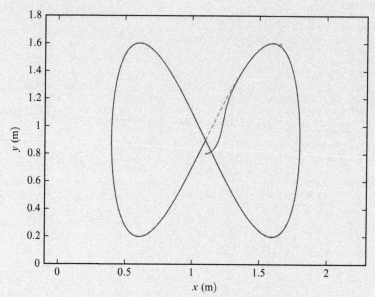

Fig. 3.26 Nonlinear trajectory tracking control of differential drive from Example 3.11: reference path (*dashed line*) and actual path (*solid line*).

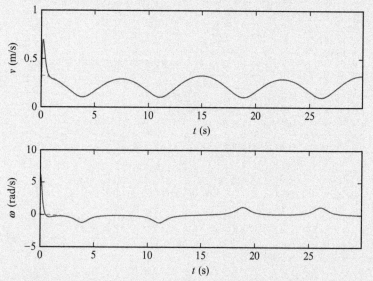

Fig. 3.27 Control signals of nonlinear trajectory tracking control of differential drive from Example 3.11.

A well-known control law by Kanayama et al. [9] can also be analyzed in the proposed framework. Choosing the Lyapunov function as

$$V = \frac{k_y}{2}\left(e_x^2 + e_y^2\right) + \frac{1}{2}\left(\frac{\sin\frac{e_\varphi}{2}}{\frac{1}{2}}\right)^2 \tag{3.60}$$

and using the control law proposed in [9] (there was also a third factor $|v_{\text{ref}}|$ in the second term of ω_{fb} that can also be included in k_φ),

$$\begin{aligned} v_{\text{fb}} &= k_x e_x \\ \omega_{\text{fb}} &= k v_{\text{ref}} e_y + k_\varphi \sin e_\varphi \end{aligned} \tag{3.61}$$

results in a stable error-system where the convergence of all the errors can be shown under the same conditions as before. Note that beside stable equilibria at $e_\varphi = 2k\pi$, $k \in \mathbb{Z}$, there is also an unstable or repelling equilibrium at $e_\varphi = (2k+1)\pi$, $k \in \mathbb{Z}$.

A framework for periodic control law design is presented in [12]. Perhaps it is worth mentioning that it is quite simple to extend the proposed techniques to the control design for symmetric vehicles that can move in the forward and the backward direction during normal operation. In this case Lyapunov functions should be periodic with a period of π on e_φ.

Four-State Error Model of the System

In this section we will tackle the same problem as in the previous section, that is, from the control point of view we often want to track any robot pose that is different from the reference one for a multiple of 360 degree. The model (3.38) does not make this problem easier because the orientation error should be usually driven to 0 using Eq. (3.38). In this section a kinematic model of the system is presented where all the poses that differ in orientation for a multiple of 360 degree are presented as one. This can be achieved by extended the state vector for one element. The variable $\varphi(t)$ from the original kinematic model (2.2) is exchanged by two new variables, $s(t) = \sin(\varphi(t))$ and $c(t) = \cos(\varphi(t))$. Their derivatives are

$$\begin{aligned} \dot{s}(t) &= \cos(\varphi(t))\dot{\varphi}(t) = c(t)\omega(t) \\ \dot{c}(t) &= -\sin(\varphi(t))\dot{\varphi}(t) = -s(t)\omega(t) \end{aligned} \tag{3.62}$$

The new kinematic model is then obtained:

$$\dot{q} = \begin{bmatrix} \dot{x} \\ \dot{y} \\ \dot{s} \\ \dot{c} \end{bmatrix} = \begin{bmatrix} c & 0 \\ s & 0 \\ 0 & c \\ 0 & -s \end{bmatrix} \begin{bmatrix} v \\ \omega \end{bmatrix} \tag{3.63}$$

The new error states are defined as follows:

$$e_x = c(x_{ref} - x) + s(y_{ref} - y)$$
$$e_y = -s(x_{ref} - x) + c(y_{ref} - y)$$
$$e_s = \sin(\varphi_{ref} - \varphi) = c \sin \varphi_{ref} - s \cos \varphi_{ref}$$
$$e_{cos} = \cos(\varphi_{ref} - \varphi) = c \cos \varphi_{ref} + s \sin \varphi_{ref}$$

(3.64)

After the differentiation of Eq. (3.64) and some manipulations, the following system is obtained:

$$\dot{e}_x = v_{ref}e_{cos} - v + e_y\omega$$
$$\dot{e}_y = v_{ref}e_s - e_x\omega$$
$$\dot{e}_s = \omega_{ref}e_{cos} - e_{cos}\omega$$
$$\dot{e}_{cos} = -\omega_{ref}e_s + e_s\omega$$

(3.65)

Like in Eq. (3.37), $v = v_{ref}e_{cos} + v_{fb}$ and $\omega = \omega_{ref} + \omega_{fb}$ will be used in the control law. The control goal is to drive e_x, e_y, and e_s to 0. The variable e_{cos} is obtained as the cosine of the error in the orientation and should be driven to 1. This is why a new error will be defined as $e_c = e_{cos} - 1$ and the final error model of the system is now

$$\dot{e}_x = \omega_{ref}e_y - v_{fb} + e_y\omega_{fb}$$
$$\dot{e}_y = -\omega_{ref}e_x + v_{ref}e_s - e_x\omega_{fb}$$
$$\dot{e}_s = -e_c\omega_{fb} - \omega_{fb}$$
$$\dot{e}_c = e_s\omega_{fb}$$

(3.66)

A controller that achieves asymptotic stability of the error model (3.66) will be developed based on a Lyapunov approach. A very straightforward idea would be to use a Lyapunov function of the type

$$V_0 = \frac{k}{2} \left(e_x^2 + e_y^2 \right) + \frac{1}{2} \left(e_s^2 + e_c^2 \right)$$

(3.67)

Interestingly, this Lyapunov function results in the control law (3.61). However, a slightly more complex function will be proposed here, which also includes the function (3.67) as a special case. The following Lyapunov-function candidate is proposed to achieve the control goal:

$$V = \frac{k}{2} \left(e_x^2 + e_y^2 \right) + \frac{1}{2 \left(1 + \frac{e_c}{a} \right)} \left(e_s^2 + e_c^2 \right)$$

(3.68)

where $k > 0$ and $a > 2$ are constants. Note that the range of the function $e_c = \cos(\varphi_{ref} - \varphi) - 1$ is $[-2, 0]$, and therefore

$$0 < \frac{a-2}{a} \le 1 + \frac{e_c}{a} \le 1$$

$$1 \le \frac{1}{1 + \frac{e_c}{a}} \le \frac{a}{a-2} \tag{3.69}$$

Due to Eq. (3.69) the function V in Eq. (3.68) is lower-bounded by the function V_0 in Eq. (3.67), and V fulfills the conditions for the Lyapunov function. The role of $(1 + \frac{e_c}{a})$ will be explained later on. The function V can be simplified by using the following:

$$e_s^2 + e_c^2 = e_s^2 + (e_{cos} - 1)^2 = 2 - 2e_{cos} = -2e_c \tag{3.70}$$

Taking into account the equations of the error model (3.66), (3.70), the derivative of V in Eq. (3.68) is

$$\dot{V} = -ke_x v_{fb} + kv_{ref}e_y e_s + \frac{1}{2\left(1 + \frac{e_c}{a}\right)}(-2e_s \omega_{fb})$$

$$+ \frac{-\frac{1}{a}e_s \omega_{fb}(-2e_c)}{2\left(1 + \frac{e_c}{a}\right)^2} = -ke_x v_{fb} + e_s\left(kv_{ref}e_y - \frac{\omega_{fb}}{\left(1 + \frac{e_c}{a}\right)^2}\right) \tag{3.71}$$

In order to make $\dot{V}$ negative semidefinite, the following control law is proposed:

$$v_{fb} = k_x e_x$$

$$\omega_{fb} = kv_{ref}e_y\left(1 + \frac{e_c}{a}\right)^2 + k_s e_s\left[\left(1 + \frac{e_c}{a}\right)^2\right]^n \tag{3.72}$$

where $k_x(t)$ and $k_s(t)$ are positive functions, while $n \in \mathbb{Z}$. For practical reasons n is a small number (usually -2, -1, 0, 1, or 2 are good choices). By taking into account the control law (3.72), the function $\dot{V}$ becomes

$$\dot{V} = -kk_x e_x^2 - k_s e_s^2\left[\left(1 + \frac{e_c}{a}\right)^2\right]^{n-1} \tag{3.73}$$

It is again very simple to show the convergence of e_x and e_s based on Eq. (3.73). The convergence of e_y and e_c is again a little more difficult [13].

EXAMPLE 3.12

A differentially driven vehicle is to be controlled to follow the reference trajectory $x_{ref} = 1.1 + 0.7\sin(\frac{2\pi t}{30})$ and $y_{ref} = 0.9 + 0.7\sin(\frac{4\pi t}{30})$. Sampling time is $T_s = 0.033\,\mathrm{s}$. The initial pose is $[x(0), y(0), \varphi(0)] = [1.1, 0.8, 0]$. Implement the control algorithm presented in this section in a Matlab code. Experiment with control gains and additional control design parameters (a and n).

Solution

The code is given in Listing 3.11. Simulation results are shown in Figs. 3.28 and 3.29 where good tracking can be observed.

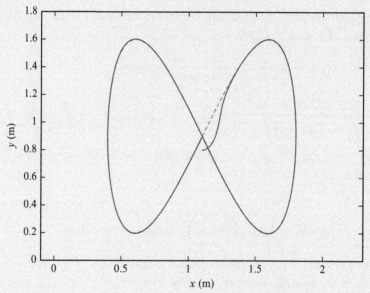

Fig. 3.28 Nonlinear trajectory tracking control of differential drive from Example 3.12: reference path (*dashed line*) and actual path (*solid line*).

EXAMPLE 3.12—cont'd

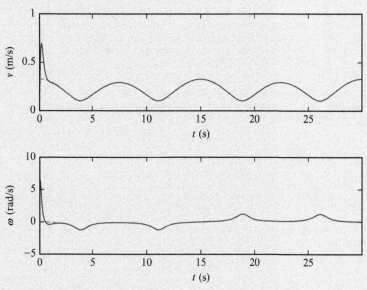

Fig. 3.29 Control signals of nonlinear trajectory tracking control of differential drive from Example 3.12.

Listing 3.11

```
1  Ts = 0.033; % Sampling time
2  t = 0:Ts:30; % Simulation time
3  q = [1.1; 0.8; 0]; % Initial robot pose
4
5  % Reference
6  freq = 2*pi/30;
7    xRef = 1.1 + 0.7*sin(freq*t);    yRef = 0.9 + 0.7*sin(2*freq*t);
8   dxRef = freq*0.7*cos(freq*t);    dyRef = 2*freq*0.7*cos(2*freq*t);
9  ddxRef =-freq^2*0.7*sin(freq*t); ddyRef =-4*freq^2*0.7*sin(2*freq*t);
10 qRef = [xRef; yRef; atan2(dyRef, dxRef)]; % Reference trajectory
11 vRef = sqrt(dxRef.^2+dyRef.^2);
12 wRef = (dxRef.*ddyRef-dyRef.*ddxRef)./(dxRef.^2+dyRef.^2);
13 uRef = [vRef; wRef]; % Reference inputs
14
15 for k = 1:length(t)
16     e = [cos(q(3)), sin(q(3)), 0; ...
17         -sin(q(3)), cos(q(3)), 0; ...
18          0,         0,         1]*(qRef(:,k) - q); % Error vector
19     eX = e(1); eY = e(2); % Two errors (due to distance)
20     eS = sin(e(3)); eCos = cos(e(3)); eC = eCos - 1; % Orientation errors
21
22     % Current reference inputs
23     vRef = uRef(1,k);
24     wRef = uRef(2,k);
25
```

(Continued)

EXAMPLE 3.12—cont'd

```
26    % Control
27    zeta = 0.9; % Experiment with this control design parameter
28    g = 85;     % Experiment with this control design parameter
29    a = 10;     % Experiment with this control design parameter
30    n = 2;      % Experiment with this parameter (integer)
31    Kx = 2*zeta*sqrt(wRef^2+g*vRef^2);
32    Ks = Kx;
33    K = g;
34    % Gains Kx and Ks could also be constant.
35    % This form is used to have the same damping of the transient
36    % irrespective of ref. velocities.
37
38    % Control: feedforward and feedback
39    v = vRef*cos(e(3)) + Kx*eX;
40    w = wRef + K*vRef*eY*(1+eC/a)^2 + Ks*eS*(1+eC/a)^(2*n);
41
42    % Robot motion simulation
43    dq = [v*cos(q(3)); v*sin(q(3)); w];
44    noise = 0.00; % Set to experiment with noise (e.g. 0.001)
45    q = q + Ts*dq + randn(3,1)*noise; % Euler integration
46    q(3) = wrapToPi(q(3)); % Map orientation angle to [-pi, pi]
47 end
```

3.3.7 Takagi-Sugeno Fuzzy Control Design in the LMI Framework

The error model (3.38) is nonlinear, as we have already stressed. Takagi-Sugeno models are known to describe the dynamics of nonlinear systems. In this section the model (3.38) will be rewritten in the form of a Takagi-Sugeno model. This form enables the control design of a parallel distributed compensation (PDC) in the linear matrix inequality (LMI) framework.

Takagi-Sugeno Fuzzy Error Model of a Differentially Driven Wheeled Mobile Robot

The Takagi-Sugeno (or TS) models have their roots in fuzzy logic where the model is given in the form of if-then rules. A TS model can also be represented in a more compact polytopic form [14]:

$$\dot{\boldsymbol{\xi}}(t) = \sum_{i=1}^{r} h_i\left(\boldsymbol{z}(t)\right)\left(A_i\boldsymbol{\xi}(t) + B_i\boldsymbol{u}(t)\right)$$

$$\boldsymbol{v}(t) = \sum_{i=1}^{r} h_i\left(\boldsymbol{z}(t)\right)\left(C_i\boldsymbol{\xi}(t)\right)$$

(3.74)

where $\boldsymbol{\xi}(t) \in \mathbb{R}^n$ is a state vector, $\boldsymbol{v}(t) \in \mathbb{R}^p$ is an output vector, and $\boldsymbol{z}(t) \in \mathbb{R}^q$ is the premise vector depending on the state vector (or some other quantity), $\boldsymbol{A}_i$, $\boldsymbol{B}_i$, and $\boldsymbol{C}_i$ are constant matrices. And finally, the nonlinear weighting functions $h_i(\boldsymbol{z}(t))$ are all nonnegative and such that $\sum_{i=1}^r h_i(\boldsymbol{z}(t)) = 1$ for an arbitrary $\boldsymbol{z}(t)$. For any nonlinear model, one can find an equivalent fuzzy TS model in a compact region of the state space variable using the sector nonlinearity approach, which consists of decomposing each bounded nonlinear term in a convex combination of its bounds [16]. The number of rules r is then related to the number of the nonlinearities of the model, as shown later.

In this section we will make use of the fact that in the wheeled robot error model the nonlinear functions are known a priori, which makes the use of the aforementioned concept possible. The nonlinear tracking error model (3.38) will therefore be rewritten in the equivalent matrix form:

$$
\begin{bmatrix} \dot{e}_x \\ \dot{e}_y \\ \dot{e}_\varphi \end{bmatrix} = \begin{bmatrix} 0 & \omega_{\text{ref}} & 0 \\ -\omega_{\text{ref}} & 0 & v_{\text{ref}}\frac{\sin e_\varphi}{e_\varphi} \\ 0 & 0 & 0 \end{bmatrix} \begin{bmatrix} e_x \\ e_y \\ e_\varphi \end{bmatrix} + \begin{bmatrix} -1 & e_y \\ 0 & -e_x \\ 0 & -1 \end{bmatrix} \begin{bmatrix} v_{\text{fb}} \\ \omega_{\text{fb}} \end{bmatrix}
$$

$$(3.75)$$

Four bounded nonlinear functions appear in this model: ω_{ref}, $v_{\text{ref}}\frac{\sin e_\varphi}{e_\varphi}$ (or with different notation $v_r sinc(\varphi)$), e_y, and e_x. The premise vector is therefore:

$$
\boldsymbol{z}(t) = \begin{bmatrix} \omega_{\text{ref}} \\ v_{\text{ref}}(t) sinc \left(e_\varphi(t)\right) \\ e_y(t) \\ e_x(t) \end{bmatrix}
$$

$$(3.76)$$

First, the controllability of Eq. (3.75) in the linearized sense will be analyzed. It is trivial to see that in the vicinity of the zero-error, the system (3.75) is controllable if v_{ref} is different from 0 and $|e_\varphi|$ is different from π, or ω_{ref} is different from 0. In practical cases ω_{ref} often crosses 0, so v_{ref} cannot become 0, and $|e_\varphi|$ cannot become π. The following assumptions are therefore needed to avoid loss of controllability and to restrict our attention to a certain compact region of the error space:

$$\underline{\omega}_{\text{ref}} \leq \omega_{\text{ref}} \leq \bar{\omega}_{\text{ref}}$$

$$|e_\varphi| \leq \bar{e}_\varphi < \pi, \quad 0 < \underline{v}_{\text{ref}} \leq v_{\text{ref}} \leq \bar{v}_{\text{ref}} \Rightarrow \underline{v}_{\text{ref}} sinc\left(\bar{e}_\varphi\right) \leq v_{\text{ref}} sinc\left(e_\varphi\right) \leq \bar{v}_{\text{ref}}$$

$$|e_y| \leq \bar{e}_y$$

$$|e_x| \leq \bar{e}_x$$

$$(3.77)$$

The bounds on v_{ref} and ω_{ref} are obtained from the actual reference trajectory, while the bounds on the tracking error are selected on the basis of any a priori knowledge available. It is important that these bounds are lower than the error due to measurement noise, initial errors, etc. The bounds from Eq. (3.77) are denoted $\underline{z}_j$ and $\bar{z}_j$, $j = 1, 2, 3, 4$. There are four nonlinearities in the system, so the number of fuzzy rules r is equal to $2^4 = 16$. The TS model of Eq. (3.75) is

$$\dot{e}(t) = A_{z(t)}e(t) + B_{z(t)}u_{fb}(t) \qquad (3.78)$$

$$A_i = \begin{bmatrix} 0 & \varepsilon_i^1 & 0 \\ -\varepsilon_i^1 & 0 & \varepsilon_i^2 \\ 0 & 0 & 0 \end{bmatrix}, \quad B_i = \begin{bmatrix} -1 & \varepsilon_i^3 \\ 0 & -\varepsilon_i^4 \\ 0 & -1 \end{bmatrix} \qquad (3.79)$$

The index i runs in an arbitrary manner through all the vertices of the hypercube defined by Eq. (3.77). The most usual way is to use binary enumeration:

$$i_1 = \begin{cases} 0 & i \le \frac{r}{2} \\ 1 & \text{else} \end{cases}$$

$$i_2 = \begin{cases} 0 & i - \frac{r}{2}i_1 \le \frac{r}{4} \\ 1 & \text{else} \end{cases}$$

$$i_3 = \begin{cases} 0 & i - \frac{r}{2}i_1 - \frac{r}{4}i_2 \le \frac{r}{8} \\ 1 & \text{else} \end{cases} \qquad (3.80)$$

$$i_4 = \begin{cases} 0 & i - \frac{r}{2}i_1 - \frac{r}{4}i_2 - \frac{r}{8}i_3 \le \frac{r}{16} \\ 1 & \text{else} \end{cases}$$

Then ε_i^j in Eq. (3.79) is defined as:

$$\varepsilon_i^j = \underline{z}_j + i_j\left(\bar{z}_j - \underline{z}_j\right), \quad i = 1, 2, \ldots, 16, \quad j = 1, 2, 3, 4 \qquad (3.81)$$

And finally the membership functions h_i need to be defined:

$$h_i(z) = w_{i_1}^1(z_1)w_{i_2}^2(z_2)w_{i_3}^3(z_3)w_{i_4}^4(z_4) \quad i = 1, 2, \ldots, 16$$

$$w_1^j(z_j) = \frac{z_j - \underline{z}_j}{\bar{z}_j - \underline{z}_j}, w_0^j(z_j) = 1 - w_1^j(z_j) \quad j = 1, 2, 3, 4 \qquad (3.82)$$

The TS model (3.78) of the tracking-error model represents the exact model of the system (3.75), that is, in this approach the TS model does not serve as an approximator but takes into account all known nonlinearities that arise in the system. The form Eq. (3.78) is very suitable for the design and analysis tasks as it will be shown in the following.

PDC Control of a Differentially Driven Wheeled Mobile Robot

In order to stabilize the TS model (3.78), a PDC control law [15],

$$u_{fb}(t) = -\sum_{i=1}^{r} h_i(z(t)) F_i e(t) = -F_{z(t)} e(t) \qquad (3.83)$$

is used. The stabilization problem by using PDC is a well-known one. Due to a specific structure where the plant model and the controller share the same membership functions it is possible to adapt certain tools for linear systems analysis and design to this nonlinear system. Particularly important is the possibility to treat system stability in a formal and straightforward manner. Roughly speaking, the system described by Eqs. (3.78), (3.83) is asymptotically stable if for each i and j the matrix $(A_i - B_i F_j)$ is Hurwitz (a matrix is Hurwitz if all the eigen values lie in the open left half-plane of the complex plane s). The number of matrices to be analyzed grows very quickly. A systematic approach is therefore needed. It was soon realized that LMIs seem to be a perfect tool for the job [16]. Given the plant parameters in the form of matrices A_i and B_i, it is possible to find the set of controller parameters F_j that asymptotically stabilize the system.

The original approach is too conservative since it does not take into account specific properties of the system, for example, the shape of the membership functions. Many relaxations of the original LMI conditions exist. The adaptation of the result due to [17] is given below:

The TS model (3.78) can be stabilized via the PDC control law (3.83) if there exist matrices M_i $(i = 1, 2, \ldots, r)$ and $X > 0$, such that the following LMI conditions hold:

$$\begin{array}{ll} \Upsilon_{ii} < 0, & i = 1, 2, \ldots, r \\ \frac{2}{r-1}\Upsilon_{ii} + \Upsilon_{ij} + \Upsilon_{ji} < 0, & i, j = 1, 2, \ldots, r, \ \ i \neq j \end{array} \qquad (3.84)$$

with $\Upsilon_{ij} = XA_i^T + A_i X - M_j^T B_i^T - B_i M_j$. The gains F_i of the PDC control law (3.83) are given by $F_i = M_i X^{-1}$.

3.3.8 Model-Based Predictive Control

Model-based predictive control (MPC) approaches are advanced methods that can be applied to various areas as well as to mobile robotics were reference trajectory is known beforehand. The use of predictive control approaches in mobile robotics usually relies on a linearized (or also nonlinear) kinematic model or dynamic model to predict future system states. A number of successful implementations to wheeled robotics have been reported such as generalized predictive control in [18], predictive control with Smith predictor to handle the system's time delay in [19], MPC based on a linear, time-varying system description in [20], nonlinear predictive controller with a multilayer neural network as a system model in [21], and many others. In most approaches the control law solutions are solved by an optimization of a cost function. Other approaches obtain control law as an analytical solution, which is computationally effective and can be easily used in fast, real-time implementations [22].

This section deals with a differentially driven mobile robot and trajectory-tracking control. Reference trajectory needs to be a smooth twice-differentiable function of time. For prediction, a linearized error dynamics model obtained around the reference trajectory is used. The controller minimizes the difference between the robot future tracking error and the reference error with defined desired dynamics.

Model-based control strategies combine a feedforward solution and a feedback action in input vector u, given as follows:

$$u = u_{ff} + u_{fb} = \begin{bmatrix} v_{ref} \cos e_\varphi + v_{fb} \\ \omega_{ref} + \omega_{fb} \end{bmatrix} \tag{3.85}$$

where the feedforward input vector $u_{ff} = [v_{ref} \cos e_\varphi \, \omega_{ref}]^T$ is calculated from the reference trajectory using relations (3.22), (3.23). The feedback input vector is $u_{fb} = [v_{fb} \, \omega_{fb}]^T$, which is the output of the MPC controller.

The control problem is given by the linear tracking error dynamics model (3.39), which in shorter notation reads

$$\dot{e} = A_c(t)e + B_c u_{fb} \tag{3.86}$$

where $A_c(t)$ and B_c are matrices of the continuous state-space model and e is the tracking error defined in local robot coordinates, defined by transformation (3.35) and shown in Fig. 3.21.

Discrete MPC

The MPC approach presented in [22] is summarized. It is designed for discrete time; therefore, model (3.86) needs to be written in discrete-time form:

$$\dot{e}(k+1) = A(k)e(k) + Bu_{fb}(k) \tag{3.87}$$

where $A(k) \in \mathbb{R}^n \times \mathbb{R}^n$, n is the number of the state variables and $B \in \mathbb{R}^n \times \mathbb{R}^m$, and m is the number of input variables. The discrete matrices $A(k)$ and B can be obtained as follows:

$$A(k) = I + A_c(t) T_s$$
$$B = B_c T_s$$
(3.88)

which is sufficient approximation for a short sampling time T_s.

The main idea of MPC is to compute optimal control actions that minimize the objective function given in prediction horizon interval h. The objective function is a quadratic cost function:

$$J(u_{fb}, k) = \sum_{i=1}^{h} \epsilon^T(k, i) Q \epsilon(k, i) + u_{fb}^T(k + i - 1) R u_{fb}(k + i - 1) \quad (3.89)$$

consisting of a future reference tracking error $e_r(k+i)$, predicted tracking error $e(k+i|k)$, difference between the two errors $\epsilon(k, i) = e_r(k+i) - e(k+i|k)$, and future actions $u_{fb}(k + i - 1)$, where i denotes the ith step-ahead prediction $(i = 1, \ldots, h)$. Q and R are the weighting matrices.

For prediction of the state $e(k + i|k)$ the error model (3.87) is applied as follows:

$$e(k + 1|k) = A(k)e(k) + Bu_{fb}(k)$$
$$e(k + 2|k) = A(k + 1)e(k + 1|k) + Bu_{fb}(k + 1)$$
$$\vdots$$
$$e(k + i|k) = A(k + i - 1)e(k + i - 1|k) + Bu_{fb}(k + i - 1)$$
$$\vdots$$
$$e(k + h|k) = A(k + h - 1)e(k + h - 1|k) + Bu_{fb}(k + h - 1)$$
(3.90)

Predictions $e(k+i|k)$ in Eq. (3.90) are rearranged so that they are dependent on current error $e(k)$, current and future inputs $u_{fb}(k + i - 1)$, and matrices $A(k + i - 1)$ and B. The model-output prediction at the time instant h can then be written as

$$e(k + h|k) = \Pi_{j=1}^{h-1} A(k + j)e(k) +$$
$$+ \sum_{i=1}^{h} \left(\Pi_{j=i}^{h-1} A(k + j) \right) Bu_{fb}(k + i - 1) + Bu_{fb}(k + h - 1)$$
(3.91)

The future reference error $(e_r(k + i))$ defines how the tracking error should decrease when the robot is not on the trajectory. Let us define the future reference error to decrease exponentially from the current tracking error $e(k)$, as follows:

$$e_r(k+i) = A_r^i e(k) \tag{3.92}$$

for $i = 1, \ldots, h$. The dynamics of the reference error is defined by the reference model matrix A_r.

According to Eqs. (3.90), (3.91) the robot-tracking prediction-error vector is defined as follows:

$$E^*(k) = \left[e(k+1|k)^T \ e(k+2|k)^T \ \ldots \ e(k+h|k)^T \right]^T$$

where E^* is given for the whole interval of the observation (h) where the control vector is

$$U_{\text{fb}}(k) = \left[u_{\text{fb}}^T(k) \ u_{\text{fb}}^T(k+1) \ldots \ u_{\text{fb}}^T(k+h-1) \right]^T \tag{3.93}$$

and

$$\Lambda(k,i) = \Pi_{j=i}^{h-1} A(k+j)$$

The robot-tracking prediction-error vector can be written in compact form:

$$E^*(k) = F(k)e(k) + G(k)U_{\text{fb}}(k) \tag{3.94}$$

where

$$F(k) = [A(k) \ A(k+1)A(k) \ \ldots \ \Lambda(k,0)]^T \tag{3.95}$$

and

$$G(k) = \begin{bmatrix} B & 0 & \cdots & 0 \\ A(k+1)B & B & \cdots & \vdots \\ \vdots & \vdots & \ddots & \vdots \\ \Lambda(k,1)B & \Lambda(k,2)B & \cdots & B \end{bmatrix} \tag{3.96}$$

where dimensions of $F(k)$ and $G(k)$ are $(n \cdot h \times n)$ and $(n \cdot h \times m \cdot h)$, respectively.

The robot reference-tracking-error vector is

$$E_r^*(k) = \left[e_r(k+1)^T \ e_r(k+2)^T \ \ldots \ e_r(k+h)^T \right]^T$$

which in compact form is computed as

$$E_r^*(k) = F_r e(k) \tag{3.97}$$

where

$$F_r = \left[A_r \; A_r A_r^2 \; \cdots \; A_r A_r^h \right]^T \qquad (3.98)$$

where F_r is the $(n \cdot h \times n)$ matrix.

The optimal control vector (3.93) is obtained by optimization of Eq. (3.89), which can be done numerically or analytically. The analytical solution is derived in the following.

The objective function (3.89) defined in matrix form reads.

$$J(U_{fb}) = \left(E_r^* - E^* \right)^T \overline{Q} \left(E_r^* - E^* \right) + U_{fb}^T \overline{R} U_{fb} \qquad (3.99)$$

The minimum of Eq. (3.99) is expressed as

$$\frac{\partial J}{\partial U_{fb}} = -2\overline{Q} G^T E_r^* + 2 G^T \overline{Q} E^* + 2 \overline{R} U_{fb} = 0 \qquad (3.100)$$

and the solution for the optimal control vector is obtained as

$$U_{fb}(k) = \left(G^T \overline{Q} G + \overline{R} \right)^{-1} G^T \overline{Q} \left(F_r - F \right) e(k) \qquad (3.101)$$

where the weighting matrices are as follows:

$$\overline{Q} = \begin{bmatrix} Q & 0 & \cdots & 0 \\ 0 & Q & \cdots & 0 \\ \vdots & \vdots & \ddots & \vdots \\ 0 & 0 & \cdots & Q \end{bmatrix} \qquad (3.102)$$

and

$$\overline{R} = \begin{bmatrix} R & 0 & \cdots & 0 \\ 0 & R & \cdots & 0 \\ \vdots & \vdots & \ddots & \vdots \\ 0 & 0 & \cdots & R \end{bmatrix} \qquad (3.103)$$

Solution (3.101) contains control vectors $u_{fb}^T(k+i-1)$ for the whole interval of prediction ($i = 1, \ldots, h$). To apply feedback control action in the current time instant k only the first vector $u_{fb}^T(k)$ (first m rows of $U_{fb}(k)$) is applied to the robot. The solution is obtained analytically; therefore, it enables fast real-time implementations that may not be possible if numeric optimization of Eq. (3.89) is used.

EXAMPLE 3.13

Implement model predictive control given in Eq. (3.101) for trajectory tracking of a differential-drive robot. Reference trajectory and robot are defined in Example 3.9.

The choice for prediction horizon is $h = 4$, the reference model matrix is $A_r = I_{3\times3} \cdot 0.65$, and the weighting matrices are

$$Q = \begin{bmatrix} 4 & 0 & 0 \\ 0 & 40 & 0 \\ 0 & 0 & 0.1 \end{bmatrix}, \quad R = I_{2\times2} \cdot 10^{-3}$$

Solution

Using MPC compute the feedback part of control signal $u_{fb}(k)$ and apply it to the robot together with the feedforward part $u_{ff}(k)$. In the following Matlab script (Listing 3.12), possible solution and trajectory tracking results are given. Obtained simulation results are shown in Figs. 3.30 and 3.31.

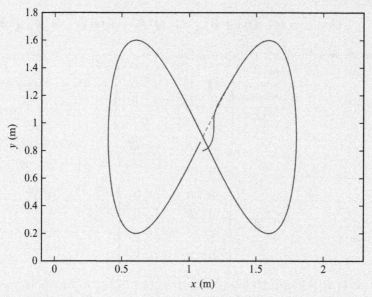

Fig. 3.30 Obtained control tracking results using explicit MPC (reference is marked as *dashed*).

EXAMPLE 3.13—cont'd

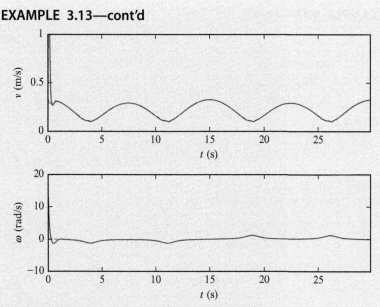

Fig. 3.31 Control action calculated using explicit MPC.

Listing 3.12

```
1  Ts = 0.033; % Sampling time
2  t = 0:Ts:30; % Simulation time
3  q = [1.1; 0.8; 0]; % Initial robot pose
4
5  % Reference
6  freq = 2*pi/30;
7    xRef = 1.1 + 0.7*sin(freq*t);    yRef = 0.9 + 0.7*sin(2*freq*t);
8   dxRef = freq*0.7*cos(freq*t);    dyRef = 2*freq*0.7*cos(2*freq*t);
9  ddxRef =-freq^2*0.7*sin(freq*t); ddyRef =-4*freq^2*0.7*sin(2*freq*t);
10 qRef = [xRef; yRef; atan2(dyRef, dxRef)]; % Reference trajectory
11 vRef = sqrt(dxRef.^2+dyRef.^2);
12 wRef = (dxRef.*ddyRef-dyRef.*ddxRef)./(dxRef.^2+dyRef.^2);
13 uRef = [vRef; wRef]; % Reference inputs
14
15 for k = 1:length(t)-4
16     e = [cos(q(3)), sin(q(3)), 0; ...
17         -sin(q(3)), cos(q(3)), 0; ...
18          0,          0,        1]*(qRef(:,k) - q); % Error vector
19     e(3) = wrapToPi(e(3)); % Correct angle
20
21     A0 = [1, Ts*uRef(2,k),   0;-Ts*uRef(2,k),   1, Ts*uRef(1,k);   0,0,1];
22     A1 = [1, Ts*uRef(2,k+1), 0;-Ts*uRef(2,k+1), 1, Ts*uRef(1,k+1); 0,0,1];
23     A2 = [1, Ts*uRef(2,k+2), 0;-Ts*uRef(2,k+2), 1, Ts*uRef(1,k+2); 0,0,1];
24     A3 = [1, Ts*uRef(2,k+3), 0;-Ts*uRef(2,k+3), 1, Ts*uRef(1,k+3); 0,0,1];
25     A4 = [1, Ts*uRef(2,k+4), 0;-Ts*uRef(2,k+4), 1, Ts*uRef(1,k+4); 0,0,1];
26     B = [Ts, 0; 0, 0; 0, Ts];
27     C = eye(3);
```

(Continued)

EXAMPLE 3.13—cont'd

```
28
29      Z = zeros(3,2);
30      Hm = [C*A0*B,          Z,          Z,          Z; ...
31            C*A0*A1*B,        C*A0*B,     Z,          Z; ...
32            C*A0*A1*A2*B,     C*A0*A1*B,  C*A0*B,     Z; ...
33            C*A0*A1*A2*A3*B, C*A0*A1*A2*B, C*A0*A1*B, C*A0*B];
34      Fm = [C*A0*A1, C*A0*A1*A2, C*A0*A1*A2*A3, C*A0*A1*A2*A3*A4].';
35
36      ar = 0.65;
37      Ar = eye(3)*ar; % Reference error dynamics
38      H = 0;
39      Fr = [Ar^(H+1), Ar^(H+2), Ar^(H+3), Ar^(H+4)].';
40
41      % Weight matrices
42      Qt = diag(repmat([1; 40; 0.1], 4, 1));
43      Rt = diag(repmat([0.001; 0.001], 4, 1));
44
45      % Optimal control calculation
46      KKgpc = (Hm.'*Qt*Hm + Rt)\(Hm.'*Qt*(-Fm));
47      KK = KKgpc(1:2,:); % Take current control gains
48
49      v = -KK*e;
50      uF = [uRef(1,k)*cos(e(3)); uRef(2,k)];
51      u = v + uF;
52
53      vMAX = 1; wMAX = 15; % Max velocities
54      if abs(u(1))>vMAX, u(1) = sign(u(1))*vMAX; end
55      if abs(u(2))>wMAX, u(2) = sign(u(2))*wMAX; end
56
57      % Robot motion simulation
58      dq = [u(1)*cos(q(3)); u(1)*sin(q(3)); u(2)];
59      noise = 0.00; % Set to experiment with noise (e.g. 0.001)
60      q = q + Ts*dq + randn(3,1)*noise; % Euler integration
61      q(3) = wrapToPi(q(3)); % Map orientation angle to [-pi, pi]
62 end
```

3.3.9 Particle Swarm Optimization-Based Control

Control of a mobile robot can also be defined as an optimization problem where the best solution among all possible solutions in a search space needs to be found in each control-loop sample time. The obtained control does not have an explicit structure, meaning that the control law cannot be given as a mapping function from system states to control actions. An optimal solution that minimizes some objective function is instead found by an iterative optimization algorithm like Newton methods, gradient descent methods, or stochastic methods such as genetic algorithms, particle swarm optimization (PSO), and the like.

If the objective function is not convex in the minimization problem then most optimization algorithms can be trapped in a local minimum, which is not the optimal solution. Using stochastic optimization the likelihood of local minimum, the solution is usually smaller because the search pattern is randomized to some extent.

PSO is inspired by social behavior of small animals such as bird flocks or fish schools [23, 24]. PSO uses a group (swarm) of particles where each particle presents its own solution hypothesis. Each particle i is characterized by parameter vector $\boldsymbol{x}_i$ defining its position in the parameter space and increment vector $\boldsymbol{v}_i$ defining velocity in the parameter space. During optimization the population of all candidate solutions are updated according to the objective function that defines a measure of quality. Each particle keeps track of its parameters and remembers the best parameters $\boldsymbol{pBest}_i$ achieved so far together with the associated objective function $J_i = f(\boldsymbol{pBest}_i)$. During optimization also the best parameter vector obtained so far for the entire swarm $\boldsymbol{gBest}$ is stored. In local version of PSO for each particle the best parameter vector $\boldsymbol{gBest}_i$ is tracked for some neighborhood of particles (e.g., k nearest particles, ring topology, and the like [25]). Local versions of PSO are less likely to be trapped in a local minimum.

Next, a basic (global) version of PSO is explained. In each control-loop iteration particles update mimic cognitive and social behavior by means of the following rules:

$$\boldsymbol{v}_i \leftarrow \omega \boldsymbol{v}_i + c_1 rand_{(0,1)}(\boldsymbol{pBest}_i - \boldsymbol{x}_i) + c_2 rand_{(0,1)}(\boldsymbol{gBest} - \boldsymbol{x}_i)$$
$$\boldsymbol{x}_i \leftarrow \boldsymbol{x}_i + \boldsymbol{v}_i$$

(3.104)

where ω is the inertia factor, c_1 is the self-cognitive constant, and c_2 is the social constant. In addition $rand_{(0,1)}$ is a vector of uniformly distributed values in the range $(0, 1)$. The dimensions of vectors in Eq. (3.104) equal the dimension of the parameter search space. The parameters ω, c_1, and c_2 are positive tuning parameters. Basic PSO code is given in Algorithm 1.

ALGORITHM 1   Particle swarm optimization
Initialization:
for each particle $i = 1, \ldots, N$ **do**
    Randomly initialize particle position $\boldsymbol{x}_i$ within boundaries of the parameter space.
    Randomly initialize particle velocity $\boldsymbol{v}_i$ or set to zero.
    Set $\boldsymbol{pBest}_i = \boldsymbol{x}_i$.
end for

Optimization:
$J_{best} = \infty$

repeat

    for each particle $i = 1, \ldots, N$ **do**

        Compute current objective function $J_i = f(x_i)$ for each particle.

        Store the best parameters:

        if $J_i < f(\pmb{pBest}_i)$ **then**

            $\pmb{pBest}_i = \pmb{x}_i$

        end if

        if $f(\pmb{pBest}_i) < J_{best}$ **then**

            $\pmb{gBest} = \pmb{pBest}_i$

            $J_{best} = f(\pmb{gBest})$

        end if

    end for

    for each particle $i = 1, \ldots, N$ **do**

        Update particle velocity and position:

        $\pmb{v}_i \leftarrow \omega \pmb{v}_i + c_1 rand_{(0,1)}(\pmb{pBest}_i - \pmb{x}_i) + c_2 rand_{(0,1)}(\pmb{gBest} - \pmb{x}_i)$

        Check if velocity $\pmb{v}_i$ is feasible.

        $\pmb{x}_i \leftarrow \pmb{x}_i + \pmb{v}_i$

    end for

until max iterations are reached or convergence criteria is satisfied

---

EXAMPLE 3.14

Write the controller using the PSO for the robot and trajectory defined in Example 3.9. PSO is applied in each sampling time $t = kT_s$ to find the best controls (translational and angular velocity) that would drive the robot as close as possible to the current reference position $x_{ref}(t)$, $y_{ref}(t)$. Next, the robot pose prediction is computed considering robot kinematics and the particles' proposed solution (control action). Optimal control minimizes the cost function consisting of the tracking error $e(t) = \left[e_x(t), e_y(t), e_\varphi(t) \right]^T$ and the feedback control effort $\pmb{u}_{fb} = [v_{fb}, \omega_{fb}]^T$. This follows the equation $J(t) = e(t)^T \pmb{Q} e(t) + \pmb{u}_{fb}^T \pmb{R} \pmb{u}_{fb}$, where $\pmb{Q}$ and $\pmb{R}$ are diagonal weighting matrices used to tune the controller behavior.

Solution

The control law includes feedforward action and feedback action similarly as in Example 3.9. Feedforward action is obtained from a known trajectory while the feedback action is computed using PSO. The tracking error is expressed in robot local coordinates (see Fig. 3.21) because this makes the optimization more efficient due to the closed-loop system being decoupled better. The error in the local x coordinate can simply be compensated by translational velocity, and errors in y and φ can be compensated by the angular velocity. However, this is not the case when using a global tracking error due to nonlinear rotation transformation in Eq. (3.35).

EXAMPLE 3.14—cont'd

In Listing 3.13 a Matlab script of a possible solution is given. Obtained simulation results are shown in Figs. 3.32 and 3.33. Similar control results as in Example 3.13 are obtained. However, the computational effort is much higher than in Example 3.13. Results of Example 3.14 are not deterministic because PSO is using randomly distributed particles in optimization to find the best control in each sampling time. Therefore, resulting trajectories of the mobile robot can vary to some extent, especially the initial transition phase, until the robot reaches the reference. However, the obtained results in the majority of simulation runs are at least comparable to the results of other presented deterministic controllers (e.g., from Example 3.9).

Listing 3.13

```
1  Ts = 0.033; % Sampling time
2  t = 0:Ts:30; % Simulation time
3  q = [1.1; 0.8; 0]; % Initial robot pose
4
5  % Reference
6  freq = 2*pi/30;
7    xRef = 1.1 + 0.7*sin(freq*t);     yRef = 0.9 + 0.7*sin(2*freq*t);
8   dxRef = freq*0.7*cos(freq*t);    dyRef = 2*freq*0.7*cos(2*freq*t);
9  ddxRef =-freq^2*0.7*sin(freq*t); ddyRef =-4*freq^2*0.7*sin(2*freq*t);
10 qRef = [xRef; yRef; atan2(dyRef, dxRef)]; % Reference trajectory
11 vRef = sqrt(dxRef.^2+dyRef.^2);
12 wRef = (dxRef.*ddyRef-dyRef.*ddxRef)./(dxRef.^2+dyRef.^2);
13 uRef = [vRef; wRef]; % Reference inputs
14
15 vMax = 1; wMax = 15; % Velocity constraints
16
17 % Swarm initialization
18 iterations = 20; % Number of iterations
19 omega = 0.5*0.5; % Inertia
20 c1 = 0.5*1; % Correction factor
21 c2 = 0.5*1; % Global correction factor
22 N = 25; % Swarm size
23 swarm = zeros([2,N,4]);
24 uBest = [0; 0];
25
26 for k = 1:length(t)-1
27     % Initial swarm position
28     swarm(:,:,1) = repmat(uBest, 1, N) + diag([0.1; 3])*randn(2,N);
29     swarm(:,:,2) = 0; % Initial particle velocity
30     swarm(1,:,4) = 1000; % Best value so far
31
32     for iter = 1:iterations % PSO iteratively find best solution
33         % Evaluate particles parameters
34         for i = 1:N
35             % Compute new predicted pose of the robot using i-th particle
36             % parameters (input velocities) and compare obtained predicted
37             % pose to the reference pose.
38             vwi = swarm(:,i,1);
39             ui = vwi + uRef(:,k); % Feedback and feedforward
40             qk = q; % Current robot pose
41             % Predict robot pose using particle parameters (velocities)
42             qk = qk + Ts*[cos(qk(3)), 0; sin(qk(3)), 0; 0, 1]*ui;
```

(Continued)

EXAMPLE 3.14—cont'd

```
43              qk(3) = wrapToPi(qk(3)); % Correct angle range
44              e = [cos(qk(3)), sin(qk(3)), 0; ...
45                   -sin(qk(3)), cos(qk(3)), 0; ...
46                    0,           0,          1]*(qRef(:,k+1)-qk); % Error
47              e(3) = wrapToPi(e(3)); % Correct angle range
48              Qt = diag([4; 80; 0.1]); Rt = diag([1; 1]*0.0001); % Weights
49              J = e.'*Qt*e + vwi.'*Rt*vwi; % Cost function
50              if J<swarm(1,i,4) % if new parameter is better, update:
51                  swarm(:,i,3) = swarm(:,i,1); % param values (v and w)
52                  swarm(1,i,4) = J;            % and best criteria value.
53              end
54          end
55          [~, gBest] = min(swarm(1,:,4)); % Global best particle parameters
56
57          % Updating parameters by velocity vectors
58          a = omega*swarm(:,:,2) + ...
59              c1*rand(2,N).*(swarm(:,:,3) - swarm(:,:,1)) + ...
60              c2*rand(2,N).*(repmat(swarm(:,gBest,3), 1, N) - swarm(:,:,1));
61          % Max param increment, acceleration: aMax=3 ==> 3*Ts=0.1
62          a(1,a(1,:)>0.1) = 0.1; a(1,a(1,:)<-0.1) = -0.1;
63          % Max param increment, angular acceleration: aMax=60 ==> 60*Ts=2
64          a(2,a(1,:)>2) = 2;     a(2,a(1,:)<-2) = -2;
65
66          v = swarm(:,:,1) + a; % Update velocity
67          % Limit velocity to preserve curvature ...
68          [m, ii] = max([v(1,:)/vMax; v(2,:)/wMax; ones(1,N)]);
69          i = ii==1; v(1,i) = sign(v(1,i))*vMax;
70                     v(2,i) = v(2,i)./m(i);
71          i = ii==2; v(2,i) = sign(v(2,i))*wMax;
72                     v(1,i) = v(1,i)./m(i);
73
74          swarm(:,:,2) = a; % Updated particle velocities (acceleration)
75          swarm(:,:,1) = v; % Updated particle positions (velocities)
76      end
77
78      % Take the best particle to get robot command velocities
79      uBest = swarm(:,gBest,1);
80      u = uBest + uRef(:,k); % Feedback and feedforward
81
82      % Velocity constraints
83      if abs(u(1))>vMax, u(1) = sign(u(1))*vMax; end
84      if abs(u(2))>wMax, u(2) = sign(u(2))*wMax; end
85
86      % Robot motion simulation
87      dq = [u(1)*cos(q(3)); u(1)*sin(q(3)); u(2)];
88      noise = 0.00; % Set to experiment with noise (e.g. 0.001)
89      q = q + Ts*dq + randn(3,1)*noise; % Euler integration
90      q(3) = wrapToPi(q(3)); % Map orientation angle to [-pi, pi]
91  end
```

EXAMPLE 3.14—cont'd

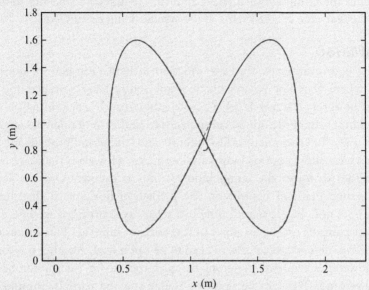

Fig. 3.32 Obtained control tracking results using particle swarm optimization approach (reference is marked as *dashed*).

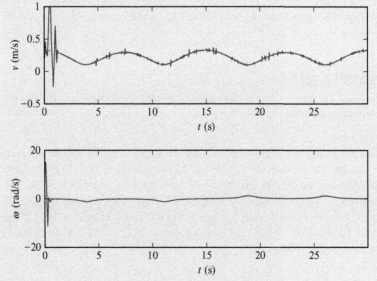

Fig. 3.33 Control action calculated using the particle swarm optimization approach.

PSO is a generic algorithm that can be applied to numerous applications as long as the computational effort and time required to find the solution meet the real time requirements of the system being controlled.

MPC With PSO

Solving optimization of objective function in MPC can also be done by PSO, where the best parameters (control actions) are searched for the prediction interval $(t, t + hT_S)$ where h is horizon. If a system has $m = 2$ control inputs then the optimization needs to find $m \cdot h$ optimal parameters, which may be computationally intense and therefore problematic for systems with short control-loop sampling times. However, there are some possibilities to lower the computational time as follows. One possibility is to assume constant controls in the prediction horizon. If the control action does not change significantly in a relatively short horizon time, then constant controls can be assumed in the horizon interval. This means that only m parameters instead of $m \cdot h$ need to be optimized. Another possibility is to lower the required iterations in optimization for each control-loop sampling time. This can be done by initializing the particles around the optimal solution from the previous time sample. Doing so, this would hopefully require less iterations for the particles to converge.

A possible solution of implementing MPC on PSO is given in Example 3.15.

EXAMPLE 3.15

Extend Example 3.14 to a model-based predictive controller with prediction horizon $h = 3$.

Solution

The presented solution supposes that the control action (feedback part) in the prediction horizon interval is constant $u_{fb}(t + (i - 1)T_S) = u_{fb}$. Feedforward action ($u_{ff}$) is obtained from known trajectory, while the feedback action is computed using PSO. From the current robot pose the h-step-ahead prediction is made using the robot kinematic model and control actions $u(t + (i - 1)T_S) = u_{fb} + u_{ff}(t + (i - 1)T_S)$ $(i = 1, \dots, h)$ are obtained as follows:

$$\hat{q}(t + iT_S) = f\left(\hat{q}(t + (i - 1)T_S), u(t + (i - 1)T_S)\right)$$

EXAMPLE 3.15—cont'd

where the initial state equals $\hat{q}(t) = q(t)$ and $i = 1, \ldots, h$. The control action is obtained by optimizing the cost function for the horizon period

$$J(t + hT_S) = \sum_{i=1}^{h} e(t + iT_S)^T Q e(t + iT_S) + u_{\text{fb}}^T R u_{\text{fb}}$$

where $e(\cdot)$ is the tracking error in local coordinates. The optimal control action is calculated using PSO as is given in the Matlab script in Listing 3.14. Simulation results are shown in Figs. 3.34 and 3.35. A smaller tracking error and a better initial transition compared to the results in Example 3.14 are obtained.

Listing 3.14

```
1  Ts = 0.033; % Sampling time
2  t = 0:Ts:30; % Simulation time
3  q = [1.1; 0.8; 0]; % Initial robot pose
4
5  % Reference
6  freq = 2*pi/30;
7    xRef = 1.1 + 0.7*sin(freq*t);      yRef = 0.9 + 0.7*sin(2*freq*t);
8   dxRef = freq*0.7*cos(freq*t);      dyRef = 2*freq*0.7*cos(2*freq*t);
9  ddxRef =-freq^2*0.7*sin(freq*t); ddyRef =-4*freq^2*0.7*sin(2*freq*t);
10 qRef = [xRef; yRef; atan2(dyRef, dxRef)]; % Reference trajectory
11 vRef = sqrt(dxRef.^2+dyRef.^2);
12 wRef = (dxRef.*ddyRef-dyRef.*ddxRef)./(dxRef.^2+dyRef.^2);
13 uRef = [vRef; wRef]; % Reference inputs
14
15 vMax = 1; wMax = 15; % Valocity constraints
16
17 % Swarm initialization
18 iterations = 20; % Number of iterations
19 omega = 0.5*0.5; % Inertia
20 c1 = 0.5*1; % Correction factor
21 c2 = 0.5*1; % Global correction factor
22 N = 25; % Swarm size
23 swarm = zeros([2,N,4]);
24 uBest = [0; 0];
25
26 H = 3; % Prediction horizon
27
28 for k = 1:length(t)-H
29     % Initial swarm position
30     swarm(:,:,1) = repmat(uBest, 1, N) + diag([0.1; 3])*randn(2,N);
31     swarm(:,:,2) = 0; % Initial particle velocity
32     swarm(1,:,4) = 1000; % Best value so far
33
34     for iter = 1:iterations % PSO iteratively find best solution
35         % Evaluate particles parameters
36         for i = 1:N
37             % Compute new predicted pose of the robot using i-th particle
38             % parameters (input velocities) and compare obtained predicted
39             % pose to the reference pose.
40             vwi = swarm(:,i,1);
```

(Continued)

EXAMPLE 3.15—cont'd

```
41          ui = vwi + uRef(:,k); % Feedback and feedforward
42          qk = q; % Current robot pose
43          % Predict robot pose using particle parameters (velocities)
44          J = 0;
45          for h = 1:H
46              qk = qk + Ts*[cos(qk(3)), 0; sin(qk(3)), 0; 0, 1]*ui;
47              qk(3) = wrapToPi(qk(3)); % Correct angle range
48              e = [cos(qk(3)), sin(qk(3)), 0; ...
49                  -sin(qk(3)), cos(qk(3)), 0; ...
50                   0,          0,          1]*(qRef(:,k+h)-qk); % Error
51              e(3) = wrapToPi(e(3)); % Correct angle range
52              Qt = diag([4; 80; 0.1]); Rt = diag([1; 1]*0.0001);% Weights
53              J = J + e.'*Qt*e + vwi.'*Rt*vwi; % Cost function
54          end
55          if J<swarm(1,i,4) % if new parameter is better, update:
56              swarm(:,i,3) = swarm(:,i,1); % param values (v and w)
57              swarm(1,i,4) = J;            % and best critiria value.
58          end
59      end
60      [~, gBest] = min(swarm(1,:,4)); % Global best particle parameters
61
62      % Updating parameters by velocity vectors
63      a = omega*swarm(:,:,2) + ...
64          c1*rand(2,N).*(swarm(:,:,3) - swarm(:,:,1)) + ...
65          c2*rand(2,N).*(repmat(swarm(:,gBest,3), 1, N) - swarm(:,:,1));
66      % Max param increment, acceleration: aMax=3 ==> 3*Ts=0.1
67      a(1,a(1,:)>0.1) = 0.1; a(1,a(1,:)<-0.1) = -0.1;
68      % Max param increment, angula acceleration: aMax=60 ==> 60*Ts=2
69      a(2,a(1,:)>2) = 2;      a(2,a(1,:)<-2) = -2;
70
71      v = swarm(:,:,1) + a; % Update velocity
72      % Limit velocity to preserve curvature ...
73      [m, ii] = max([v(1,:)/vMax; v(2,:)/wMax; ones(1,N)]);
74      i = ii==1; v(1,i) = sign(v(1,i))*vMax;
75                 v(2,i) = v(2,i)./m(i);
76      i = ii==2; v(2,i) = sign(v(2,i))*wMax;
77                 v(1,i) = v(1,i)./m(i);
78
79      swarm(:,:,2) = a; % Updated particle velocities (acceleration)
80      swarm(:,:,1) = v; % Updated particle positions (velocities)
81  end
82
83  % Take the best particle to get robot command velocities
84  uBest = swarm(:,gBest,1);
85  u = uBest + uRef(:,k); % Feedback and feedforward
86
87  % Velocity constraints
88  if abs(u(1))>vMax, u(1) = sign(u(1))*vMax; end
89  if abs(u(2))>wMax, u(2) = sign(u(2))*wMax; end
90
91  % Robot motion simulation
92  dq = [u(1)*cos(q(3)); u(1)*sin(q(3)); u(2)];
93  noise = 0.00; % Set to experiment with noise (e.g. 0.001)
94  q = q + Ts*dq + randn(3,1)*noise; % Euler integration
95  q(3) = wrapToPi(q(3)); % Map orientation angle to [-pi, pi]
96 end
```

EXAMPLE 3.15—cont'd

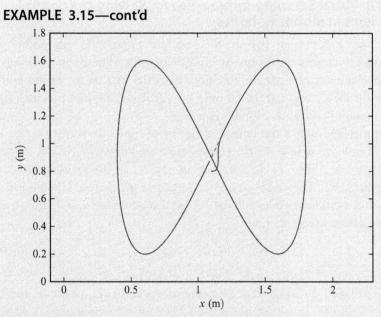

Fig. 3.34 Obtained control tracking results using MPC on PSO (reference is marked as *dashed*).

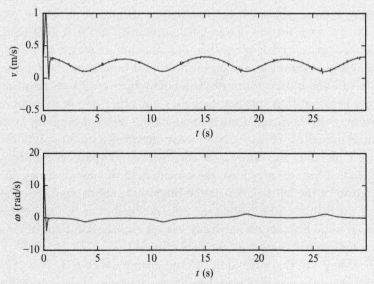

Fig. 3.35 Control action calculated using MPC on PSO.

3.3.10 Visual Servoing Approaches for Solving the Control Problems in Mobile Robotics

This part will present the use of visual servoing (VS) approaches for solving the control problems in mobile robotics. The main emphasis is on development of a mobile robotic system that is able to autonomously complete the desired task based only on visual information provided by a camera mounted on the mobile system.

In the field of VS the estimated signals obtained from images that are used in order to determine the control error are known as *features*. Features can be points, lines, circles, area of an object, angles between lines, or something else. The control error of a general VS control scheme can be written as a difference between the desired feature vector $x_{\text{ref}}(t)$ and current feature vector $x(y(t), \zeta)$ [26]:

$$e(t) = x_{\text{ref}}(t) - x(y(t), \zeta) \tag{3.105}$$

The K-element feature state vector $x(y(t), \zeta) \in \mathbb{R}^K$ in Eq. (3.105) is obtained based on image-based measurements $y(t)$ (e.g., position, area, or shape of patterns in image) and some additional time-invariant knowledge about the system ζ (e.g., intrinsic camera parameters, known properties of the observed scene, system constraints). One of the crucial challenges in designing a visual control is the appropriate definition of feature vector, since this vector is used to determine control error. If the feature vector is not defined appropriately more than one situation may exist in which the control error is minimal—the problem of local minimums. Some features may only be appropriate for some classes of motions (e.g., only translations) and may be completely inappropriate for some other class of motions (e.g., translation with rotation). If the features are not selected carefully, in some situations undesired, unnecessary, or even unexpected control motions can occur—a problem also known as *camera retreat*. During visual servoing some tracked features may leave the camera field of view that may disable completion of the VS task. Therefore the visual control algorithm should prevent the features from leaving the camera's field of view, use alternative features if some features are inevitably lost, or estimate locations of features that are temporarily outside of the field of view.

According to the definition of the feature vector x, VS control approaches can be classified into three main categories [27]: position-based visual servoing (PBVS), image-based visual servoing (IBVS), and hybrid VS. In the case of PBVS the control error is defined as the error between the desired and current robot pose in the 3D task space. The camera in the PBVS control scheme is used in order to estimate the 3D positions of objects and robot in the environment from the image, which is a projection of the environment. Therefore all the control approaches that have already been presented in the rest of this chapter can be used directly in this class of control approaches. Normally with the PBVS approach optimum control actions can be achieved; however, the approach requires accurate camera calibration or else the true control error cannot be eliminated. On the other hand, the control error in the IBVS control scheme is defined directly in the 2D image space (e.g., as a difference between two image points). Since the control error is defined directly in image, IBVS is less susceptible to accuracy of camera calibration. However, IBVS normally achieves less optimum trajectories than can be achieved with PBVS approach, and the IBVS approach is more susceptible to the camera retreat problem if the image features are not chosen appropriately. IBVS approaches are especially interesting, since they enable in-image task definition and the control approach known as *teach-by-showing*. HVS approaches try to combine good properties of PBVS and IBVS control schemes. Some control schemes only switch between PBVS and IBVS controllers according to some switching criteria that selects the most appropriate control approach based on the current system states [28]. In some other hybrid approaches the control error is composed of features from 2D image space and features from 3D task space [29, 30]. Visual control of nonholonomic systems is considered a difficult problem [31–34]. Visual control schemes are sometimes used in combination with some additional sensors [35] that may provide additional data or ease image processing.

In [26, 36] the authors have presented the general control design approach that can be used regardless of the class of VS scheme. Let us consider the case where the reference feature state vector in Eq. (3.105) is time invariant ($x_{ref}(t) = x_{ref} = $ const.). Commonly a velocity controller is used. The velocity control vector can be written as a combination of a translational velocity vector $v(t)$ and an angular velocity vector $\omega(t)$ in a combined vector $u^T(t) = [v^T(t), \omega^T(t)] \in \mathbb{R}^M$. In general the velocity vector and angular velocity vector of an object in a 3D space both consist of three velocities, that is, $v^T(t) = [v_x(t), v_y(t), v_z(t)]$ and $\omega^T(t) = [\omega_x(t), \omega_y(t), \omega_z(t)]$. However, in the case of wheeled mobile robotics these two control vectors normally have some zero elements, since the mobile robot in normal driving conditions is assumed to move on a plane, and no tumbling or lifting is expected. The relation between the velocity control vector $u^T(t)$ and the rate of feature states $\dot{s}^T(t)$ can be written as follows:

$$\dot{x}(t) = L(t)u(t) \tag{3.106}$$

where the matrix $L(t) \in \mathbb{R}^K \times \mathbb{R}^M$ is known as the *interaction matrix*. From relations (3.105), (3.106) the rate of error can be determined:

$$\dot{e}(t) = -L(t)u(t) \tag{3.107}$$

The control action $u(t)$ should minimize the error e. If exponential decay of error e is desired in the form $\dot{e}(t) = -ge(t), g > 0$, the following control law can be derived:

$$u(t) = gL^{\dagger}(t)e(t) \tag{3.108}$$

where $L^{\dagger}(t)$ is the Moore-Penrose pseudoinverse of the interaction matrix $L(t)$. If the matrix $L(t)$ is of full rank, its pseudoinverse can be computed as $L^{\dagger}(t) = (L^T(t)L(t))^{-1}L^T(t)$. If the interaction matrix is of square shape ($K = M$) and it is not singular ($\det(L(t)) \neq 0$), the calculation of pseudoinverse in Eq. (3.108) simplifies to the ordinary matrix inverse $L^{-1}(t)$.

In practice only an approximate value of the true interaction matrix can be obtained. Therefore in control scheme (3.108) only the interaction matrix estimate $\hat{L}(t)$ or estimate of the interaction matrix inverse $\hat{L}^{\dagger}(t)$ can be used:

$$u(t) = g\hat{L}^{\dagger}(t)e(t) \tag{3.109}$$

The interaction matrix inverse estimate $\hat{L}^{\dagger}(t)$ can be determined from linearization of the system around the current system state ($\hat{L}^{\dagger}(t) = \hat{L}_t^{\dagger}(t)$) or around the desired (reference) system state ($\hat{L}^{\dagger}(t) = \hat{L}_{\text{ref}}^{\dagger}(t)$). Sometimes a combination in the form $\hat{L}^{\dagger}(t) = \frac{1}{2}(\hat{L}_{\text{ref}}(t) + \hat{L}_t(t))^{\dagger}$ is used [36].

If the control (3.109) is inserted into Eq. (3.107), the differential equation of the closed-loop system is obtained:

$$\dot{e}(t) = -gL(t)\hat{L}^{\dagger}(t)e(t) \tag{3.110}$$

A Lyapunov function $V(t) = \frac{1}{2}e^T(t)e(t)$ can be used to check stability of the closed-loop system (3.110). The derivative of the Lyapunov function is

$$\dot{V}(t) = -ge^T(t)L(t)\hat{L}^{\dagger}(t)e(t) \tag{3.111}$$

A sufficient condition for global stability of the system (3.110) is satisfied if the matrix $L(t)\hat{L}^{\dagger}(t)$ is semipositive definite. In the case the number of observed features K equals to the number of control inputs M (i.e., $K = M$) and the matrices $L(t)$ and $\hat{L}^{\dagger}(t)$ both have full rank; the closed-loop system (3.110) is stable if the estimate of the matrix $\hat{L}^{\dagger}(t)$ is not too coarse [26]. However, it should be noted that in the case of IBVS it is not simple to select appropriate features in the image space that eliminate the desired control problem in the task space.

EXAMPLE 3.16

A camera, C, is mounted on the wheeled mobile robot R with differential drive at $t_C^R = [0, 0, 0.5]^T$ with orientation $R_C^R = R_x(90 \text{ degree})$ $R_y(-90 \text{ degree}) R_x(45 \text{ degree})$. The intrinsic camera parameters are $\alpha_x f = \alpha_y f = 300$, no skewness ($\gamma = 0$), and the image center is in the middle of the image with a dimension of 1024×768. Design an IBVS control that drives the mobile robot from initial pose $[x(0), y(0), \varphi(0)] = [1 \text{ m}, 0, 100 \text{ degree}]$ to the goal position $x_{\text{ref}} = 4$ m and $y_{\text{ref}} = 4$ m. For the sake of simplicity assume that the rotation center of the mobile robot is visible in the image.

Solution

The camera is observing the scene in front of the mobile robot. Since the camera is mounted without any roll around the optical axis center, a decoupled control can be designed directly based on image error between the observed goal position and image of the robot's center of rotation. A possible implementation of the solution in Matlab is shown in Listing 3.15. The obtained path of the mobile robot is shown in Fig. 3.36 and the path of the observed goal in the image is shown in Fig. 3.37. The control signals are shown in Fig. 3.38. The results confirm the applicability of the IBVS control scheme. The robot reaches the goal pose, since the observed feature (goal) is visible during the entire time of VS (Fig. 3.37).

Listing 3.15

```
1  Ts = 0.03; % Sampling time
2  t = 0:Ts:15; % Simulation1 time
3  r = 0.5; % Distance parameter for the intermediate point
4  dTol = 1; % Tolerance distance (to the intermediate point) for switch
5  qRef = [4; 4; 0]; % Reference pose
6  q = [1; 0; 100/180*pi]; % Initial pose
7
8  % Camera
9  alphaF = 300; % alpha*f, in px/m
10 s = [1024; 768]; % screen size, in px
11 c = s/2; % image centre, in px
12 S = [alphaF, 0, c(1); 0, alphaF, c(2); 0, 0, 1]; % Internal camera model
13 RL2C = rotX(pi/2)*rotY(-pi/2)*rotX(pi/4); tL2C=[0;0;0.5]; % Camera mounting
```

(Continued)

EXAMPLE 3.16—cont'd

```
14 % Camera simulation
15 pOP = S*RL2C.'*([0; 0; 0]-tL2C); pOP = pOP/pOP(3);
16 RW2L = rotZ(-q(3)); tW2L = [q(1:2); 0];
17 pP = S*RL2C.'*(RW2L.'*([qRef(1:2); 0]-tW2L)-tL2C); pP = pP/pP(3);
18
19 u = [0; 0];
20 for k = 1:length(t)
21     if pP(1)<0 || pP(2)<0 || pP(1)>s(1) || pP(2)>s(2) % Invisible feature
22         u = [0; 0]; % Lost tracked feature
23     else
24         D = sqrt(sum((pP(1:2)-pOP(1:2)).^2));
25         if D<dTol % Stop when close to the goal
26             u = [0; 0];
27         else
28             u = [0, 0.002; 0.005, 0]*(pOP(1:2)-pP(1:2));
29         end
30     end
31
32     % Robot motion simulation
33     dq = [u(1)*cos(q(3)); u(1)*sin(q(3)); u(2)];
34     noise = 0.00; % Set to experiment with noise (e.g. 0.001)
35     q = q + Ts*dq + randn(3,1)*noise; % Euler integration
36     q(3) = wrapToPi(q(3)); % Map orientation angle to [-pi, pi]
37
38     % Camera simulation
39     RW2L = rotZ(-q(3)); tW2L = [q(1:2); 0];
40     pP = S*RL2C.'*(RW2L.'*([qRef(1:2); 0]-tW2L)-tL2C); pP = pP/pP(3);
41
42 end
```

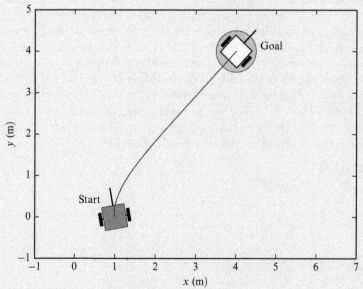

Fig. 3.36 Obtained path with IBVS control from Example 3.16.

EXAMPLE 3.16—cont'd

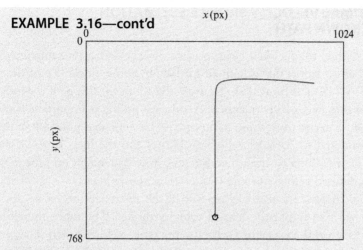

Fig. 3.37 Path of the observed feature (goal) in the image space from Example 3.16.

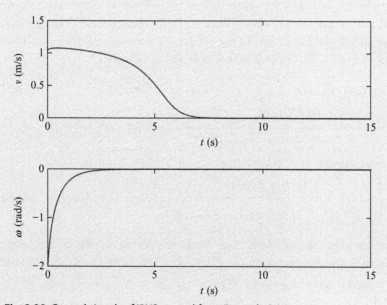

Fig. 3.38 Control signals of IBVS control from Example 3.16.

3.4 OPTIMAL VELOCITY PROFILE ESTIMATION FOR A KNOWN PATH

Wheeled mobile robots often need to drive on an existing predefined path (e.g., road or corridor) that is defined spatially by some schedule parameter u as $x_p(u)$, $y_p(u)$, $u \in [u_{SP}, u_{EP}]$. To drive the robot on this path a desired velocity profile needs to be planned. A velocity profile is important if the robot needs to arrive from some start point SP to some end point EP in the shortest time and to consider robot capabilities and the like. To drive the robot on such path its position therefore becomes dependent on time $x(t)$, $y(t)$, with defined velocity profile $v(t)$, $\omega(t)$, as shown next.

Suppose there is the special case $u = t$; the reference path becomes a trajectory with an implicitly defined velocity profile. The robot therefore needs to drive on the trajectory with the reference velocities $v(t) = v_{ref}(t)$ and $\omega(t) = \omega_{ref}(t)$ calculated from the reference trajectory using relations (3.22), (3.23).

To plan the desired velocity profile for the spatially given path, one needs to find the schedule $u = u(t)$. The planned velocity profile must be consistent with the robot constraints (maximum velocity and maximum acceleration that motors can produce and which assures safe driving without longitudinal and lateral slip of the wheels). Reference velocities are similar to those in Eqs. (3.22), (3.23) are expressed as follows:

$$v(t) = \sqrt{x_p'(u(t))^2 + y_p'(u(t))^2}\, \dot{u}(t) = v_p(u)\, \dot{u}(t) \tag{3.112}$$

$$\omega(t) = \frac{x_p'(u(t))y_p''(u(t)) - y_p'(u(t))x_p''(u(t))}{x_p'(u(t))^2 + y_p'(u(t))^2}\, \dot{u}(t) = \omega_p(u)\, \dot{u}(t) \tag{3.113}$$

and curvature is

$$\kappa(t) = \frac{x_p'(u(t))y_p''(u(t)) - y_p'(u(t))x_p''(u(t))}{\left(x_p'(u(t))^2 + y_p'(u(t))^2\right)^{3/2}} = \kappa_p(u) \tag{3.114}$$

where primes denote derivatives with respect to u, while dots denote derivatives with respect to t. Time derivatives of the path therefore consider the schedule $u(t)$ as follows: $\frac{dx(t)}{dt} = \frac{dx_p}{du}\frac{du}{dt} = x_p'\dot{u}(t)$, $\frac{dy(t)}{dt} = \frac{dy_p}{du}\frac{du}{dt} = y_p'\dot{u}(t)$.

The main idea of velocity profile planning is adopted from [37] and is as follows. For a given path a velocity profile is determined considering pure rolling conditions, meaning that command velocities are equal to the real robot velocities (no slipping of the wheels). This is archived by limitation of the overall allowable acceleration,

$$a = \sqrt{a_t^2 + a_r^2} \qquad (3.115)$$

where

$$a_t = \frac{dv}{dt}, \quad a_r = v\omega = v^2\kappa \qquad (3.116)$$

are the tangential and radial acceleration, respectively. Maximum acceleration that prevents sliding is defined by friction force F_{fric}:

$$a_{\text{MAX}} = \frac{F_{\text{fric}}}{m} = \frac{mgc_{\text{fric}}}{m} = gc_{\text{fric}} \qquad (3.117)$$

where m is the vehicle mass, g is the gravity acceleration, and c_{fric} is the friction coefficient. Due to vehicle construction, maximal tangential acceleration $a_{\text{MAX}t}$ and radial acceleration $a_{\text{MAX}r}$ are usually different and can be estimated by experiment. Therefore the overall acceleration should be somewhere inside the ellipse

$$\frac{a_t^2}{a_{\text{MAX}t}^2} + \frac{a_r^2}{a_{\text{MAX}r}^2} \leq 1 \qquad (3.118)$$

or always on the boundary in the case of time-optimal planning. In path turns the robot needs to drive slower due to higher radial acceleration. Therefore to estimate the schedule $u(t)$, first the turning points (TPs) of the path (where the absolute value of the curvature is locally maximal) are located where u is in the interval $[u_{\text{SP}}, u_{\text{EP}}]$. Positions of TPs are denoted $u = u_{\text{TP}i}$, where $i = 1, \ldots, n_{\text{TP}}$ and n_{TP} is the number of TPs. In TPs translational velocity is locally the lowest, tangential acceleration is supposed to be 0, and radial acceleration is maximal. Tangential velocity in TPs can then be calculated considering Eq. (3.116)

$$v_p(u_{\text{TP}i}) = \sqrt{\frac{a_{\text{MAX}r}}{|\kappa(u_{\text{TP}i})|}} \qquad (3.119)$$

Before and after the TP, the robot can move faster, because the curvature is lower than in the TP. Therefore the robot must decelerate before each TP ($u < u_{\text{TP}i}$) and accelerate after the TP ($u > u_{\text{TP}i}$) according to the acceleration constraint (3.118).

From Eq. (3.112) it follows that in each fixed point of the path $v(t)$ and $v_p(u)$ are proportional with proportional factor $\dot{u}(t)$ defined in time. A feasible time-minimum velocity profile is therefore defined by the derivative of the schedule $\dot{u}(t)$, where the minmax solution is computed as described next (minimizing maximum velocity profile candidates to comply with acceleration constrains). The radial and the tangential acceleration are expressed from Eq. (3.116) considering Eq. (3.112) as follows:

$$a_r(t) = \left(x_p'(u)^2 + y_p'(u)^2\right)\kappa_p(u)\,\dot{u}^2(t) = v_p^2(u)\,\kappa_p(u)\,\dot{u}^2(t) \qquad (3.120)$$

$$a_t(t) = \frac{x_p'(u)x_p''(u) + y_p'(u)y_p''(u)}{\sqrt{x_p'(u)^2 + y_p'(u)^2}}\dot{u}^2(t) + \sqrt{x_p'(u)^2 + y_p'(u)^2}\ddot{u}^2(t)$$

$$= \frac{dv_p(u)}{du}\dot{u}^2(t) + v_p(u)\ddot{u}(t) \qquad (3.121)$$

where dependence on time for $u(t)$ is omitted due to shorter notation. Taking the boundary case of Eqs. (3.118), (3.120), Eq. (3.121) results in an optimal schedule differential equation

$$\ddot{u} = \pm a_{\mathrm{MAX}t}\sqrt{\frac{1}{x_p'^2 + y_p'^2} - \frac{(x_p'^2 + y_p'^2)\kappa_p^2\dot{u}^4}{a_{\mathrm{MAX}r}^2} - \frac{x_p'x_p'' + y_p'y_p''}{x_p'^2 + y_p'^2}\dot{u}^2} \qquad (3.122)$$

A solution of a differential equation is found by means of explicit simulation integrating from the TPs forward and backward in time. When accelerating a positive sign is used, and a negative sign is used when decelerating. The initial condition $u(t)$ and $\dot{u}$ are defined by the position of TPs $u_{\mathrm{TP}i}$ and from Eq. (3.120), knowing that the radial acceleration is maximal allowable in the TPs:

$$\dot{u}|_{\mathrm{TP}i} = \sqrt{\frac{a_{\mathrm{MAX}r}}{\left(x_p'(u_{\mathrm{TP}i})^2 + y_p'(u_{\mathrm{TP}i})^2\right)\kappa_p(u_{\mathrm{TP}i})}} \qquad (3.123)$$

Only a positive initial condition (3.123) is shown because $u(t)$ has to be a strictly increasing function. The differential equations are solved until the violation of the acceleration constraint is found or u leaves the interval $[u_{\mathrm{SP}}, u_{\mathrm{EP}}]$.

The violation usually appears when moving (accelerating) from the current TPi toward the next TP ($i - 1$ or $i + 1$), and the translational velocity becomes much higher than allowed by the trajectory. The solution of the differential equation (3.122) therefore consists of the segments of $\dot{u}$ around each turning point:

$$\dot{u}_l = \dot{u}_l(u), \quad u \in [\underline{u}_l, \overline{u}_l], \quad l = 1, \ldots, n_{\mathrm{TP}} \qquad (3.124)$$

where $\dot{u}_l = \dot{u}_l(u(t))$ is the derivative of the schedule depending on u, and $\underline{u}_l$, $\overline{u}_l$ are the lth segment borders. Here the segments in Eq. (3.124) are given as a function of u although the simulation of Eq. (3.122) is done with respect to time. This is because time offset (time needed to arrive in TP) is not yet known; what is known at this point is the relative time interval corresponding to each segment solution $\dot{u}_l$.

The solution of the time-minimum velocity profile obtained by accelerating on the slipping edge throughout the trajectory is possible if the union of the intervals covers the whole interval of interest $[u_{SP}, u_{EP}] \subseteq \bigcup_{l=1}^{n_{TP}} [\underline{u}_l, \overline{u}_l]$. Actual $\dot{u}$ is found by minimizing the individual ones:

$$\dot{u} = \min_{1 \leq l \leq n_{TP}} \dot{u}_l(u) \qquad (3.125)$$

Absolute time corresponding to the schedule $u(t)$ is obtained from inverting $\dot{u}(u(t)) = \frac{du}{dt}$ to obtain $dt = \dot{u}^{-1}(u)du$ and its integration:

$$t = \int_{u_{SP}}^{u_{EP}} \frac{du}{\dot{u}(u)} = t(u) \qquad (3.126)$$

where $\dot{u}(u)$ must not become 0. In our case the time-optimal solution is searched; therefore, the velocity (and also $\dot{u}(u)$) is always higher than 0. $\dot{u}(u) = 0$ would imply that the velocity is zero, meaning the system would stop, and that cannot lead to the time-optimal solution.

Also, initial (in SP) and terminal (in EP) velocity requirements can be considered as follows. Treat SP and EP as normal TPs where the initial velocities v_{SP}, $v_p(u_{SP})$, v_{EP}, $v_p(u_{EP})$ are known so the initial condition for $\dot{u}_{SP} = \frac{v_{SP}}{v_p(u_{SP})}$, $\dot{u}_{EP} = \frac{v_{EP}}{v_p(u_{EP})}$ can be computed. If these initial conditions for SP or EP are larger than the solution for $\dot{u}$ (obtained from TPs), then the solution does not exist because it would be impossible to make it through the first turning point, even if the robot brakes maximally. If the solution for given SP and EP velocities exist then also include segments of $\dot{u}$ computed for SP and EP to Eqs. (3.124), (3.125).

From the computed schedule $u(t)$, $\dot{u}(t)$, the reference trajectory is given by $x(t) = x_p(u(t))$ and $y(t) = y_p(u(t))$, and the reference velocity is given by Eqs. (3.22), (3.23).

EXAMPLE 3.17

Compute the optimal velocity profile that will result in the shortest traveling time for a given path $x_p(u) = \cos(u)$, $y_p(u) = \sin(2u)$ where $u \in [0, 2\pi]$ and considering maximal tangential accelerations $a_{MAXt} = 2\,\text{m/s}^2$ and radial acceleration $a_{MAXr} = 4\,\text{m/s}^2$.

Compute the schedule $u(t)$ and velocity profiles $v(u)$, $v(t)$.

Solution

To compute the optimal schedule $u(t)$ the algorithm presented in this chapter needs to be implemented. First the TPs need to be computed with the initial conditions given in relations (3.119), (3.123); then the solutions for each TP are simulated using Eq. (3.122). Finally, the time derivative of the schedule is minimized according to Eq. (3.125).

A possible implementation of the solution is given in Listing 3.16 (neglect code clocks that could be evaluated if the variable velCnstr would be set to true). Optimum schedule determination of Example 3.17 is given in Figs. 3.39–3.41. In Fig. 3.39 it can be seen that solution for each TP is integrated up to the point where the acceleration constraints are violated, which is marked with a thin line. The final solution is marked with a thick line.

Listing 3.16

```
1  % Define trajectory with function handles
2    x = @(u) cos(u);     y = @(u) sin(2*u);     % Path
3    dx = @(u) -sin(u);  dy = @(u) 2*cos(2*u);  % First derivative
4    ddx = @(u) -cos(u); ddy = @(u) -4*sin(2*u); % Second derivative
5    v = @(u) sqrt(dx(u).^2 + dy(u).^2); % Tangential velocity
6    w = @(u) (dx(u).*ddy(u)-dy(u).*ddx(u))./(dx(u).^2+dy(u).^2); % Angular vel.
7    kappa = @(u) w(u)./v(u); % Curvature
8
9    u = 0:0.001:2*pi; % Time
10   arMax = 4; atMax = 2; % Acceleration constraints
11   vSP = 0.2; vEP = 0.1; % Initial and final velocity requirements
12   uSP = u(1); uEP = u(end); % Start point and end point
13   uTP = []; % Turn points
14   for i = 2:length(u)-1 % Determine turn points
15       if all(abs(kappa(u(i))) > abs(kappa(u([i-1, i+1]))))
16           uTP = [uTP, u(i)];
17       end
18   end
19   up0 = sqrt(arMax./abs(v(uTP).*w(uTP))) ; % Derivative in turn points
20
21   velCnstr = false; % Enable velocity contraints (disabled)
22   if velCnstr
23       vMax = 1.5*inf; %Velocity constraints
24       for i = 1:length(uTP) % Make uTP in accordance with velocity constraint
25           vvu = v(uTP(i)); vvt = vvu*up0(i);
26           if abs(vvt) > vMax, up0(i) = abs(vMax/vvu); end
27       end
28       % Add requirements for initial and final velocity
29       uTP  = [uSP, uTP, uEP]; up0 = [vSP/v(uSP), up0, vEP/v(uEP)];
```

EXAMPLE 3.17—cont'd

```
30   end
31
32   Ts = 0.001; % Simulation sampling time
33   N = length(uTP); ts = cell(1,N); us = cell(1,N); ups = cell(1,N);
34   for i = 1:N % Loop through all turn points
35       uB = uTP(i); upB = up0(i); tB = 0;
36       uF = uTP(i); upF = up0(i); tF = 0;
37       uBs =[]; upBs = []; tBs = []; uFs = []; upFs =[]; tFs = []; % Storage
38       goB = true; goF = true;
39
40       while goB || goF
41           % Integrate back from the turn point
42           if uB > uSP && goB
43               dxT = dx(uB);    dyT = dy(uB);
44               ddxT = ddx(uB); ddyT = ddy(uB);
45               vT = v(uB)*upB; wT = w(uB)*upB; kappaT = kappa(uB);
46               arT = vT*wT; atT = atMax*sqrt(1 - (arT/arMax)^2);
47
48               if velCnstr && abs(vT) > vMax
49                   upB = vMax/v(uB); upp = 0;
50               elseif abs(arT)-arMax > 0.001
51                   arT = arMax; atT = 0; upp = 0; goB = false;
52               else
53                   atT = -real(atT);
54                   upp = real(-atMax*sqrt(1/(dxT^2 + dyT^2) - ...
55                       (dxT^2 + dyT^2)*kappaT^2*upB^4/arMax^2) - ...
56                       (dxT*ddxT + dyT*ddyT)/(dxT^2 + dyT^2) * upB^2);
57               end
58
59               uBs = [uBs; uB]; upBs = [upBs; upB]; tBs = [tBs; tB]; % Store
60               tB = tB + Ts;
61               uB = uB - upB*Ts; % Euler integration
62               upB = upB - upp*Ts; % Euler integration
63           else
64               goB = false;
65           end
66
67           % Integrate forward from the turn point
68           if uF < uEP && goF
69               dxT = dx(uF);    dyT = dy(uF);
70               ddxT = ddx(uF); ddyT = ddy(uF);
71               vT = v(uF)*upF; wT = w(uF)*upF; kappaT = kappa(uF);
72               arT = vT*wT; atT = atMax*sqrt(1 - (arT/arMax)^2);
73
74               if velCnstr && abs(vT) > vMax
75                   upF = vMax/v(uF); upp = 0;
76               elseif abs(arT)-arMax > 0.001
77                   arT = arMax; atT = 0; upp = 0; goF = false;
78               else
79                   atT = real(atT);
80                   upp = real(+atMax*sqrt(1/(dxT^2 + dyT^2) - ...
81                       (dxT^2 + dyT^2)*kappaT^2*upF^4/arMax^2) - ...
82                       (dxT*ddxT + dyT*ddyT)/(dxT^2 + dyT^2) * upF^2);
83               end
84
85               uFs = [uFs; uF]; upFs = [upFs; upF]; tFs = [tFs; tF]; % Store
86               tF = tF + Ts;
```

(Continued)

EXAMPLE 3.17—cont'd

```
87              uF = uF + upF*Ts; % Euler integration
88              upF = upF + upp*Ts; % Euler integration
89          else
90              goF = false;
91          end
92      end
93
94      ts{i} = [tBs; tB+tFs(2:end)];
95      us{i} = [flipud(uBs); uFs(2:end)];
96      ups{i} = [flipud(upBs); upFs(2:end)];
97  end
98
99  % Find minimum of all profiles ups (schedule derivative)
100 usOrig = us;
101 for i = 1:N-1
102     d = ups{i+1} - interp1(us{i}, ups{i}, us{i+1});
103     j = find(d(1:end-1).*d(2:end)<0, 1); % Where ups{i} is approx. ups{i+1}
104     % Find more exact u where profiles ups{i} and ups{i+1} are equal
105     uj = us{i+1}(j) + (us{i+1}(j+1)-us{i+1}(j))/(d(j+1)-d(j))*(0-d(j));
106     rob = interp1(us{i}, ups{i}, uj);
107
108     keep = us{i} < uj;
109     us{i} = [us{i}(keep); uj];      ups{i} = [ups{i}(keep); rob];
110     keep = us{i+1} > uj;
111     us{i+1} = [uj; us{i+1}(keep)]; ups{i+1} = [rob; ups{i+1}(keep)];
112 end
113
114 % Construct final solution profile
115 tt = interp1(usOrig{1}, ts{1}, us{1}); uu = us{1}; uup = ups{1};
116 for i = 2:N
117     ti = interp1(usOrig{i},ts{i},us{i});
118     tt = [tt; ti + tt(end) - ti(1)];
119     uu = [uu; us{i} + uu(end) - us{i}(1)];
120     uup = [uup; ups{i}];
121 end
122 vv = v(uu).*uup;
```

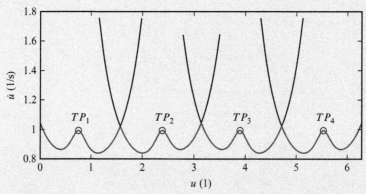

Fig. 3.39 Optimum schedule determination finding the minimum profile $\dot{u}$ of all the turn points.

EXAMPLE 3.17—cont'd

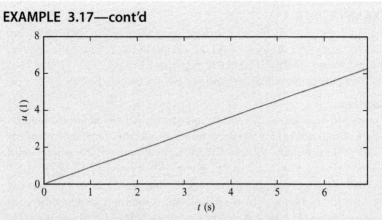

Fig. 3.40 Optimum schedule $u(t)$, which is the nonlinear function, although for this case it looks close to linear.

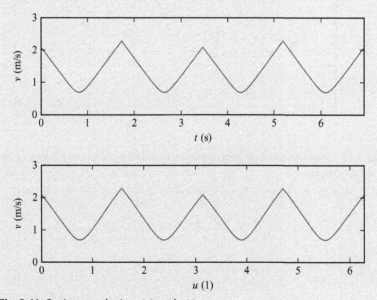

Fig. 3.41 Optimum velocity $v(u)$ and $v(t)$.

EXAMPLE 3.18

Extend Example 3.17 to also include requirements for initial and final velocity, $v_{SP} = 0.2$ and $v_{EP} = 0.1\,\text{m/s}$, respectively. Additionally consider that maximum velocity is limited to $v_{MAX} = 1.5\,\text{m/s}$.

Compute the schedule $u(t)$ and velocity profiles $v(u)$, $v(t)$.

Solution

Code from Example 3.17 can be modified to include additional requirements. Requirements for initial and terminal velocities are handled similarly as other TPs. Simply, SP and EP are treated as new TPs whose initial conditions are $u_{SP} = 0$, $u_{EP} = 2\pi$, $\dot{u}_{SP} = \frac{v_{SP}}{v_p(u_{SP})}$, and $\dot{u}_{EP} = \frac{v_{EP}}{v_p(u_{EP})}$.

The velocity constraints are taken into account if the variable velCnstr in Listing 3.16 is set to true. Optimum schedule determination of Example 3.18 is shown in Figs. 3.42–3.44.

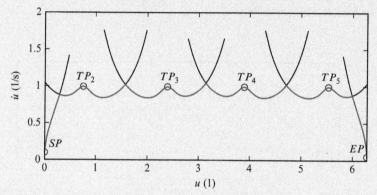

Fig. 3.42 Optimum schedule determination finding the minimum profile $\dot{u}$ of all the turn points.

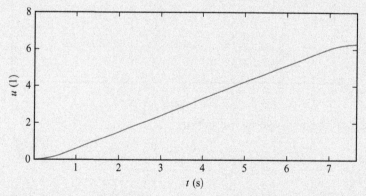

Fig. 3.43 Optimum schedule $u(t)$, which is a nonlinear function, although for this case it looks close to linear.

EXAMPLE 3.18—cont'd

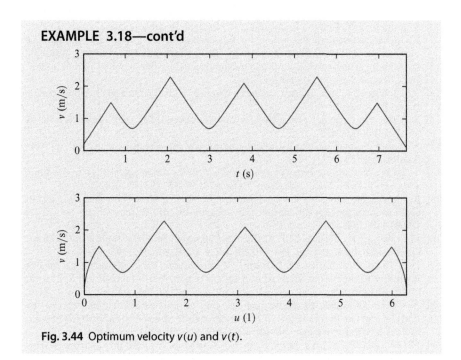

Fig. 3.44 Optimum velocity $v(u)$ and $v(t)$.

REFERENCES

[1] R.W. Brockett, Asymptotic stability and feedback stabilization, in: R.S. Millman, H.J. Sussmann (Eds.), Differential Geometric Control Theory, Birkhuser, Boston, MA, 1983, pp. 181–191.

[2] M. Bowling, M. Veloso, Motion control in dynamic multi-robot environments, in: IEEE International Symposium on CIRA'99, IEEE, Monterey, CA, 1999, pp. 168–173.

[3] L.E. Dubins, On curves of minimal length with a constraint on average curvature, and with prescribed initial and terminal positions and tangents, Am. J. Math. 79 (3) (1957) 497–516.

[4] J.A. Reeds, L.A. Shepp, Optimal paths for a car that goes both forward and backward, Pac. J. Math. 145 (2) (1990) 367–393.

[5] C.C. de Wit, O.J. Sordalen, Exponential stabilization of mobile robots with nonholonomic constraints, in: Proceedings of the 30th IEEE Conference on Decision and Control, vol. 37, 1991, pp. 1791–1797.

[6] A. De Luca, G. Oriolo, M. Vendittelli, Control of wheeled mobile robots: an experimental overview, in: S. Nicosia, B. Siciliano, A. Bicchi, P. Valigi (Eds.), Ramsete, Lecture Notes in Control and Information Sciences, vol. 270, Springer, Berlin, 2001, pp. 181–226.

[7] A. Zdešar, Simulacijsko okolje za analizo in sintezo algoritmov vodenja na osnovi digitalne slike, Bsc, Univerza v Ljubljani, 2010.

[8] Y. Kanayama, A. Nilipour, C.A. Lelm, A locomotion control method for autonomous vehicles, in: Proceedings of the 1988 IEEE International Conference on Robotics and Automation, 1988, vol. 2, April, 1988, pp. 1315–1317.

[9] Y. Kanayama, Y. Kimura, F. Miyazaki, T. Noguchi, A stable tracking control method for an autonomous mobile robot, in: Proceedings of the 1990 IEEE International Conference on Robotics and Automation, IEEE, Cincinnati, OH, 1990, pp. 384–389.

[10] C. Samson, Time-varying feedback stabilization of car-like wheeled mobile robots, Int. J. Robot. Res. 12 (1) (1993) 55–64.

[11] P.A. Ioannou, J. Sun, Robust Adaptive Control, PTR Prentice-Hall, Upper Saddle River, NJ, USA, 1996.

[12] S. Blažič, On periodic control laws for mobile robots, IEEE Trans. Ind. Electron. 61 (7) (2014) 3660–3670.

[13] S. Blažič, A novel trajectory-tracking control law for wheeled mobile robots, Rob. Autom. Syst. 59 (2011) 1001–1007.

[14] T. Takagi, M. Sugeno, Fuzzy identification of systems and its applications to modeling and control, IEEE Trans. Syst. Man. Cybern. SMC-15 (1985) 116–132.

[15] K. Tanaka, M. Sano, A robust stabilization problem of fuzzy control systems and its application to backing up control of a truck-trailer, IEEE Trans. Fuzzy Syst. 2 (2) (1994) 119–134.

[16] H.O. Wang, K. Tanaka, M.F. Griffin, An approach to fuzzy control of nonlinear systems: stability and design issues, IEEE Trans. Fuzzy Syst. 4 (1996) 14–23.

[17] H.D. Tuan, P. Apkarian, T. Narikiyo, Y. Yamamoto, Parameterized linear matrix inequality techniques in fuzzy control system design, IEEE Trans. Fuzzy Syst. 9 (2) (2001) 324–332.

[18] A. Ollero, O. Amidi, Predictive path tracking of mobile robots. application to the CMU navlab, in: Proceedings of 5th International Conference on Advanced Robotics, Robots in Unstructured Environments, ICAR, vol. 91, 1991, pp. 1081–1086.

[19] J.E. Normey-Rico, J. Gómez-Ortega, E.F. Camacho, A Smith-predictor-based generalised predictive controller for mobile robot path-tracking, Control. Eng. Pract. 7 (1999) 729–740.

[20] F. Kuhne, W.F. Lages, J.M.G. da Silva Jr, Model predictive control of a mobile robot using linearization, in: Proceedings of Mechatronics and Robotics, 2004, pp. 525–530.

[21] D. Gu, H. Hu, Neural predictive control for a car-like mobile robot, Rob. Autom. Syst. 39 (2) (2002) 73–86.

[22] G. Klančar, I. Škrjanc, Tracking-error model-based predictive control for mobile robots in real time, Rob. Autom. Syst. 55 (2007) 460–469.

[23] J. Kennedy, R. Eberhart, Particle swarm optimization, in: Proceedings of IEEE International Conference on Neural Networks, IEEE, 1995, pp. 1942–1948.

[24] J. Kennedy, R. Eberhart, Swarm Intelligence, Morgan Kaufmann, San Francisco, CA, 2001.

[25] J. Kennedy, R. Mendes, Neighborhood topologies in fully informed and best-of-neighborhood particle swarms, IEEE Trans. Syst. Man Cybern. C Appl. Rev. 36 (4) (2006) 515–519.

[26] F. Chaumette, S. Hutchinson, Visual servo control, part I: basic approaches, IEEE Robot. Autom. Mag. 13 (2006) 82–90.

[27] P.I. Corke, Visual Control of Robots: High-Performance Visual Servoing, Wiley, New York, 1996.

[28] N.R. Gans, S.A. Hutchinson, Stable visual servoing through hybrid switched-system control, IEEE Trans. Robot. 23 (2007) 530–540.

[29] E. Malis, F. Chaumette, S. Boudet, 2-1/2-D visual servoing, IEEE Trans. Robot. Autom. 15 (1999) 238–250.

[30] E. Malis, F. Chaumette, 2-1/2-D visual servoing with respect to unknown objects through a new estimation scheme of camera displacement, Int. J. Comp. Vis. 37 (1) (2000) 79–97.

[31] D. Fontanelli, P. Salaris, F.A.W. Belo, A. Bicchi, Unicycle-like robots with eye-in-hand monocular cameras: from PBVS towards IBVS, in: G. Chesi, K. Hashimoto (Eds.), Lecture Notes in Control and Information Sciences, Springer, London, 2010, pp. 335–360.

[32] A. De Luca, G. Oriolo, P.R. Giordano, Image-based visual servoing schemes for nonholonomic mobile manipulators, Robotica 25 (2007) 131–145.

[33] G.L. Mariottini, Image-based visual servoing for mobile robots: the multiple view geometry approach, PhD thesis, University of Siena, 2005.

[34] G. Chesi, K. Hashimoto, Visual Servoing via Advanced Numerical Methods, Springer-Verlag, Berlin, 2010.

[35] A. Cherubini, F. Chaumette, Visual navigation of a mobile robot with laser-based collision avoidance, Int. J. Robot. Res. 32 (2012) 189–205.

[36] F. Chaumette, S. Hutchinson, Visual servo control, part II: advanced approaches, IEEE Robot. Autom. Mag. 14 (2007) 109–118.

[37] M. Lepetič, I. Škrjanc, H.G. Chiacchiarini, D. Matko, Predictive functional control based on fuzzy model: comparison with linear predictive functional control and PID control, J. Intell. Robot. Syst. 36 (4) (2003) 467–480.

CHAPTER 4

Path Planning

4.1 INTRODUCTION

Path planning from location A to location B, simultaneous obstacle,
avoidance and reacting to environment changes are simple tasks for humans
but not so straightforward for an autonomous vehicle. These tasks present
challenges that each wheeled mobile robot needs to overcome to become
autonomous. A robot uses sensors to perceive the environment (up to some
degree of uncertainty) and to build or update its environment map. In
order to determine appropriate motion actions that lead to the desired
goal location, it can use different decision and planning algorithms. In the
process of path planning, the robot's kinematic and dynamic constraints
should be considered.

Path planning is used to solve problems in different fields, from simple
spatial route planning to selection of an appropriate action sequence that
is required to reach a certain goal. Since the environment is not always
known in advance, this type of planning is often limited to the environments
designed in advance and environments that we can describe accurately
enough before the planning process. Path planning can be used in fully
known or partly known environments, as well as in entirely unknown
environments where sensed information defines the desired robot motion.

Path planning in known environments is an active research area and
presents a foundation for more complex cases where the environment is not
known a priori. This chapter presents an overview of the most often used
path planning approaches applicable to wheeled mobile robots. First, some
basic definitions of path planning are given. For more detailed information
and further reading see also [1–3].

4.1.1 Robot Environment

The environment in which the robot moves consists of *free space* and space
occupied by the obstacles (Fig. 4.1). The start and goal configuration are located
in free space. Configuration is a complete set of parameters that define
the robot in space. These parameters usually include robot position and

Wheeled Mobile Robotics
http://dx.doi.org/10.1016/B978-0-12-804204-5.00004-4

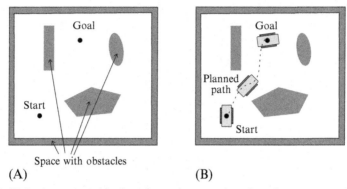

Fig. 4.1 (A) Environment with obstacles and start and goal configurations, and (B) one out of many possible paths from the start to the goal configuration.

orientation or also its joints angles. The number of these parameters equals to the number of degrees of freedom (DOF).

The environment that includes moving obstacles is the *dynamic environment*, while the *static environment* does not change with time. The environment with known positions of the obstacles is called known the environment; otherwise, the environment is unknown.

4.1.2 Path Planning

Path planning is the task of finding a continuous path that will drive the robot from the start to the goal configuration. The entire path must lie in the free space (as shown in Fig. 4.1). In path planning the mobile system uses known environment map, which is stored in the robot's memory.

The state or configuration gives a possible pose of the robot in the environment, and it can be represented as a point in the configuration space that includes all possible robot states (configurations). The robot can move from one state to the other by implementing different actions. A suitable path is therefore described as a sequence of actions that guide the robot from the start configuration (state) through some intermediate configurations to the goal configuration. Which action is chosen in the current state and which state will be next depends on the used path planning algorithm and used criteria. The algorithm chooses the next most suitable state from the set of all possible states that can be visited from the current state. This decision is made according to some criteria function, usually defined with one of the distance measures, such as the shortest Euclidean distance to the goal state.

Between certain start and goal states, there may be one or more paths, or there is no path at all that connects the states. Usually there are several feasible paths (i.e., paths that do not collide with obstacles). To narrow the selection choice additional demands or criteria are introduced that define desired optimality:

- the path length must be the shortest,
- the suitable path is the one in which the robot can pass in the shortest time,
- the path should be as far as possible from the obstacles,
- the path must be smooth without sharp turns,
- the path must consider motion constraints (e.g., nonholonomic constraints, where at the current time not all driving directions are possible).

4.1.3 Configuration and Configuration Space

Configuration is the state of the system in the environment space spanned in n dimensions. The state vector is described by $q = [q_1, \ldots, q_n]^T$. The state q is a point in n-dimensional space called *configuration space* Q, which presents all possible configurations of the mobile system according to its kinematic model.

Part of the configuration space that is occupied by the obstacles O_i is denoted $Q_{\text{obst}} = \bigcup_i O_i$. So the free part of the environment without the obstacles is defined by

$$Q_{\text{free}} = Q - Q_{\text{obst}}$$

Q_{free} therefore contains space where the mobile system can plan its motion.

Assume there is a circular robot that can only execute translatory motions in a plane (i.e., it has two DOF $q = [x, y]^T$) then its configuration can be defined by a point (x, y) presenting the robot center. In this case the configuration space Q is determined by moving the robot around obstacles while keeping it in contact with the obstacle, as shown in Fig. 4.2. In doing so the robot center point describes the border between Q_{free} and Q_{obs}. In other words, the obstacles are enlarged by the known robot dimension (radius) in order to treat the robot as a point in the path planning problem.

An additional example of configuration space for a triangular robot and rectangular obstacle is shown in Fig. 4.3. The robot can only move in x and y directions ($q = [x, y]^T$).

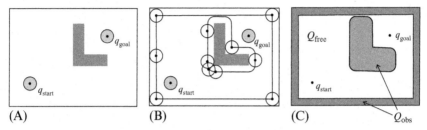

Fig. 4.2 (A) Robot with a circular shape in the environment containing an obstacle with marked start and goal configurations, (B) configuration space definition, and (C) transformed environment from (A) to a configuration space where the robot is presented simply by its center point.

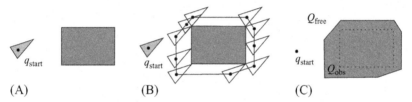

Fig. 4.3 (A) Rectangular obstacle and triangular robot with a point defining its configuration q, (B) configuration space determination, (C) free configuration Q_{free}, and space occupied by the obstacle Q_{obs}.

If the robot from Fig. 4.3 could also rotate then its configuration would have three dimensions $q = [x, y, \varphi]^T$ and the configuration space would be more complex. By simplification we could determine the configuration space using a circle that outlines the robot shape and has a center in the robot center. Obtained configuration Q_{free} is then smaller than the true free configuration because the outlined circle area is bigger than the robot area. However, this significantly simplifies the path planning problem.

4.1.4 Mathematical Description of Obstacle Shape and Pose in the Environment

The mathematical description of obstacle shape and position is required to estimate robot configuration space and additional environment simplifications. The two most common approaches to describe the obstacles are border presentation by recording vertices and presentation given with half-planes.

Border-Based Description of Obstacles

The obstacle ground plan is an m-sided polygon and has m vertices. The obstacle border (shape) can be described by its vertices that are listed in counterclockwise order. The holes in the obstacles and surrounding walls are described by vertices in the clockwise order (Fig. 4.4). Such an obstacle description is valid for convex and nonconvex polygons.

Obstacle Description Using Intersecting Half-Planes

A convex polygon with m vertices can be described by a union of m half planes, where the half plane is defined as $f(x, y) \leq 0$ for straight lines or $f(x, y, z) \leq 0$ for planes (3D obstacles). An example of how to describe a polygon with five vertices is given in Fig. 4.5. Nonconvex shapes or shapes with holes can be described with the use of operations over sets such as union, intersections, difference among sets, and the like.

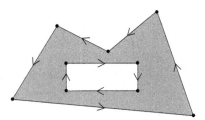

Fig. 4.4 Example of an obstacle description with listed vertices: counterclockwise for the outer obstacle boundary and clockwise for its holes. The left-hand side of each arrowed straight line segment belongs to the obstacle (*shaded area*).

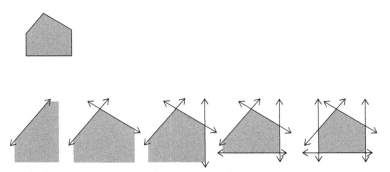

Fig. 4.5 Example of an obstacle description by half planes.

4.2 ENVIRONMENT PRESENTATION FOR PATH PLANNING PURPOSES

Before path planning, the environment needs to be presented in a unified mathematical manner that is suitable for processing in path searching algorithms.

4.2.1 Graph Representation

Configuration space consists of free space, which contains all possible configurations (states) of the mobile system, and space occupied by obstacles. If free space is reduced and presented by a subset of configurations (e.g., centers of cells) that include start and goal configurations, the desired number of intermediate configurations and transitions among them, then the *state transition graph* (also graph) is obtained. States are presented by circles and called nodes, and connections among them are given by lines. The connections (lines) therefore present actions required for the system to move among the states.

In a *weighted graph* each connection has some prescribed weight or cost that is needed for action execution and transition among the states of this connection. In a *directed graph* the connections are also marked with possible directions. In a directed graph the cost depends on the direction of transition, while in an undirected graph the transition is possible in both directions. Fig. 4.6 shows an example of a weighed and directed weighted graph. Path searching in a graph is possible with different algorithms such as A*, Dijkstra's algorithm, and the like.

4.2.2 Cell Decomposition

The environment can be partitioned to *cells*, which are usually defined as simple geometric shapes. Cells must be convex because each straight line segment connecting any two configurations inside the cell must lie entirely

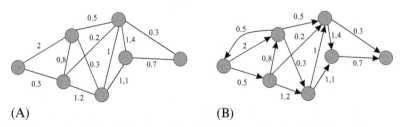

Fig. 4.6 Example of (A) weighted graph, and (B) directed and weighted graph.

in that cell. An environment partitioned to cells can be presented with a state transition graph where states are certain points inside the cells (e.g., centers) and connections among the states (nodes) are possible only between neighbor cells (e.g., the ones with a common edge).

Accurate Decomposition to Cells

Decomposition to cells is accurate if all the cells lie entirely in the free space or entirely in the space occupied by the obstacles. *Accurate decomposition* is therefore without losses because the union of all free cells equals the free configuration space Q_{free}.

An example of accurate decomposition to cells is vertical decomposition, shown in Fig. 4.7. This decomposition can be obtained by using an imaginary vertical line traveling from the left environment border to the right border. Whenever an obstacle corner appears (vertices of some polygon) a vertical border between the cells is made, traveling from the corner only up or only down, or up and down. The complexity of this approach depends on the geometry of the environment. For simple environments the number of cells and the connection among them is small; while with the increasing number of polygons (obstacles) and their vertices, the number of cells also increases.

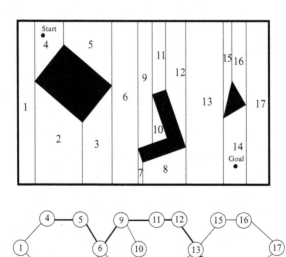

Fig. 4.7 Vertical cell decomposition (*top*) and the associated graph (*bottom*) with a marked path among the start and the goal configuration.

Accurate decomposition to cells can be presented with a state transition graph where graph nodes are, for example, the cells' centers and the transitions among the cells go, for example, through the middle points of the borders between the cells.

Approximate Decomposition to Cells

Environment decomposition to cells is approximate when it is possible that certain cells contain free configuration space as well as an obstacle or only part of some obstacle. All cells containing at least part of the obstacle are therefore normally marked as occupied, while the remaining cells are marked as free. If decomposition to the cells is made of the same size, then an *occupancy grid* (Fig. 4.8) is obtained.

The center of each cell (in Fig. 4.8, the free cell is marked with white) is presented as a node in a graph. Transitions among the cells are in four or eight directions if diagonal transitions are allowed. This approach is very simple for application, but due to the constant cell size some information about the environment might get lost (this is not lossless decomposition); for example, obstacles are enlarged and narrow passages among them might get lost. The main drawback of this approach is the memory usage, which increases either by enlarging the size of the environment or by lowering the cell size.

A smaller loss of the environment information at lower memory usage is obtained if the variable cells size is used. A representative example of the latter are quadtrees where initially a single cell is put over the environment. If entire cell is in free space or the entire cell is occupied by the obstacle then the cell remains as is. Otherwise, if this cell is only partly occupied by an obstacle, then it is split into smaller cells. This procedure is repeated until

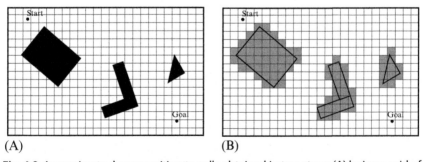

(A)   (B)

Fig. 4.8 Approximate decomposition to cells obtained in two steps: (A) laying a grid of equal cells on the environment and (B) marking cells as free or occupied (*shaded cells*).

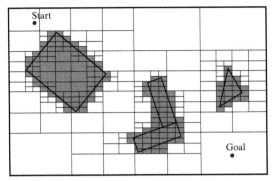

Fig. 4.9 Approximate decomposition to cells using variable cell size: quadtree. Free cells are marked as *white* and occupied cells are *shaded*.

there is a desired resolution. An example of a quadtree is given in Fig. 4.9, and its presentation can be described with a state graph. Approximate decomposition to cells is simpler than the accurate decomposition to cells; however, it can happen that due to the loss of information a path planning algorithm may not be able to find a solution although it does exist in the environment.

EXAMPLE 4.1

Write a program that performs a quadtree decomposition of some environment with a rectangular shape of the obstacles. A framework that generates a random environment with obstacles is given in Listing 4.1, where a function for the calculation of quadtree is already declared. The function accepts the obstacles, environment dimension, and desired tree depth, and it returns a quadtree in the form of a Matlab structure.

Listing 4.1 Framework for implementation of quadtree decomposition

```
1  bb = [0, 16, 0, 12]; % Environment dimension: xa, xb, ya, yb
2  N = 10; % Number of obstacles
3  minDim = [0.1; 0.1]; % Min dimension of an obstacle: xMin and yMin
4  maxDim = [2; 2]; % Max dimension of an obstacle: xMax and yMax
5  % Random obstacle map, vertices in a column: x1, y1, x2, y2, x3, y3, x4, y4
6  obst = zeros(8, N);
7  for i = 1:N
8      p = [bb(1); bb(3)] + [diff(bb(1:2)); diff(bb(3:4))].*rand(2,1);
9      phi = 2*pi*rand(); d = minDim/2 + (maxDim-minDim)/2*rand();
10     R = [cos(phi),-sin(phi); sin(phi), cos(phi)];
11     v = repmat(p, 1, 4) + R*([-1, 1, 1,-1; -1,-1, 1, 1].*repmat(d, 1, 4));
12     obst(:,i) = reshape(v, [], 1);
13 end
14
15 tree = quadTree(obst, bb, 4); % Create quad tree with depth-level 4
```

(Continued)

EXAMPLE 4.1—cont'd
Solution

A possible implementation of a quadtree algorithm in Matlab is given in Listing 4.2. The result of the algorithm on a randomly generated map is visualized in Fig. 4.10.

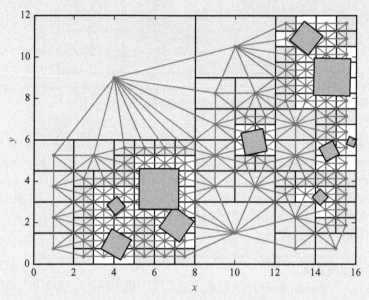

Fig. 4.10 Quadtree decomposition of the environment with random obstacles. The quadtree is overlaid with a mesh that connects the centers of all the neighboring cells.

Listing 4.2 Quadtree-decomposition algorithm

```
1  function tree = quadTree(obst, bb, level)
2      % Generate quad $tree$ of depth $level>=0$ around obstacles $obst$
3      % (every column contains the obstacle vertices: x1, y1, x2, y2, ...) in
4      % an environment of dimension $bb$ (xa, xb, ya, yb).
5      minDim = [diff(bb(1:2)); diff(bb(3:4))]/2^level; % Min cell dimension
6      % Root node
7      tree(1).leaf = true; % % Is the cell a leaf?
8      tree(1).free = false; % Is the cell occupied?
9      tree(1).bounds = bb; % Cell boundary: xa, xb, ya, yb
10     tree(1).center = [mean(bb(1:2)); mean(bb(3:4))]; % Cell center
11
12     id = 1; k = 2;
13     while id < k
14         occupied = isOccupied(tree(id).bounds, obst);
15
16         d = [diff(tree(id).bounds(1:2)), diff(tree(id).bounds(3:4))]/2;
17         if occupied && d(1) > minDim(1)/2 % Split the cell into 4 new cells
18             tree(id).leaf = false;
```

EXAMPLE 4.1—cont'd

```
19              tree(id).free = false;
20
21              b = tree(id).bounds;
22              bs = [b(1),         b(1)+d(1), b(3),        b(3)+d(2); ...
23                   b(1)+d(1), b(2),        b(3),        b(3)+d(2); ...
24                   b(1),         b(1)+d(1), b(3)+d(2), b(4); ...
25                   b(1)+d(1), b(2),        b(3)+d(2), b(4)];
26              for i = 1:4
27                  tree(k).leaf = true;
28                  tree(k).free = false;
29                  tree(k).bounds = bs(i,:);
30                  tree(k).center = [mean(bs(i,1:2)); mean(bs(i,3:4))];
31                  k = k + 1;
32              end
33          elseif ~occupied
34              tree(id).free = true;
35          end
36          id = id + 1;
37      end
38
39      % Create visibility graph
40      a = zeros(2,length(tree)*4); leafs = zeros(1,length(tree));
41      for i = 1:length(tree)
42          a(:,i*4-3:i*4) = tree(i).bounds([1, 2, 2, 1; 3, 3, 4, 4]);
43          leafs(i) = tree(i).leaf;
44      end
45      offset = [-1, 1, 1,-1; -1,-1, 1, 1]/2.*repmat(minDim, 1, 4);
46      for i = 1:length(tree)
47          tree(i).neighbours = [];
48          if tree(i).leaf
49              b = tree(i).bounds([1, 2, 2, 1; 3, 3, 4, 4]) + offset;
50              c = reshape(inpolygon(a(1,:), a(2,:), b(1,:), b(2,:)), 4, []);
51              tree(i).neighbours = setdiff(find(any(c).*leafs), i);
52          end
53      end
54 end
55
56 function occupied = isOccupied(bounds, obst)
57      occupied = false;
58      pb = bounds([1, 2, 2, 1, 1; 3, 3, 4, 4, 3]);
59      for j = 1:size(obst, 2) % Loop through the obstacles
60          pa = reshape(obst(:,j), 2, []); N = size(pa, 2);
61          ina = inpolygon(pa(1,:), pa(2,:), pb(1,:), pb(2,:));
62          inb = inpolygon(pb(1,:), pb(2,:), pa(1,:), pa(2,:));
63          if any(ina) || any(inb) % Are any vertices in the obstacle or cell?
64              occupied = true; break;
65          else % Check if there are any intersecting edges.
66              for k = 1:size(pb, 2)-1 % Loop through the boundary edges
67                  for i = 1:N % Loop through the obstacle edges
68                      a1 = [pa(:,i); 1]; a2 = [pa(:,mod(i,N)+1); 1];
69                      b1 = [pb(:,k); 1]; b2 = [pb(:,k+1); 1];
70                      pc = cross(cross(a1, a2), cross(b1, b2));% Intersection
71                      if abs(pc(3))>eps
72                          pc = pc/pc(3);
73                          da = a2-a1; ca = pc-a1; ea = (ca.'*da)/(da.'*da);
74                          db = b2-b1; cb = pc-b1; eb = (cb.'*db)/(db.'*db);
75                          if eb>eps && eb<1 && ea>eps && ea<1
```

(Continued)

EXAMPLE 4.1—cont'd

```
76                              occupied = true; break;
77                          end
78                      end
79                  end
80              if occupied, break; end
81          end
82      end
83      if occupied, break; end
84  end
85 end
```

4.2.3 Roadmap

A roadmap (map that contains roads), which consists of lines, curves, and their points of intersection, gives possible connections between points in the free space. The task of path planning is to connect the start point and the goal point with an existing road connection in the map to find a connecting sequence of roads. A roadmap depends on environment geometry. The challenge is to find a minimum number of roads that enable a mobile robot to access any free part of the environment. In the following, two different roadmap building algorithms are presented: visibility map and Voronoi graph.

Visibility Graph

A *visibility graph* consists of all possible connections among any two vertices that lie entirely in the free space of the environment. This means that for each vertex connections are made to all the other vertices that can be seen from it. The start point and the goal point are treated as vertices. Connections are also made between neighboring vertices of the same polygon. An example of a visibility graph is given in Fig. 4.11A. The path obtained using a visibility graph tends to be as close as possible to the obstacles; therefore, it is the shortest path possible. To prevent collision of the robot with the obstacles the obstacles can be enlarged by the robot diameter as described in Section 4.1.3.

Visibility graphs are simple to use but the number of road connections increases with the number of obstacles, which can result in higher complexity and therefore lower efficiency. A visibility graph can be simplified

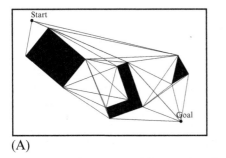 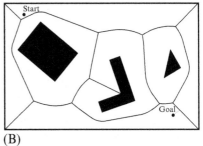

(A)   (B)

Fig. 4.11 Roadmap: (A) visibility graph and (B) Voronoi graph.

by removing redundant connections that can be replaced by existing shorter connections.

Voronoi Graph

A Voronoi graph (Fig. 4.11B) consists of road sections that have the largest distance from the obstacles. This also means that the road between any two obstacles has equal distance to both obstacles.

A basic Veronoi graph (diagram) is defined for a plane with n points (e.g., point obstacle). It partitions the plane into n regions whose borders define the roadmap. Each region has exactly one generating point and any point in a certain region is closer to its generating point than to any other generating point. An example of a general Voronoi graph for the plane environment consisting of obstacles of any shape (square, straight line, etc.) is shown in Fig. 4.11. Here, space is partitioned to regions where each region has exactly one generating obstacle. Any point in a certain region is closer to the generating obstacle than to any other obstacle. Borders between the regions define the roadmap.

Driving on such a road minimizes the risk of colliding with obstacles, which can be desired when the robot pose is known with some uncertainty (due to measurement noise or control). In environments constructed from polygons the roadmap consists of three typical *Voronoi curves* as shown in Fig. 4.12. A Voronoi curve that has equal distance between
- two vertices is a straight line,
- two edges is also a straight line, and
- a vertex and a straight line segment is a parabola.

In a roadmap the start and the goal configuration are also connected, and the obtained graph is used to find the solution.

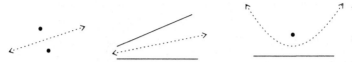

Fig. 4.12 Typical Voronoi curves.

As already mentioned this approach maximizes robot distance to the obstacles. However, the obtained path length is far from being the optimal (shortest) one. A robot with distance sensors (such as an ultrasonic or laser range finder) can apply a control to drive equally away from all the surrounding obstacles, which means that it follows roads from the Voronoi graph. Although robots using only touch or vicinity sensors cannot follow Voronoi roads because they may have problems with localization, they can easily follow roads in the visibility graph.

EXAMPLE 4.2

Compute the Voronoi graph for the environment in Fig. 4.11 using the function voronoi in Matlab. The environment (o) and object (o1, o2, o3) coordinates are defined in Listing 4.3.

Listing 4.3 Vertices of the obstacles in the environment

```
1  o  = 1000*[0.0149, 0.0693; ...
2              1.6228, 0.0679; ...
3              1.6241, 1.0867; ...
4              0.0112, 1.0854];
5  o1 = 1000*[0.4263, 0.4569; ...
6              0.6144, 0.6857; ...
7              0.3097, 0.9414; ...
8              0.1190, 0.7126];
9  o2 = 1000*[0.8151, 0.2079; ...
10             1.0885, 0.3008; ...
11             0.9644, 0.6574; ...
12             0.8753, 0.6278; ...
13             0.9706, 0.3573; ...
14             0.7838, 0.2927];
15 o3 = 1000*[1.3319, 0.4865; ...
16             1.4723, 0.5659; ...
17             1.3845, 0.7112];
18 % Free space is on the right-hand side of the obstacle
19 obstacles = {flipud(o), o1, o2, o3};
```

Solution

In Matlab the Voronoi graph for a list of points can be drawn using a function voronoi. The environment in Fig. 4.11 also contains polygons with defined edges; therefore, the voronoi function cannot be used

EXAMPLE 4.2—cont'd

directly. But if each edge is presented by a number of equidistant (assistant) points, an approximation of the Voronoi graph can be computed (using function `voronoi`) as shown in Fig. 4.13, where 20 points are used for each obstacle edge.

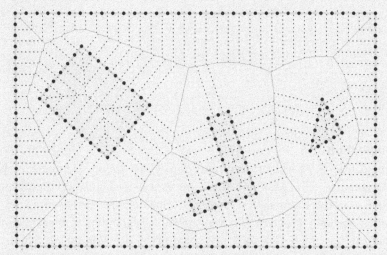

Fig. 4.13 Approximation of Voronoi graph using Matlab function `voronoi` and assistant points for the borders' presentation.

To obtain the final Voronoi graph the obtained cell borders between assistant points belonging to the same edge need to be removed; better yet, they do not need to be computed.

Listing 4.4 Approximation of Voronoi graph

```
1  % Break down the line segments into intermediate points
2  dMin = 50;
3  points = []; obst = [];
4  B = length(obstacles);
5  for i = 1:B
6      ob = obstacles{i};
7      M = size(ob, 1);
8      for j = 1:M
9          k = mod(j, M)+1; % j+1
10         d = sqrt((ob(j,1)-ob(k,1))^2 + (ob(j,2)-ob(k,2))^2);
11         n = ceil(d/dMin)+1;
12         x = linspace(ob(j,1), ob(k,1), n).';
13         y = linspace(ob(j,2), ob(k,2), n).';
14         points = [points; x(1:end-1) y(1:end-1)];
15     end
16     obst = [obst; obstacles{i}([1:end,1],:); nan(1,2)];
17 end
18
19 % Calculate voronoi segments on a set of intermediate points
```

(Continued)

EXAMPLE 4.2—cont'd

```
20 [vx, vy] = voronoi(points(:,1), points(:,2));
21
22 % Eliminate auxiliary (not in the free space) voronoi segments
23 s = false(1, size(vx, 2));
24 for j = 1:size(vx, 2)
25     in = inpolygon(vx(:,j), vy(:,j), obst(:,1), obst(:,2));
26     s(j) = all(in==1);
27 end
28 ux = vx(:,s); uy = vy(:,s); % Approximated Voronoi segments
29
30 % Plot
31 voronoi(points(:,1), points(:,2), 'b:'); hold on;
32 plot([ux;nan(1,size(ux,2))], [uy;nan(1,size(uy,2))], 'g-', 'LineWidth', 1);
33 axis equal tight;
```

Triangulation

Triangulation is a form of decomposition where the environment is split into triangular cells. Although there are various triangulation algorithms, finding a good triangulation approach that avoids narrow triangles is an open research problem [2]. One possible algorithm is the Delaunay triangulation, which is a dual presentation of the Voronoi graph. In a Delaunay graph the center of each triangle (center of the circumscribed circle) coincides with each vertex of the Voronoi polygon. An example of the latter is given in Fig. 4.14.

From accurate environment decomposition approaches, path searching algorithms can be used in order to determine if the path exists or if it does not. These approaches are therefore complete.

EXAMPLE 4.3

Obtain Delaunay triangulation for the environment given in Fig. 4.11. Points of the obstacles borders are given in Listing 4.3.

Solution

In Matlab, the Delaunay triangulation can be computed for known vertices using function DelaunayTri and plotted using function triplot, as shown in Listing 4.5.

Listing 4.5 Delaunay triangulation

```
1 points = cell2mat(obstacles(:));
2 dt = delaunayTriangulation(points);
3 triplot(dt, 'b-'); axis equal tight;
```

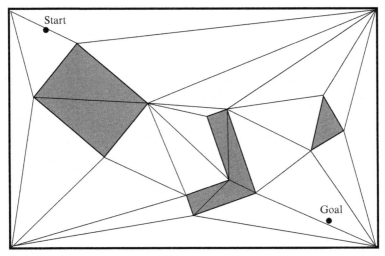

Fig. 4.14 Delaunay triangulation where some triangle edges coincide with the edges of the obstacles.

4.2.4 Potential Field

The potential field method describes the environment with the *potential field*, which can be thought of as an imaginary height. The goal point is in the bottom, and the height increases with the distance to the goal point and is even higher at the obstacles. A path planning procedure can then be explained as the motion of the ball that rolls downhill to the goal, as shown in Fig. 4.15.

A potential field is expressed as a sum of the attractive field due to the goal point $U_{attr}(q)$ and a repulsive field $U_{rep}(q)$ due to the obstacles:

$$U(q) = U_{attr}(q) + U_{rep}(q) \qquad (4.1)$$

The goal point is the global minimum of the potential field. Attractive potential $U_{attr}(q)$ in Eq. (4.1) can be defined to be proportional to the squared Euclidean distance to the goal point $D(q, q_{goal}) = \sqrt{(x - x_{goal})^2 + (y - y_{goal})^2}$, as follows:

$$U_{attr}(q) = k_{attr}\frac{1}{2}D^2(q, q_{goal}) \qquad (4.2)$$

where k_{attr} is a positive constant.

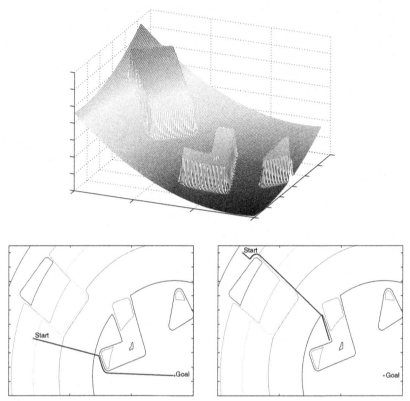

Fig. 4.15 Potential field for known goal point (*upper*) and contours with equal potential and calculated path for two start points (*lower*), where the calculated path reaches the goal (*left*) and where the path is trapped in the concave obstacle (*right*).

Repulsive potential $U_{\text{rep}}(q)$ should be very high near the obstacles and with a decreasing potential as the distance from the obstacles increases $D(q, q_{\text{obst}})$. It should be zero when $D(q, q_{\text{obst}})$ is higher than some threshold value D_0. The repulsive potential can be expressed as follows:

$$U_{\text{rep}}(q) = \begin{cases} \frac{1}{2} k_{\text{rep}} \left(\frac{1}{D(q, q_{\text{obst}})} - \frac{1}{D_0} \right)^2 ; & D(q) \le D_0 \\ 0; & D(q) > D_0 \end{cases} \tag{4.3}$$

where k_{rep} is a positive constant and $D(q, q_{\text{obst}})$ is the distance to the nearest point on the nearest obstacle.

To obtain the path from the start to the goal point we need to follow the negative gradient of potential field $(-\nabla U(q))$.

EXAMPLE 4.4

Calculate the negative gradient of the potential field (4.1).

Solution

The negative gradient of attractive field (4.2) equals to

$$-\nabla U_{attr}(q) = -k_{attr}\frac{1}{2}\begin{bmatrix} 2(x - x_{goal}) \\ 2(y - y_{goal}) \end{bmatrix} = k_{attr}(q_{goal} - q)$$

and it is in the direction from the robot pose q to the goal q_{goal} where the magnitude (norm) is proportional to the distance between the points q and q_{goal}.

The negative gradient of the repulsive field (4.3) when $D_{obst} \leq D_0$ equals

$$-\nabla U_{rep}(q) = -k_{rep}\left(\frac{1}{D_{obst}} - \frac{1}{D_0}\right)\frac{-1}{D_{obst}^2}\nabla D_{obst}$$

$$= k_{rep}\left(\frac{1}{D_{obst}} - \frac{1}{D_0}\right)\frac{1}{D_{obst}^3}(q - q_{obst}),$$

where $D_{obst} = D(q, q_{obst}) = \sqrt{(x - x_{obst})^2 + (y - y_{obst})^2}$. The direction of the repulsive field is always away from the obstacle and its strength decreases with the distance to the obstacle. For situations where $D_{obst} > D_0$, the repulsive gradient is $-\nabla U_{rep}(q) = 0$.

By using a potential field for environment presentation the robot can reach the goal point simply by following the negative gradient of the potential field. The negative gradient of the potential field is explicitly calculated from the known robot position. The main disadvantage of this approach is the possibility that the robot can become trapped (jittering behavior) in the local minimum, which can happen if the environment contains any obstacle of concave shape (Fig. 4.15) or when the robot starts to oscillate between more equally distanced points to the obstacle. The path planning using a potential field can be used to calculate the reference path that the robot needs to follow, or it can also be used online in motion control to steer the robot in the current negative gradient direction.

EXAMPLE 4.5
Calculate the potential field shown in Fig. 4.15 for the environment with obstacle coordinates given in Listing 4.3.

Solution
The potential field is calculated using relations (4.1)–(4.3), while the path is calculated by integration of the negative gradient, calculated in Example 4.4.

4.2.5 Sampling-Based Path-Planning

Up to now the presented path planning methods required explicit presentation of free environment configuration space. By increasing the dimension of the configuration space these methods tend to become time consuming. In such cases, sampling-based methods could be used for environment presentation and path planning.

In sampling-based path planning, random points (robot configurations) are sampled; then collision detection methods are applied in order to verify if these points belong to the free space [4, 5]. From a set of such sampled points and connections among them (connections must also lie in the free space) a path between the known start and the desired goal point is searched.

Sampling-based methods do not require calculation of free configuration space Q_{free}, which is a time-consuming operation for complex obstacle shapes and for many DOF. Instead, the random samples are used to present the configuration space, and independently from the environment geometry a path planning solution can be obtained for a wide variety of problems. In comparison to decomposition of the environment to cells, the sampling-based approaches do not require a high number of cells and time–consuming computation that is normally required to achieve accurate decomposition to cells. Due to inclusion of stochastic mechanisms (random walk) such as in random path planner the problem of being trapped in some local minimum (e.g., as in the potential field method) is unlikely, because motion is defined between randomly chosen samples from the free space.

To save processing time the collision detection is checked only for obstacles that are close enough and present potential danger of colliding with the robot. Additionally, the robot and obstacles can be encircled by

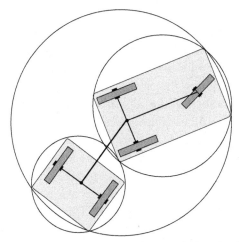

Fig. 4.16 Presenting a complex robot shape with an enclosing simple shape (*circle*) that can be split into two smaller simple shapes while performing hierarchical detection of collision of the robot.

simple shapes so that complex collision detection (between true robot shape and obstacle shapes) is performed only when those enclosing shapes are overlapping. Furthermore, collision detection can be performed hierarchically by exchanging bigger shapes that encircle the whole robot for two smaller shapes that encircle two halves of the robot, as shown in Fig. 4.16. If any of the two shapes are overlapping with the obstacle it is again split in two appropriate smaller shapes. This is continued until the collision is confirmed or rejected or until the desired resolution is reached.

Sampling-based approaches can be divided into the ones that are appropriate for a single path search and those that can be used for many path search queries. In the former approaches the path from a single start to a single goal point needs to be found as quickly as possible; therefore, the algorithms normally focus only on parts of the environment that are more promising to find the solution. New points and connections to the graph are included at runtime until the solution is found. In the latter approaches, the entire procedure of presenting the whole free environment space by an undirected graph or roadmap is performed first. The path planning solution can then be found in the obtained graph for an arbitrary pair of start and goal points. In the following, an example of both approaches is given.

RRT Method

A rapidly exploring random tree (RRT) is the method for searching the path from a certain known start point to one goal point [4]. In each iteration, the method adds a new connection in the direction from the randomly sampled point to the closest point in the existing graph.

In the first iteration of the algorithm the starting configuration q_i presents the graph tree. In each next iteration a configuration q_{rand} is chosen randomly and the closest node q_{near} from the existing graph is searched for. In the direction from q_{near} to q_{rand} at some predefined distance ε a candidate for a new node q_{new} is calculated. If q_{new} and the connection from q_{near} to q_{new} are in the free space then q_{new} is selected as the new node and its connection to q_{near} is added to the graph. The procedure is illustrated in Fig. 4.17.

The search is complete after some number of iterations are evaluated (e.g., 100 iterations) or when some probability is reached (e.g., 10%). Once the algorithm termination criterion is satisfied, the goal point is selected instead of taking a new random sample, and a check is made to determine whether the goal point can be connected to the graph [6]. Such a graph tree is quickly extending to unexplored areas as shown in Fig. 4.18. This method has only two parameters: the size of the step ε and the desired resolution or number of iterations that define the algorithm termination conditions. Therefore, the behavior of the RRT algorithm is consistent and its analysis is simple.

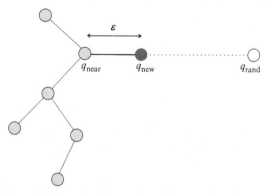

Fig. 4.17 Illustration of RRT method. Extension of the graph with a new point q_{new} in the direction of randomly sampled point q_{rand}.

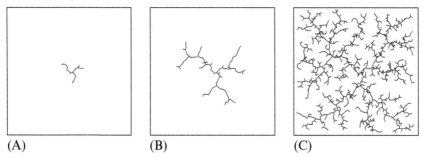

Fig. 4.18 The tree built by RRT method quickly progress to the unexplored free space. From the left to the right figure the trees contain (A) 20, (B) 100, and (C) 1000 nodes, respectively.

EXAMPLE 4.6

Implement an RRT algorithm that will produce a graph tree for a 2D free space of size 10 by 10 squared meters where the parameter $\varepsilon = 0.2$ m.

Solution

A Matlab script that produces results similar as in Fig. 4.18 could be implemented as shown in Listing 4.6.

Listing 4.6 RRT algorithm

```
1  xi = [5, 5]; % Initial configuration
2  D = 0.2; % Distance to new node
3  maxIter = 1000;
4  M = [xi]; % Map
5
6  j = 1;
7  while j < maxIter
8      xRand = 10*rand(1,2); % Random configuration
9      dMin = 100; iMin = 1; % Search for the closest point in the map M
10     for i = 1:size(M,1)
11         d = norm(M(i,:)-xRand);
12         if d<dMin
13             dMin = d;
14             iMin = i;
15         end
16     end
17
18     xNear = M(iMin,:);
19     v = xRand - xNear;
20     xNew = xNear + v/norm(v)*D; % Calculate new point
21
22     con = [xNear; xNew];
23     M = [M; xNew];
24     j = j + 1;
25
26     line(con(:,1), con(:,2), 'Color', 'b');
27  end
```

EXAMPLE 4.7

Extend Example 4.6 to also include simple obstacles (e.g., circular-shaped obstacles).

Solution

Assume there is an environment with simple obstacles, such as circular objects with known positions and diameters. Check if the candidate for a new node q_{new} is in the free area and if the straight line segment connecting q_{near} and q_{new} also lies in the free space.

PRM Method

A probabilistic roadmap (PRM) is the method used for path searching between more start points and more goal points [7]. The algorithm has two steps. The first step is the learning phase where a roadmap or undirected graph of the free space is constructed as shown in Fig. 4.19A. In the second step the current start and goal point are connected to the graph and some graph path searching algorithm is used to find the optimum path, as shown in Fig. 4.19B.

In the learning phase a roadmap is constructed. The roadmap is initially an empty set and it is later filled with nodes by repeating the following steps. A randomly chosen configuration q_{rand} from the free space is included in the map, and nodes Q_n that are needed to expand the map are found. These nodes can be found by choosing K nearest neighbor nodes (Q_n) or all neighbor nodes whose distance to q_{rand} is smaller than some predefined parameter D, as shown in Fig. 4.20. Note that in the first iteration or at

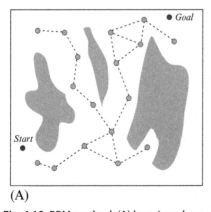
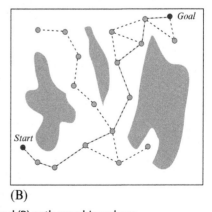

(A)   (B)

Fig. 4.19 PRM method: (A) learning phase and (B) path searching phase.

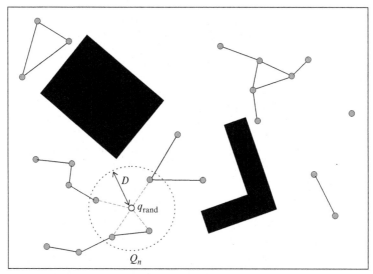

Fig. 4.20 In PRM method all possible connections from the random point q_{rand} to neighbor nodes Q_n are included in the map.

first few iterations neighbor nodes probably cannot be found. All simple connections from q_{rand} to the nodes Q_n that entirely lie in the free space are included in the map. This procedure is repeated until the map contains the desired number of N nodes.

In the path searching phase the desired start and the goal point are connected through free space with the closest possible nodes from the map. Then a path searching algorithm is applied to find the path from the start to the goal point.

These two phases do not need to be performed separately, as they can be repeated iteratively until the current number of nodes is high enough to find the solution. If the solution cannot be found, the map is further expanded with new nodes and connections until the solution is feasible. In this way we iteratively approach toward the most appropriate presentation of the free space.

This method is very effective for robots with many DOF; however, there can be problems finding the path between two regions that are connected with a narrow passage. This problem can be solved by adding additional nodes using the *bridge test*, where three nearby random points on some straight line is selected, as shown in Fig. 4.21. If the outermost points are in collision with the obstacle and the middle one is not, then the latter is included as a node in the map. We try to connect this node with neighbor

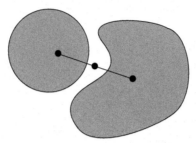

Fig. 4.21 In the bridge test two random nearby points that define a line are selected. The middle point is added as a node to the roadmap if it is in the free space and the outer points are inside obstacles.

nodes in a similar way as other connections and add these connections to the map. By combining a bridge test with uniform sampling into some hybrid sampling strategy [8], smaller roadmaps that cover free space efficiently and still have good connectively in narrow passages can be obtained.

EXAMPLE 4.8

Implement a PRM algorithm for 2D free space of size 10 by 10 squared meters. Try to extend the obtained algorithm for an environment with obstacles.

Solution

A solution is given in Listing 4.7.

Listing 4.7 PRM algorithm

```
1  D = 1; % Distance parameter
2  maxIter = 200;
3  M = []; % Map
4
5  j = 1;
6  while j <= maxIter
7      xRand = 10*rand(1,2); % Random configuration
8      M = [M; xRand];
9      con = []; % Connections
10     for i = 1:size(M,1) % Search connections to the neighbor nodes
11         d = norm(M(i,:)-xRand);
12         if d<D && d>eps % Add connections from xRand to neighbor
13             con = [con; xRand, M(i,:)];
14         end
15     end
16     j = j + 1;
17
18     line(xRand(1), xRand(2), 'Color', 'r', 'Marker', '.');
19     for i = 1:size(con,1)
20         line(con(i,[1,3]), con(i,[2,4]), 'Color', 'b');
21     end
22  end
```

4.3 SIMPLE PATH PLANNING ALGORITHMS: BUG ALGORITHMS

Bug algorithms are the simplest path planning algorithms that assume only local knowledge and do not need a map of the environment. Therefore, they are appropriate in situations where an environment map is unknown or it is changing rapidly and also when the mobile platform has very limited computational power. These algorithms use local information obtained from their sensors (e.g., distance sensor) and global goal information. Their operation consists of two simple behaviors: motion in a straight line toward the goal and obstacle boundary following.

Mobile robots that use these algorithms can avoid obstacles and move toward the goal. These algorithms require low memory usage and the obtained path is usually far from being optimal. Bug algorithms were first implemented in [9] and several improvements followed such as in [10–12].

In the following three basic versions bug algorithms are described.

4.3.1 Bug0 Algorithm

A Bug0 algorithm has two basic behaviors:
1. Move toward the goal until an obstacle is detected or the goal is reached.
2. If an obstacle is detected, then turn left (or right, but always in the same direction) and follow the contour of the obstacle until motion in a straight line toward the goal is again possible.

A example of Bug0 algorithm performance is given in Fig. 4.22.

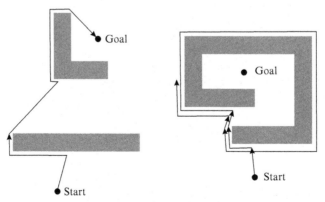

Fig. 4.22 Bug0 algorithm successfully finds a path to the goal in the environment on the *left* while it is unsuccessful in the environment on the *right*.

4.3.2 Bug1 Algorithm

A Bug1 algorithm in comparison to Bug0 uses more memory and requires several more computations. In each iteration it needs to calculate Euclidean distance to the goal and remember the closest point on the obstacle circumference to the goal. Its operation is described by the following:

1. Move on the straight line toward the goal until an obstacle is hit or the goal is reached.
2. If an obstacle is detected, then turn left and follow the entire contour of the obstacle and measure the Euclidean distance to the goal. When the point where the obstacle was initially detected is reached again, follow the contour of the obstacle in the direction that is shortest to the point on the contour that is closest to the goal. Then resume moving toward the goal in a straight line.

A example of a Bug1 algorithm operation is given in Fig. 4.23.

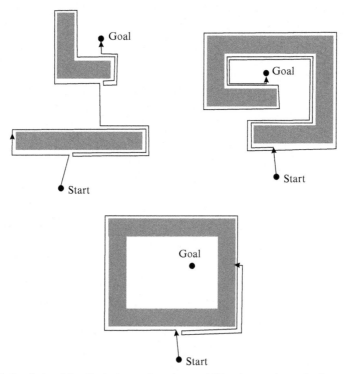

Fig. 4.23 Bug1 algorithm finds the path to the goal if it exists as shown in the *upper two figures*. In the worst case its path is for $\frac{3}{2}$ of the length of all obstacle contours longer than the Euclidean distance from the start to the goal configuration. The algorithm can identify unreachable goal.

The obtained path is not optimal and is in the worst case for $\frac{3}{2}$ of the length of all obstacle contours longer than the Euclidean distance from the start to the goal configuration. For each obstacle that is detected on its path from the start to the goal, the algorithm finds only one entry point and one leaving point from the obstacle contour. So it never detects the obstacle more than once and therefore never circles among the same obstacles. When the algorithm detects the same obstacle more than once it knows that either the start or the goal point is captured inside the obstacle and the path searching can be terminated, since no feasible path to the goal exists (bottom situation in Fig. 4.23).

4.3.3 Bug2 Algorithm

The Bug2 algorithm always tries to move on the main line that is defined as a straight line connecting the start point and the goal point. It operates by repeating the following steps:

1. Move on the main line until an obstacle is hit or the goal point is reached. In the latter case the path searching is finished.
2. If an obstacle is detected, follow the obstacle contour until the main line is reached where the Euclidean distance to the goal point is shorter than the Euclidean distance from the point where the obstacle was first detected.

Although the Bug2 algorithm in general seems much more effective than Bug1 (see left part of Fig. 4.24), it does not guarantee that the robot detects certain obstacles only once. At some obstacle configurations the robot with Bug2 could make unnecessary repeating obstacle encirclements

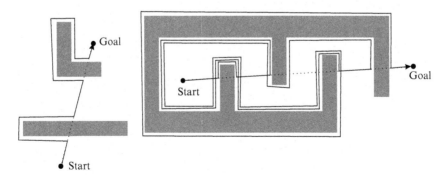

Fig. 4.24 The Bug2 algorithm follows the main line to the goal point. It can encircle the same obstacle more than once; therefore, unnecessary circling can occur. The Bug2 algorithm can identify an unreachable goal.

until it reaches the goal point as illustrated in the right part of Fig. 4.24. The algorithm can identify that the goal is unreachable if it comes across (detects) the same obstacle in the same point more than once.

From comparison of Bug1 and Bug2 algorithms the following can be concluded:

- Bug1 is the more thorough search algorithm, because it evaluates all the possibilities before making a decision.
- Bug2 is the greedy algorithm, because it selects the first option that looks promising.
- In most cases the Bug2 algorithm is more efficient than Bug1.
- However, the operation of the Bug1 algorithm is easier to predict.

EXAMPLE 4.9

Implement path planning using Bug0 algorithm for a mobile robot with differential drive. Assume that the environment map is not available and the robot only knows its current pose, goal point, and current distance to the goal (measurement from sensor).

According to the Bug0 algorithm the robot should drive toward the goal if it is far enough from any obstacle (e.g., more than 0.2 m), and it should follow the obstacle if it is close to the obstacle. Code your implementation of the algorithm in the simulation framework given in Listing 4.8, which already enables simulation of robot motion and sensor measurements. The environment and an example of an obtained robot path are shown in Fig. 4.25.

Listing 4.8 Simulation framework for Bug0 algorithm

```
1  Ts = 0.03; % Sampling time
2  t = 0:Ts:30; % Simulation1 time
3  q = [0; 0; 0];% Initial pose
4  goal = [4; 4]; % Goal location
5  % Obstacles
6  obstacles{1} = flipud([-1 -1; 7 -1; 7 5; -1 5]);
7  obstacles{2} = [0.5 1; 4 1];
8  obstacles{3} = [3 3.5; 3 2.5; 5 2.5; 3 2.5];
9  obst = [];
10 for i = 1:length(obstacles)
11     obst = [obst; obstacles{i}([1:end,1],:); nan(1,2)];
12 end
13
14 for k = 1:length(t)
15     % Distance to the nearest obstacle and orientation of the segment
16     [dObst, ~, z] = nearestSegment(q(1:2).', obst);
17     phiObst = atan2(obst(z+1,2)-obst(z,2), obst(z+1,1)-obst(z,1));
18
19     % Control algorithm goes here ...
20
21     % Robot motion simulation
22     dq = [v*cos(q(3)); v*sin(q(3)); w];
```

EXAMPLE 4.9—cont'd

```
23    noise = 0.00; % Set to experiment with noise (e.g. 0.001)
24    q = q + Ts*dq + randn(3,1)*noise; % Euler integration
25    q(3) = wrapToPi(q(3)); % Map orientation angle to [-pi, pi]
26 end
```

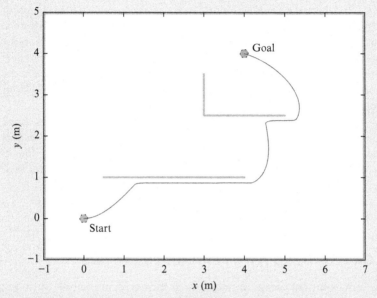

Fig. 4.25 Path planning and control example using the Bug0 algorithm to drive the robot to the goal while avoiding obstacles.

Solution

A possible implementation of the solution that produces the path shown in Fig. 4.25 is given in Listing 4.9; the code needs to be inserted into the marked line in Listing 4.8.

Listing 4.9 Bug0 algorithm

```
1  % Control based on distance to the obstacle
2  if dObst>0.2 % Drive toward the goal
3      phiRef = atan2(goal(2)-q(2), goal(1)-q(1));
4      ePhi = wrapToPi(phiRef - q(3));
5      dGoal = sqrt(sum((goal-q(1:2)).^2));
6      g = [dGoal/2, 1]; % Control gains
7  else % Drive right arround the obstacle
8      phiRef = wrapToPi(phiObst + pi*0);  % Add pi to go always left
9      ePhi = wrapToPi(phiRef-q(3));
10     g = [0.4, 5]; % Control gains
11 end
12 % Simple control of differential drive
13 v = g(1)*abs(cos(ePhi));
14 w = g(2)*ePhi;
15 v = min([v, 0.5]);
```

EXAMPLE 4.10

Also implement Bug1 and Bug2 algorithms by extending Example 4.9.

Solution

To implement the Bug1 algorithm you can adapt Listing 4.9 in Example 4.25 where the main behavior consists of two parts. The first behavior (driving toward the goal) remains unchanged, while the second needs to be changed as follows. Store the start position where the robot first detects the obstacle. Drive around the obstacle and measure the distance to the goal and remember the closest point. Perform this until the robot arrives in the stored start position or sufficiently close. Return to the remembered closest point.

Similarly you can adapt Listing 4.9 in Example 4.25 to the Bug2 algorithm as follows. Store the start position where the robot first detects the obstacle. Do the obstacle encircling until the main straight line is crossed. If this cross point is closer to the goal than the starting point then drive toward the goal; otherwise, continue encircling the obstacle.

4.4 GRAPH-BASED PATH PLANNING METHODS

If the environment with obstacles is sufficiently presented by a graph, one can use path searching algorithms to find the path from the start to goal configuration. In the following, a few well-known path graph-based searching algorithms are presented.

In general, path searching algorithms proceed as follows. The starting node is first checked to determine if it is also the goal node. Normally this is not the case; therefore, the search is extended to other nodes in the neighborhood of the current node. One of the neighbor nodes is selected (how the nodes are selected depends on the used algorithm and its cost function), and if it is not the goal node, then the search is also extended to the neighbor nodes of this new node. This procedure is continued until the solution is found or until all graph nodes have been investigated.

When performing a search in a graph, a list of nodes that have already been visited is made with the main purpose to prevent visiting the same node more than once. Nodes that can still extend the search area to other neighbor nodes (also known as alive nodes) are kept in an *open list*. Nodes that have no successors or have been checked already (also known as dead nodes) are kept in a *closed list*.

The evolution of path searching depends on the strategy of selecting new nodes for extending the search area. A list of open nodes is sorted according to some criteria and when the search is extended the first node from that sorted list is taken; this node best fits the used sorting criteria (has the smallest value of the criteria).

In the beginning, the open list Q only contains the start node. Nodes in the neighborhood of the start node are calculated and put in the open list, and the starting node is put in the closed list. The search is then extended with the first node in the open list and successors (of the chosen node from the open list) are calculated. The open list then contains the remaining previous successors (except the first one that has been already extended) and currently calculated successors of the chosen node. This procedure is shown in Fig. 4.26, where nodes in the open list are colored with light gray, nodes in the closed list are colored black and unvisited nodes are white. In Fig. 4.26 it can be seen why the nodes from the open list are called leaf nodes.

Graph search algorithms can be classified as *informed* or *noninformed*. Noninformed algorithms use no additional information beside the problem definition. These approaches search the graph systematically and do not distinguish more promising nodes from the less promising ones. Informed search, also known as heuristic search, contains additional information about the nodes. Therefore, it is possible to select more promising nodes and in that way searching for the final solution can be more efficient.

4.4.1 Breadth-First Search

Breadth-first search belongs to a class of noninformed graph search algorithms. It first explores the shallow nodes, that is, the nodes that are closest to the starting node. All nodes that can be accessed in k steps are visited before any of the nodes that require $k + 1$ steps (Fig. 4.26).

Fig. 4.26 Breadth-first search algorithm first explores the closest nodes. The currently explored node is marked with an *arrow*, nodes in open list are *colored gray*, nodes in the closed list are *black*, and unvisited nodes are *white*.

In the open list Q, the nodes are sorted using a FIFO approach (first-in first-out). The newly opened nodes are added to the end of the list Q and the nodes for extending the search are taken from the beginning of the list Q.

The algorithm is *complete* because it finds solution if it exists. If there exist several solutions, it finds the one that has the least steps from the starting node. In general, the found solution is not optimal because it is not necessary that all transitions between nodes have equal cost.

The algorithm generally requires high memory usage and long computing times. Computational time and memory consumption both increase exponentially as graph branching progresses.

4.4.2 Depth-First Search

Depth-first search is a noninformed graph search algorithm where the search is first extended in the depth. The node that is the farthest away from the starting node is used to extend the search. The search continues in the depth until the current node has no further successors. The search is then continued with the next deepest node whose successors have not been explored yet, as shown in Fig. 4.27.

The list of open nodes Q is a stack sorted by the LIFO (last-in first-out) method. The newly opened nodes are added to the beginning of the list Q and from the same end; also, the nodes for extending the search area are taken.

The depth-first search is not complete. In the case of an infinite graph (with infinite branch that does not end), it can get trapped in one branch of the graph; or in the case of a loop branch (loop in finite graph depth), it can get stuck in cycles. To avoid this problem the search can be limited to a certain depth only, but then the solution can have a higher depth than the maximum depth limitation. The algorithm is also not optimal because the found path is not necessarily the shortest one also.

This method has low memory usage as it only stores the path from the start node to the current node and intermediate nodes that have not been explored yet. When some nodes and all of their successors are explored, this node no longer needs to be stored in the memory.

4.4.3 Iterative Deepening Depth-First Search

This algorithm combines advantages of the breadth-first search and depth-first search algorithms. It iteratively increases the search depth limit and explores nodes using the depth-first search algorithm until the solution is

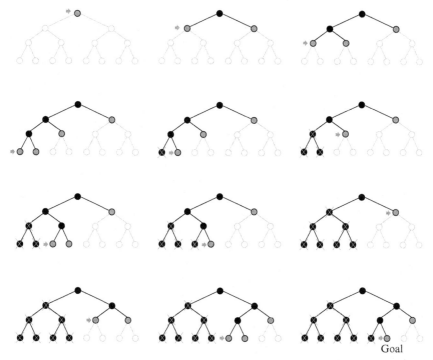

Fig. 4.27 Depth-first search has low memory usage because it stores only leaf nodes (*light gray*) and extended nodes (*gray*). Nodes removed from the memory are marked with *crosses*.

found (Fig. 4.28). First, the depth-first search is performed for the nodes that are zero steps away from the starting node. If the solution is not found, the search in the depth is repeated for nodes that are up to one step away from the starting one and so on.

This algorithm is complete (the solution is found if it exists), and it has small memory usage and is optimal if the cost of all transitions are equal or if transition costs increase with the node depth. If all the nodes have approximately the same rate of branching, then the repeated calculation of nodes is also not a big computational burden because the majority of the nodes are in the bottom of the tree, and those nodes are visited only once or a few times.

4.4.4 Dijkstra's Algorithm

Dijkstra's algorithm is a noninformed algorithm that finds the shortest paths from one node (source node) to all the other nodes in the graph [13].

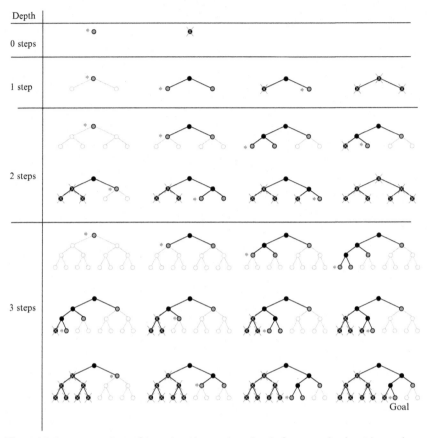

Fig. 4.28 Demonstration of iterative deepening depth-first search algorithm, where search is iterated to the depth of tree steps. Nodes removed from the memory are marked with *crosses*. The algorithm terminates when the goal node in the third step is found.

The result is therefore the shortest path tree from the source node to any other node. Originally it was proposed by Dijkstra [14] and later extended with many modifications. In problems where the shortest path between one starting node and one goal node must be found Dijkstra's algorithm may seem inefficient due to calculation of optimum paths to all the other nodes. In this case the algorithm can be terminated as soon the shortest desired path is calculated.

The algorithm finds the shortest path between the starting (source) node and the goal node because it calculates the cost function for the path between the start node to the current node, which we name cost-to-here.

The cost of the path to the current node (cost-to-here) is calculated as the sum of the path cost to the previous node (from which we came to the current node) and the cost of the transition between the previous node and the current node. In the case of several shortest paths (paths with the same cost) the algorithm returns only one, and it is not important which one.

To run the algorithm the transitions between the nodes and their costs need to be defined. For each visited node the algorithm stores the cost of currently shortest path to it (cost-to-here) and the connection from where we arrived to the current node. During the search, the lists of open nodes and closed nodes are adapted.

The operation of the algorithm is as follows. At the beginning, the open list contains only the start node, which has zero cost-to-here and it is without a connection to a previous node. The list of closed nodes is empty. Then the following steps (also shown in Fig. 4.29) are performed.

1. From the open list take the first node; call it the current node. The open list is sorted according to the cost-to-here in increasing order, where the first node is the one with the smallest cost-to-here.
2. For all nodes that can be reached from the current node calculate the cost-to-here as a sum of the current node's cost-to-here and transition cost.
3. Calculate and store cost-to-here and appropriate connection to the current node for all the nodes that do not already have these information stored.
4. If in the previous step some nodes already have cost-to-here and appropriate connection from some previous iteration, then compare both costs-to-here and keep in storage the lower cost-to-here and its connection.
5. Nodes are added to the open list and sorted in increasing order of the costs-to-here. Such a list is called a priority queue and enables that the node with the lowest price-to-here is found faster than in an unsorted list. The current node is moved to the closed list.

Originally, Dijkstra's algorithm terminates when the open nodes list is empty and the result are the shortest paths from the starting node to all the other nodes. If only the shortest path from the starting node and some goal node is needed then the algorithm terminates when the goal node is added to the closed list.

The resulting path can be obtained by back-tracking the connections that brought us to the goal node. At the goal node the connection to

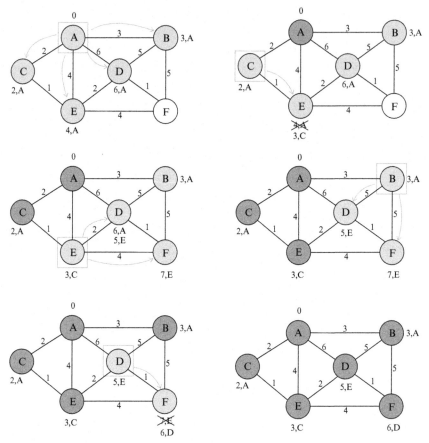

Fig. 4.29 Dijkstra's algorithm for searching the shortest paths from the starting node A to all the other nodes. The current node is surrounded with a *gray square* and its successors are marked with *arrows*. Costs for transitions between the nodes are marked at connections. Along nodes the cost-to-here and connection to the previous node are given. Nodes in the open list are colored with *light gray* while the nodes in the closed list are colored with *dark gray*. Example: If we need to obtain the shortest path between the nodes A and F its cost reads $Cost_{F-D-E-C-A} = 6$, while the path goes through nodes $A \rightarrow C \rightarrow E \rightarrow D \rightarrow F$.

the previous node is obtained then the connection to its previous node is read and so on until the start node is reached. Finally, the obtained list of connections is inverted.

Dijkstra's algorithm is complete and optimal algorithm if all the connections have costs higher than zero.

4.4.5 A* Algorithm

A* (A star) is an informed algorithm because it includes additional information or heuristic. Heuristic is the cost estimate of the path from the current node to the goal that is for the part of the graph tree that has not been explored yet. This enables the algorithm to distinguish between more or less promising nodes, and consequentially it can find the solution more efficiently. For each node the algorithm computes the heuristic function that is the cost estimate for the path from this node to the goal, and it is called cost-to-goal. This heuristic function can be Euclidean distance or Manhattan distance (sum of vertical and horizontal moves) from the current node to the goal node. The heuristic can be computed also by some other appropriate function.

During algorithm execution for each node the cost of the whole path (cost-of-the-whole-path) is calculated to consist of cost-to-here and cost-to-goal. During the path search, the open nodes list and closed nodes list are also used.

The algorithm operation is as follows. In the beginning, the open list contains only the starting node, which has zero cost-to-here and is without a connection to a previous node. Then the following steps (also illustrated in Fig. 4.30) are repeated.

1. From the open list the first node is taken; it is called the current node. The open list is sorted according to the cost-of-the-whole-path in increasing order, where the first node is the one with the smallest cost-of-the-whole-path.
2. For all nodes that can be reached from the current node we calculate the following:
   - cost-to-goal,
   - cost-to-here as the sum of the cost-to-here of the current node and the transition cost, and
   - cost-of-the-whole-path as the sum of cost-to-here and cost-to-goal.
3. For each of those nodes that do not already have stored cost-to-here, the values of the connection from the current node, cost-to-goal, and cost-of-the-whole-path are stored.
4. If in the previous step some node already has stored the cost values from some previous iteration, then both costs-to-here (current and the previous one) are compared and the one with a lower value is stored, alongside with the corresponding connection and cost-of-the-whole-path.

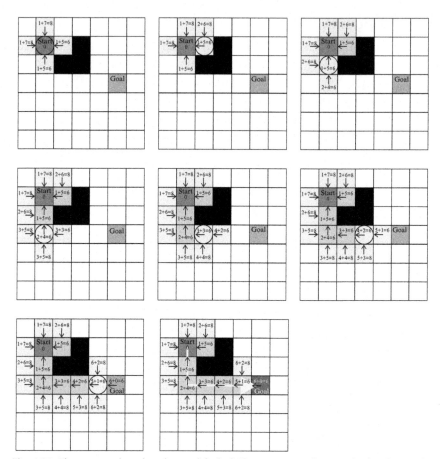

Fig. 4.30 The start and goal nodes are labeled. The current node is marked with a *circle*, nodes in the open list are *light gray*, and nodes in the closed list are *dark gray*, while obstacles are *black*. Transitions are possible in four directions (left, right, up, and down). In each visited cell (node) the direction to the parent node is denoted with an *arrow*. Each visited cell contains the cost of the path, which is the sum of cost-to-here and cost-to-goal. For the cost function the Manhattan distance is used. The found path can be tracked by following the connections marked with *arrows*.

5. Nodes whose cost values have been calculated for the first time are added to the open list. The nodes that have already been in the open-list and were updated are kept in the open list. The nodes that have been updated and were in the closed list are moved to the open-list. The open list is sorted according to the cost-of-the-whole-path in increasing order. The current node is transferred to the closed list.

In the first subfigure (in Fig. 4.30) the current node is the start node and its successor nodes (accessed in four directions: left, right, up, and down from the current node) are selected. For all the successor nodes, the cost-to-here is one as they are only one step away from the star node and cost-to-goal is the Manhattan distance (heuristic) from the successor node to the goal node, which can also be measured through the obstacle. The sum of both costs is the cost-of-the-whole path. The successors nodes also have a connection marked by an arrow (direction) to its parent node (current node marked by a circle). The open list now contains these four successor nodes and closed list contains the starting node.

In the second subfigure (in Fig. 4.30) the node with the lowest cost-of-the-whole path (with cost 6) is selected from the open list as the current node. The current node has only one successor (other cells are blocked by the obstacle or in the closed list). Cost-to-here for this successor node is 2 as it is two steps away (Manhattan distance) from the start node and cost-to-goal is 8. The current node is put to the closed list and the successor node to the open list. In the third subfigure (in Fig. 4.30) the current node becomes the node from the open list with the lowest cost-of-the-whole path, which is 6. Then the algorithm continues similarly as in the first two steps.

The A* algorithm is guaranteed to find the optimal path in graph if the heuristic for calculation of the cost-to-goal is admissible (or optimistic), which means that the estimated cost-to-goal is smaller or equal to true cost-to-goal. The algorithm finishes when the goal node is added to the closed list.

The A* algorithm is a complete algorithm because it finds the path if it exists, and as already mentioned, it is optimal if the heuristic is admissible (optimistic). Its drawback is large memory usage. If all costs-to-goal are set to zero the A* operation equals Dijkstra's algorithm. In Fig. 4.31 a comparison of Dijkstra's and A* algorithm performance is shown.

4.4.6 Greedy Best-First Search

Greedy best-first search is the informed algorithm. The open list is sorted in the increasing cost-to-goal. Therefore, the search in each iteration is extended to the open node that is closest to the goal (has the smallest cost-to-goal) assuming it will reach the goal quickly. The found path is not guaranteed to be optimal as shown in Fig. 4.32. The algorithm only considers cost from the current node to the goal and ignores the cost required to come to the current node; therefore, the overall path may become longer than the optimum one. Because the algorithm is not optimal it is also not necessary that the heuristic is admissible as it is important in A*.

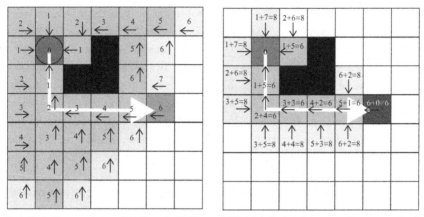

Fig. 4.31 Comparison of Dijkstra's (*left*) and A* (*right*) algorithm performance. Both algorithms find the shortest path to the goal; however, A* algorithm requires less iterations due to the used heuristics when searching the graph.

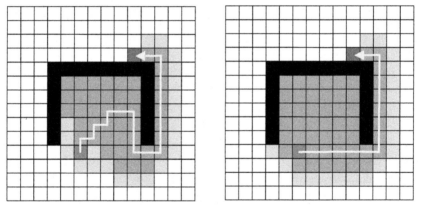

Fig. 4.32 Comparison of greedy best-first search (*left*) and A* (*right*) algorithm. The path found by the greedy best-first search algorithm is not optimal. In most cases it can find the solution faster but it is not always optimal. *Dark gray* nodes are the ones from the closed list, while the *light gray* nodes are from the open list.

EXAMPLE 4.11

Implement path planning using A* algorithm for the environment with cell decomposition in Fig. 4.32. Compare the obtained results with the results in Fig. 4.32. For calculation of the distance use Manhattan distance, while for the heuristic (cost-to-goal) use Manhattan distance or Euclidean distance, and also compare results obtained by both distances.

EXAMPLE 4.11—cont'd

Modify obtained code to also implement the greedy best-first search algorithm.

Solution

Although the reader is encouraged to implement the A* algorithm, a possible implementation of the algorithm is shown in Listing 4.10, where Manhattan distance is used to calculate the cost-to-here and Euclidean distance is used to calculate the heuristic.

Listing 4.10 Implementation of A* algorithm

```
1  classdef AStarBase < handle
2  properties
3      map = []; % Map: 0 - free space, 1 - obstacles
4      open = []; closed = []; start = []; goal = []; act = []; path = [];
5  end
6
7  methods
8      function path = find(obj, start, goal) % start=[is; js], goal=[ig; jg]
9          obj.start = start; obj.goal = goal; obj.path = [];
10         obj.closed = []; % Empty closed list
11         obj.open = struct('id', start, 'src', [0; 0], 'g', 0, ...
12                           'h', obj.heuristic(start)); % Initial open list
13
14         if obj.map(start(1), start(2))~=0 || obj.map(goal(1), goal(2))~=0
15             path = []; return; % Path not feasible!
16         end
17
18         while true % Search loop
19             if isempty(obj.open), break; end % No path found :(
20
21             obj.act = obj.open(1); % Get node from the ordered open list,
22             obj.closed = [obj.closed, obj.act]; % add it to the closed list
23             obj.open = obj.open(2:end); % and remove it from the open list.
24
25             if obj.act.id(1)==obj.goal(1) && obj.act.id(2)==obj.goal(2)
26                 % Path found :) Get the path from the closed list ...
27                 p = obj.act.id; obj.path = [p]; ids = [obj.closed.id];
28                 while sum(abs(p-start))~=0 % Follow src nodes to the start
29                     p = obj.closed(ids(1,:)==p(1) & ids(2,:)==p(2)).src;
30                     obj.path = [p, obj.path];
31                 end
32                 break;
33             end
34
35             neighbours = obj.getNodeNeighbours(obj.act.id);
36             for i = 1:size(neighbours, 2)
37                 n = neighbours(:,i);
38                 % Add neighbour to the open list if it is not in the closed
39                 % list and it is not an obstacle.
40                 ids = [obj.closed.id]; z = ids(1,:)==n(1) & ids(2,:)==n(2);
41                 if isempty(find(z, 1)) && ~obj.map(n(1), n(2))
42                     obj.addNodeToOpenList(n);
43                 end
44             end
45         end
```

(Continued)

EXAMPLE 4.11—cont'd

```
46         path = obj.path;
47     end
48     function addNodeToOpenList(obj, i)
49         g = obj.act.g + obj.cost(i); % Path cost
50         % Check if the node is in already the open list
51         ids = [obj.open.id]; s = [];
52         if ~isempty(ids)
53             s = sum(abs(ids-repmat(i, 1, size(ids, 2))))==0;
54         end
55         if isempty(find(s, 1)) % Add new node to the open list
56             node = struct('id', i, 'src', obj.act.id, ...
57                          'g', g, 'h', obj.heuristic(i));
58             obj.open = [obj.open, node];
59         else % Update the node in the open list if it has better score
60             if g<obj.open(s).g
61                 obj.open(s).g = g;
62                 obj.open(s).src = obj.act.id;
63             end
64         end
65         % Sort open list
66         [~,i] = sortrows([[obj.open.g]+[obj.open.h]; obj.open.h].', [1,2]);
67         obj.open = obj.open(i);
68     end
69
70     function n = getNodeNeighbours(obj, a)
71         n = [a(1)-1, a(1), a(1), a(1)+1; a(2), a(2)-1, a(2)+1, a(2)];
72         [h, w] = size(obj.map);
73         n = n(:, n(1,:)>=1 & n(1,:)<=h & n(2,:)>=1 & n(2,:)<=w); % Bounds
74     end
75
76     function g = cost(obj, a)
77         g = sum(abs(a-obj.act.id)); % Manhattan distance
78     end
79
80     function h = heuristic(obj, a)
81         h = sqrt(sum((a-obj.goal).^2)); % Euclidean distance
82     end
83 end
84 end
```

The usage of the A* algorithm given in Listing 4.10 is shown in Listing 4.11. The columns of output variable path represent an ordered list of cells that lead from the start to goal. A minimum modification of the algorithm is needed in order to use different heuristic (e.g., Manhattan distance as it was used to obtain the path shown in Fig. 4.32).

Listing 4.11 Usage of A* algorithm

```
1 map = zeros(14, 14); % Map
2 map(5:10,[4 11]) = 1; map(5,4:11) = 1; % Obstacles
3
4 astar = AStarBase();
5 astar.map = map;
6 path = astar.find([11; 6], [4; 10])
```

EXAMPLE 4.11—cont'd

```
path =
  Columns 1 through 13
    11    11    11    11    11    11    11    10     9     8     7     6     5
     6     7     8     9    10    11    12    12    12    12    12    12    12
  Columns 14 through 16
     4     4     4
    12    11    10
```

Also the A^* algorithm can be converted to the greedy best-first search algorithm if the cost-to-here is set to zero.

EXAMPLE 4.12

Extend Example 4.11 also with the quadtree environment decomposition and optimal path solution using the A^* algorithm.

Solution

In order to extend the A^* algorithm in Listing 4.10 with the quadtree map, only some minor modifications are required. In this case all the cells are not of the same size, therefore, the distance between the cells varies, and also the number of neighbors varies for each cell. Hence, the way neighboring cells are determined needs to be modified. Implementation of a quadtree-decomposition algorithm in Listing 4.2 already creates a visibility map; that is, it finds all the neighboring cells for each cell in the quadtree. In order to determine the cost-to-here, the distance between the cells can be defined as the Euclidean distance between the cell centers (see Fig. 4.10). The reader is encouraged to implement these modifications.

REFERENCES

[1] H. Choset, K. Lynch, S. Hutchinson, G. Kantor, W. Burgard, L. Kavraki, S. Thrun, Principles of Robot Motion: Theory, Algorithms, and Implementations, MIT Press, Boston, MA, 2005.

[2] S.M. LaValle, Planning Algorithms, Cambridge University Press, Cambridge, NY, 2006.

[3] B. Siciliano, O. Khatib, Springer Handbook of Robotics, Springer, Berlin, 2008.

[4] S.M. LaValle, Rapidly-exploring random trees: a new tool for path planning, Technical Report TR 98-11, Computer Science Department, Iowa State University, Iowa, 2013.

[5] S.M. LaValle, J.J. Kuffner, Randomized kinodynamic planning, Int. J. Robot. Res. 20 (5) (2001) 278–400.

[6] J.J. Kuffner, S.M. LaValle, RRT-connect: an efficient approach to single-query path planning, in: IEEE International Conference on Robotics and Automation (ICRA 2000), IEEE, 2000, pp. 1–7.

[7] L.E. Kavraki, P. Svestka, J.-C. Latombe, M.H. Overmars, Probabilistic roadmaps for path planning in high-dimensional configuration spaces, IEEE Trans. Robot. Autom. 12 (4) (1996) 566–580.

[8] Z. Sun, D. Hsu, T. Jiang, H. Kurniawati, J.H. Reif, Narrow passage sampling for probabilistic roadmap planning, IEEE Trans. Robot. 21 (6) (2005) 1105–1115.

[9] V. Lumelsky, P. Stepanov, Dynamic path planning for a mobile automaton with limited information on the environment, IEEE Trans. Autom. Control 31 (11) (1986) 1058–1063.

[10] A. Sankaranarayanan, M. Vidyasagar, A new path planning algorithm for moving a point object amidst unknown obstacles in a plane, in: IEEE Conference on Robotics and Automation, IEEE, 1990, pp. 1930–1936.

[11] I. Kamon, E. Rivlin, Sensory-based motion planning with global proofs, IEEE Trans. Robot. 13 (6) (1997) 814–821.

[12] S. Laubach, J. Burdick, An autonomous sensor-based path-planner for planetary microrovers, in: IEEE Conference on Robotics and Automation, IEEE, 1999, pp. 347–354.

[13] K. Mehlhorn, P. Sanders, Algorithms and Data Structures: The Basic Toolbox, Springer, Berlin, 2008.

[14] E.W. Dijkstra, A note on two problems in connexion with graphs, Numer. Math. 1 (1959) 269–271.

CHAPTER 5

Sensors Used in Mobile Systems

5.1 INTRODUCTION

Wheeled mobile robots need to sense the environment using sensors in order to autonomously perform their mission. Sensors are used to cope with uncertainties and disturbances that are always present in the environment and in all robot subsystems. Mobile robots do not have exact knowledge of the environment, and they also have imperfect knowledge about their motion models (uncertainty of the map, unknown motion models, unknown dynamics, etc.). The outcomes of the actions are also uncertain due to nonideal actuators. The main purpose of the sensors is therefore to lower these uncertainties and to enable estimation of robot states as well as the states of the environment.

Usually there is no elegant single sensor solution (especially for indoor use) that would be accurate enough and robust in measuring desired information such as the robot pose. The pose is necessary to localize the robot, which is one of the most important challenges in mobile robotics. The developers therefore need to rely on more sensors and on a fusion of their measured information. Such approaches benefit in higher quality and more robust information. Robot pose estimation usually combines relative sensors and absolute sensors. *Relative sensors* provide information that is given relatively to the robot coordinates, while the information of *absolute sensors* is defined in some global coordinate system (e.g., Earth coordinates).

Using sensors, a mobile robot can sense the states of the environment. The measured sensor information needs to be analyzed and interpreted adequately. In the real world measurements are changing dynamically (e.g., change of illumination and different sound or light absorption on surfaces). The measurement error is often modeled statistically by a probability density function for which symmetric distribution is usually supposed or even normal distribution. However, this assumption may not always be correct (e.g., ultrasonic sensor distance measurement can be larger than the true distance due to multiple reflections of the ultrasound path from the sensor transmitter to the sensor receiver).

Wheeled Mobile Robotics
http://dx.doi.org/10.1016/B978-0-12-804204-5.00005-6
207

First, the coordinate system transformations are briefly described. They are needed to correctly map sensor measurements to the robot and to estimate relevant information in robot-related coordinates. Then, the main localization methods for estimation of robot pose in the environment using specific sensors are explained. Finally, a short overview of sensors used in mobile robotics is given.

5.2 COORDINATE FRAME TRANSFORMATIONS

Sensors that are mounted on the robot are usually not in the robot's center or in the origin of the robot's coordinate frame. Their position and orientation on the robot is described by a translation vector and rotation according to the robot's frame. Those transformations are needed to relate measured quantities in the sensor frame to robot coordinates.

With these transformations we can describe how the sensed direction vector (e.g., accelerometer, magnetometer) or sensed position coordinates (e.g., laser range scanner or camera) are expressed in the robot coordinates. Furthermore, mobile robots are moving in space, and therefore, their poses or movements can be described by appropriate transformations.

Here, a short general overview for 3D space is given, although in wheeled mobile robots two dimensions are usually sufficient (e.g., motion on the plane described by two translations and one rotation). First some notations for rotation transformation will be described, and later a translation will be included.

5.2.1 Orientation and Rotation

Orientation of some local reference frame (e.g., sensor) according to the reference frame (e.g., robot) is described by a rotation matrix R:

$$R = \begin{bmatrix} u_1 & u_2 & u_3 \\ v_1 & v_2 & v_3 \\ w_1 & w_2 & w_3 \end{bmatrix} \tag{5.1}$$

where unit vectors of a local coordinate system u, v, and w are expressed in the reference frame by $u = [u_1, u_2, u_3]^T$, $v = [v_1, v_2, v_3]^T$, $w = [w_1, w_2, w_3]^T$, where $u \times v = w$. The rows of R are components of body unit vectors along the reference coordinate unit vectors x, y, and z. The elements of matrix R are cosine of the angles among the axis of both coordinate systems; therefore, matrix R is also called *the direction cosine matrix* or DCM. Because vectors u, v, and w are orthonormal the inverse of R

equals to the transpose of $\boldsymbol{R}$ (det $\boldsymbol{R} = 1$ and $\boldsymbol{R}^{-1} = \boldsymbol{R}^T$). The DCM has nine parameters to describe three degrees of freedom, but those parameters are not mutually independent but are defined by six constraint relations (sum of squared elements of each line in $\boldsymbol{R}$ is one and a dot product of any couple of the lines in $\boldsymbol{R}$ is zero).

The vector in a local (body) coordinate frame $\boldsymbol{v}_L$ is expressed from a global coordinate (reference) using the following transformation

$$\boldsymbol{v}_L = \boldsymbol{R}_G^L \boldsymbol{v}_G \tag{5.2}$$

Matrix $\boldsymbol{R}_G^L$ therefore transforms vector description (given by its components) in the global coordinates to the vector description in the local coordinates.

Basic rotation transformations are obtained by rotation around axis x, y, and z by elementary rotation matrices:

$$\boldsymbol{R}_x(\varphi) = \begin{bmatrix} 1 & 0 & 0 \\ 0 & \cos(\varphi) & \sin(\varphi) \\ 0 & -\sin(\varphi) & \cos(\varphi) \end{bmatrix} \tag{5.3}$$

$$\boldsymbol{R}_y(\theta) = \begin{bmatrix} \cos(\theta) & 0 & -\sin(\theta) \\ 0 & 1 & 0 \\ \sin(\theta) & 0 & \cos(\theta) \end{bmatrix} \tag{5.4}$$

$$\boldsymbol{R}_z(\psi) = \begin{bmatrix} \cos(\psi) & \sin(\psi) & 0 \\ -\sin(\psi) & \cos(\psi) & 0 \\ 0 & 0 & 1 \end{bmatrix} \tag{5.5}$$

where $\boldsymbol{R}_{axis}(angle)$ is rotation around the axis for a given angle. Successive rotations are described by the product of the rotation matrix where the order of rotations is important. DCM is the elementary presentation of the rigid body orientation; however, in some cases other presentations are more appropriate. Therefore two additional methods will be presented in the following.

Euler Angles
Euler angles describe orientation of some rigid body by rotation around axes x, y, and z. Those angles are marked as
- φ: roll angle (around x-axis),
- θ: pitch angle (around y-axis), and
- ψ: yaw or heading angle (around z-axis).

There are 12 possible rotation combinations around axes x, y, and z [1]. The most often used is the combination 3–2–1, where the orientation of somebody coordinate frame according to the reference coordinate frame is obtained from the initial pose where both coordinate frames are aligned and then the body frame is rotated in the following order:

1. First is rotation around z-axis for yaw angle ψ,
2. then rotation around newly obtained y for a pitch angle θ, and
3. finally rotation around newly obtained x for a roll angle φ.

The DCM of this rotation transformation is

$$R = R_x(\varphi)R_y(\theta)R_z(\psi)$$

$$= \begin{bmatrix} \cos\theta\cos\psi & \cos\theta\sin\psi & -\sin\theta \\ \sin\varphi\sin\theta\cos\psi - \cos\varphi\sin\psi & \sin\varphi\sin\theta\sin\psi + \cos\varphi\cos\psi & \sin\varphi\cos\theta \\ \cos\varphi\sin\theta\cos\psi + \sin\varphi\sin\psi & \cos\varphi\sin\theta\sin\psi - \sin\varphi\cos\psi & \cos\varphi\cos\theta \end{bmatrix}$$

and the Euler presentation is as follows:

$$\varphi = \arctan\left(\frac{R_{23}}{R_{33}}\right)$$
$$\theta = -\arcsin(R_{13}) \tag{5.6}$$
$$\psi = \arctan\left(\frac{R_{12}}{R_{11}}\right)$$

Parameterization using Euler angles is not redundant (three parameters for three degree of freedom). However, its main drawback is that it becomes singular at rotation $\theta = \pi/2$ where rotation around z- and x-axes have the same effect (they coincide). This effect is known as *Gimbal lock* in classic airplane gyroscopes. In rotation simulation using Euler angles notation, this singularity appears due to division by $\cos\theta$ (see Eq. 5.47 in Section 5.2.3).

Quaternions

Quaternions present orientation in 3D space using four parameters and one constraint equation without singularity problems such as in Euler angles presentation. Mathematically the quaternions are a noncommutative extension of the complex numbers defined by

$$q = q_0 + q_1 i + q_2 j + q_3 k \tag{5.7}$$

where complex elements i, j in k are related by expression $i^2 = j^2 = k^2 = ijk = -1$. q_0 is the scalar part of the quaternion and $q_1 i + q_2 j + q_3 k$ is termed the vector part. The quaternion norm is defined by

$$\|q\| = \sqrt{qq^*} = \sqrt{q_0^2 + q_1^2 + q_2^2 + q_3^2} \tag{5.8}$$

where q^* is the conjugated quaternion obtained by $q^* = q_0 - q_1i - q_2j - q_3k$. The inverse quaternion is calculated using its conjugate and norm value as follows:

$$q^{-1} = \frac{q^*}{\|q\|^2} \tag{5.9}$$

Rotation in space is parameterized using unit quaternions by

$$
\begin{aligned}
q_0 &= \cos \Delta\varphi/2 \\
q_1 &= e_1 \sin \Delta\varphi/2 \\
q_2 &= e_2 \sin \Delta\varphi/2 \\
q_3 &= e_3 \sin \Delta\varphi/2
\end{aligned} \tag{5.10}
$$

where vector $e^T = [e_1, e_2, e_3]$ is the unit vector of the rotation axis and $\Delta\varphi$ rotation angle around that axis. For unit quaternion it holds $q_0^2 + q_1^2 + q_2^2 + q_3^2 = 1$.

Transformation

$$q_{v'} = q^{-1} \circ q_v \circ q \tag{5.11}$$

rotates vector $v = [x, y, z]^T$ expressed by quaternion

$$q_v = xi + yj + zk \tag{5.12}$$

(or equivalently $q_v = [0, x, y, z]^T$) around axis e for angle $\Delta\varphi$ to the vector $v' = [x', y', z']^T$ expressed by quaternion

$$q_{v'} = x'i + y'j + z'k \tag{5.13}$$

where $\circ$ denotes the product of quaternions defined in Eqs. (5.16), (5.17).

An additional advantage of quaternions is the relatively easy combination of successive rotations. Product of two DCM can be equivalently written by the product of two quaternions [1]. Having two quaternions,

$$q = q_0 + q_1i + q_2j + q_3k \tag{5.14}$$

and

$$q' = q_0' + q_1'i + q_2'j + q_3'k \tag{5.15}$$

and rotating some vector from its initial orientation first by rotation defined by q' and then for rotation defined by q, we obtain the following:

$$
\begin{aligned}
q'' = q' \circ q = &(q_0'q_0 - q_1'q_1 - q_2'q_2 - q_3'q_3) \\
&+ i(q_1'q_0 + q_2'q_3 - q_3'q_2 + q_0'q_1) \\
&+ j(-q_1'q_3 + q_2'q_0 + q_3'q_1 + q_0'q_2) \\
&+ k(q_1'q_2 - q_2'q_1 + q_3'q_0 + q_0'q_3)
\end{aligned} \tag{5.16}
$$

or in vector-matrix form

$$
\begin{bmatrix} q_0'' \\ q_1'' \\ q_2'' \\ q_3'' \end{bmatrix} = \begin{bmatrix} q_0 & -q_1 & -q_2 & -q_3 \\ q_1 & q_0 & q_3 & -q_2 \\ q_2 & -q_3 & q_0 & q_1 \\ q_3 & q_2 & -q_1 & q_0 \end{bmatrix} \begin{bmatrix} q_0' \\ q_1' \\ q_2' \\ q_3' \end{bmatrix}
\tag{5.17}
$$

The relation between the quaternion and DCM presentation is given by

$$
R(q) = \begin{bmatrix} q_0^2 + q_1^2 - q_2^2 - q_3^2 & 2(q_0 q_3 + q_1 q_2) & 2(q_1 q_3 - q_0 q_2) \\ 2(q_1 q_2 - q_3 q_0) & q_0^2 - q_1^2 + q_2^2 - q_3^2 & 2(q_0 q_1 + q_2 q_3) \\ 2(q_0 q_2 + q_1 q_3) & 2(q_2 q_3 - q_0 q_1) & q_0^2 - q_1^2 - q_2^2 + q_3^2 \end{bmatrix}
\tag{5.18}
$$

or in opposite direction

$$
q_0 = \frac{1}{2}\sqrt{1 + R_{11} + R_{22} + R_{33}}
\tag{5.19}
$$

$$
q_1 = \frac{1}{4q_0}(R_{23} - R_{32})
\tag{5.20}
$$

$$
q_2 = \frac{1}{4q_0}(R_{31} - R_{13})
\tag{5.21}
$$

$$
q_1 = \frac{1}{4q_0}(R_{12} - R_{21})
\tag{5.22}
$$

If q_0 in Eq. (5.19) is close to zero then the transform from DCM to quaternion (relations 5.19–5.22) is singular. In this case we can calculate the quaternion using equivalent form (relations 5.23–5.26) without numerical problems:

$$
q_0 = \frac{1}{4T}(R_{32} - R_{23})
\tag{5.23}
$$

$$
q_1 = T
\tag{5.24}
$$

$$
q_2 = \frac{1}{4T}(R_{12} + R_{21})
\tag{5.25}
$$

$$
q_3 = \frac{1}{4T}(R_{13} + R_{31})
\tag{5.26}
$$

where $T = \frac{1}{2}\sqrt{1 + R_{11} - R_{22} - R_{33}}$

The relation between quaternions and Euler angles (notation 3–2–1) is obtained by matrices $R_x(\varphi)$, $R_y(\theta)$, and $R_z(\psi)$, which suits to quaternions $[\cos(\varphi/2) + i\sin(\varphi/2)]$, $[\cos(\theta/2) + j\sin(\theta/2)]$ in $[\cos(\psi/2) + k\sin(\psi/2)]$. The quaternion for rotation 3–2–1 is

$$
q = [\cos(\psi/2) + k\sin(\psi/2)][\cos(\theta/2) + j\sin(\theta/2)][\cos(\varphi/2) + i\sin(\varphi/2)]
\tag{5.27}
$$

or in vector form

$$q = \begin{bmatrix} \cos(\varphi/2)\cos(\theta/2)\cos(\psi/2) + \sin(\varphi/2)\sin(\theta/2)\sin(\psi/2) \\ \sin(\varphi/2)\cos(\theta/2)\cos(\psi/2) - \cos(\varphi/2)\sin(\theta/2)\sin(\psi/2) \\ \cos(\varphi/2)\sin(\theta/2)\cos(\psi/2) + \sin(\varphi/2)\cos(\theta/2)\sin(\psi/2) \\ \cos(\varphi/2)\cos(\theta/2)\sin(\psi/2) - \sin(\varphi/2)\sin(\theta/2)\cos(\psi/2) \end{bmatrix} \quad (5.28)$$

The opposite transformation is

$$\varphi = \arctan\left(\frac{2(q_1 q_0 + q_2 q_3)}{q_0^2 - q_1^2 - q_2^2 + q_3^2}\right) \quad (5.29)$$

$$\theta = -\arcsin\left(2(q_1 q_3 - q_2 q_0)\right) \quad (5.30)$$

$$\psi = \arctan\left(\frac{2(q_3 q_0 + q_1 q_2)}{q_0^2 + q_1^2 - q_2^2 - q_3^2}\right) \quad (5.31)$$

EXAMPLE 5.1

A sensor (magnetometer) coordinate system is rotated according to a robot coordinate system. Lets mark the sensor coordinate system as local (L) and the robot coordinate system as global (G). The orientation of the sensor according to the robot is described by two rotations. Suppose that initially (L) is aligned to (G) then rotating (L) around x-axis for $\alpha_x = 90$ degree and finally around new y-axis for $\alpha_y = 45$ degree. Answer the following:
1. What is the orientation of the sensor expressed by the DCM (R_G^L) according to robot coordinates?
2. Determine Euler angles (notation 3–2–1) for this transformation.
3. Determine quaternion q_G^L that describes this transformation.
4. The magnetometer measures direction vector $v = [0, 0, 1]^T$ of the magnetic field. What is the presentation of this vector in robot coordinates?

Solution

1. Final orientation is described by the rotation matrix obtained from successive rotations where multiplication order is important.

$$R_G^L = R_y(\alpha_y)R_x(\alpha_x)$$

$$= \begin{bmatrix} \cos(\alpha_y) & 0 & -\sin(\alpha_y) \\ 0 & 1 & 0 \\ \sin(\alpha_y) & 0 & \cos(\alpha_y) \end{bmatrix} \begin{bmatrix} 1 & 0 & 0 \\ 0 & \cos(\alpha_x) & \sin(\alpha_x) \\ 0 & -\sin(\alpha_x) & \cos(\alpha_x) \end{bmatrix}$$

$$= \begin{bmatrix} 0.7071 & 0.7071 & 0 \\ 0 & 0 & 1 \\ 0.7071 & -0.7071 & 0 \end{bmatrix}$$

(Continued)

EXAMPLE 5.1—cont'd

Rows of matrix R_G^L are components of the sensor axis expressed in robot coordinates, which is seen from a graphical presentation of this rotation.

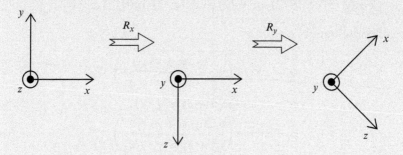

2. Euler angles for 3–2–1 notation are calculated from matrix $R = R_G^L$:

$$\varphi = \arctan\left(\frac{R_{23}}{R_{33}}\right) = 90 \text{ degree}$$

$$\theta = -\arcsin(R_{13}) = -0 \text{ degree}$$

$$\psi = \arctan\left(\frac{R_{12}}{R_{11}}\right) = 45 \text{ degree}$$

where the rotation matrix is obtained by $R_G^L = R_x(\varphi)R_y(\theta)R_z(\psi)$ and where elementary rotations $R_x(\varphi)$, $R_y(\theta)$, and $R_z(\psi)$ are defined in Eq. (5.5).

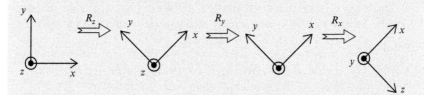

3. Quaternion q_G^L is obtained by the product of quaternions for successive rotations:

$$q_G^L = q_x \circ q_y$$

where q_x is according to Eq. (5.10) defined by rotation angle $\Delta\varphi_x = 90$ degree around rotation axis $e_x = [1, 0, 0]^T$ and q_y with rotation angle $\Delta\varphi_y = 45$ degree around rotation axis $e_y = [0, 1, 0]^T$

EXAMPLE 5.1—cont'd

$$q_x = \begin{bmatrix} \cos \Delta\varphi_x/2 \\ e_{x_1} \sin \Delta\varphi_x/2 \\ e_{x_2} \sin \Delta\varphi_x/2 \\ e_{x_3} \sin \Delta\varphi_x/2 \end{bmatrix} = \begin{bmatrix} 0.7071 \\ 0.7071 \\ 0 \\ 0 \end{bmatrix}$$

$$q_y = \begin{bmatrix} \cos \Delta\varphi_z/2 \\ e_{y_1} \sin \Delta\varphi_y/2 \\ e_{y_2} \sin \Delta\varphi_y/2 \\ e_{y_3} \sin \Delta\varphi_y/2 \end{bmatrix} = \begin{bmatrix} 0.9239 \\ 0 \\ 0.3827 \\ 0 \end{bmatrix}$$

The final quaternion (see quaternion product definition in Eq. 5.17) reads

$$q_G^L = q_x \circ q_y = \begin{bmatrix} 0.6533 \\ 0.6533 \\ 0.2706 \\ 0.2706 \end{bmatrix}$$

which suits to the rotation angle $\Delta\varphi = 2\arccos(q_0) = 2\arccos(0.6533) = 98.41$ degree around the rotation axis $e = \frac{1}{\sin\frac{\Delta\varphi}{2}}[q_1, q_2, q_3]^T = [0.8630, 0.3574, 0.3574]^T$

4. The measured direction vector $v_L = [0, 0, 1]^T$ is in robot coordinates:

$$v_G = R_L^G v_L = [0.70711, -0.70711, 0]^T$$

where $R_L^G = (R_G^L)^{-1} = R_G^{L\,T}$.

The same is obtained by quaternions where the vector after rotation is

$$q_{v_G} = (q_L^G)^{-1} \circ q_{v_L} \circ q_L^G$$

where $q_{v_L} = [0, v_L^T]^T$ and $q_L^G = (q_G^L)^{-1}$ (see Eq. 5.9). Products of quaternions are defined in Eq. (5.17). The solution is

$$q_{v_G} = \begin{bmatrix} 0.6533 \\ 0.6533 \\ 0.2706 \\ 0.2706 \end{bmatrix} \circ \begin{bmatrix} 0 \\ 0 \\ 0 \\ 1 \end{bmatrix} \circ \begin{bmatrix} 0.6533 \\ -0.6533 \\ -0.2706 \\ -0.2706 \end{bmatrix} = \begin{bmatrix} 0 \\ 0.7071 \\ -0.7071 \\ 0 \end{bmatrix}$$

which is vector $v_G = [0.7071, -0.7071, 0]^T$ (only the vector part of quaternion is considered).

EXAMPLE 5.2

Initially global coordinate frame G and local coordinate frame L are aligned; then frame L is rotated around x-axis for angle $\alpha_x = 90$ degree and then again around newly obtained z-axis for angle $\alpha_z = 90$ degree. Answer the following:

1. What is the orientation of frame L expressed by the DCM ($\boldsymbol{R}_G^L$) according to frame G?
2. Determine Euler angles (notation 3–2–1) for this transformation.
3. Determine quaternion $\boldsymbol{q}_G^L$ that describes this transformation.
4. The vector in global coordinates is $\boldsymbol{v}_G = [0, 0, 1]^T$. What is the presentation of this vector in local coordinates?

Solution

1. The DCM and graphical presentation for this rotation are

$$\boldsymbol{R}_G^L = \begin{bmatrix} 0 & 0 & 1 \\ -1 & 0 & 0 \\ 0 & -1 & 0 \end{bmatrix}$$

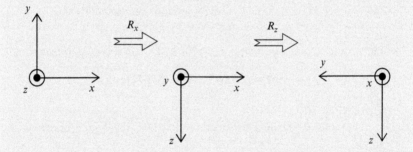

2. Euler angles (3–2–1) are

$$\varphi = \arctan\left(\frac{R_{23}}{R_{33}}\right) = \text{undefined}$$

$$\theta = -\arcsin(R_{13}) = -90 \text{ degree}$$

$$\psi = \arctan\left(\frac{R_{12}}{R_{11}}\right) = \text{undefined}$$

Notice that $\theta = -90$ degree, which means that parameterization using Euler angles is singular, and therefore φ and ψ are not defined. Using Euler angles this orientation (rotation) cannot be described.

EXAMPLE 5.2—cont'd

3. Quaternion q_G^L is

$$q_G^L = q_x \circ q_z$$

where

$$q_x = \begin{bmatrix} 0.7071 \\ 0.7071 \\ 0 \\ 0 \end{bmatrix}$$

$$q_z = \begin{bmatrix} 0.7071 \\ 0 \\ 0 \\ 0.7071 \end{bmatrix}$$

and

$$q_G^L = q_x \circ q_z = \begin{bmatrix} 0.5 \\ 0.5 \\ -0.5 \\ 0.5 \end{bmatrix}$$

which result in the rotation angle $\Delta\varphi = 2\arccos(0.5) = 120$ degree around axis of rotation $e = \frac{1}{\sin\frac{\Delta\varphi}{2}}[q_1, q_2, q_3]^T = [0.5774, -0.5774, 0.5774]^T$.

4. Vector $v_G = [0, 0, 1]^T$ reads in local coordinates as

$$v_L = R_G^L v_G = [1, 0, 0]^T$$

or with quaternions

$$q_{v_L} = (q_G^L)^{-1} \circ q_{v_G} \circ q_G^L = \begin{bmatrix} 0 \\ 1 \\ 0 \\ 0 \end{bmatrix}$$

where the vector part is $v_L = [1, 0, 0]^T$.

5.2.2 Translation and Rotation

To make presentation more general lets mark sensor coordinates by L (as local coordinates) and robot coordinates by G (as global coordinates). Sensor location according to robot coordinates is described by translation vector $t_L^G = [t_x, t_y, t_z]$ and rotation matrix R_G^L. Translation t_L^G describes the position of the local coordinates' origin in global coordinates, and rotation matrix R_G^L describes the local coordinate frame orientation according to global (robot) coordinates. A point p_G given in global coordinates can be described by local coordinates using transformation

$$p_L = R_G^L \left(p_G - t_L^G \right) \tag{5.32}$$

and its inverse transformation is given by

$$p_G = (R_G^L)^{-1} p_L + t_L^G = \left(R_G^L \right)^T p_L + t_L^G \tag{5.33}$$

EXAMPLE 5.3

A robot has a laser range finder (LRF) sensor that measures the position of the closest obstacle point $p_L = [1, 0.5, 0.4]^T$ m in sensor coordinates. It also has a magnetometer that measures Earth's magnetic field vector $v_L^T = [22, 1, 42]$ nT.

The laser range sensor is translated according to a robot (global) coordinate frame by $t_1 = [0.1, 0, 0.25]^T$ and rotated (according to a robot coordinate frame) around z-axis for 30 degree. The magnetometer translation is $t_2 = [0, 0.1, 0.2]$, and rotation is described by roll, pitch, and yaw angles $\varphi = 0$ degree, $\theta = 10$ degree, and $\psi = 20$ degree (Euler notation 3–2–1), respectively.

Answer the following:

1. What are the closest obstacle point coordinates in the robot coordinate system?
2. How is Earth's magnetic field vector presented by robot coordinates?

Solution

1. The DCM is

$$R_G^L = \begin{bmatrix} \cos(30°) & \sin(30°) & 0 \\ -\sin(30°) & \cos(30°) & 0 \\ 0 & 0 & 1 \end{bmatrix} = \begin{bmatrix} 0.866 & 0.5 & 0 \\ -0.5 & 0.8660 & 0 \\ 0 & 0 & 1 \end{bmatrix}$$

The point expressed in robot coordinates reads

$$p_G = (R_G^L)^T p_L + t_1 = [0.716, 0.933, 0.65]^T$$

2. For vector transformations only rotation matters so the magnetic vector components in robot coordinates are obtained by rotation transformation

$$R_G^L = R_x(0°)R_y(10°)R_z(20°) = \begin{bmatrix} 0.9254 & 0.3368 & -0.1736 \\ -0.3420 & 0.9397 & 0 \\ 0.1632 & 0.0594 & 0.9848 \end{bmatrix}$$

$$v_G = (R_G^L)^T v_L = [26.8705, 10.8443, 37.5417]^T$$

5.2.3 Kinematics of Rotating Frames

This section will introduce how rigid body orientation presented by quaternions or the DCM is related to angular rates around local axes of the rigid body. The rigid body is rotating around its x-, y-, and z-axes with angular velocities ω_x, ω_y, and ω_z, respectively. Therefore, the orientation of the rigid body (e.g., robot or sensor unit) is changing according to the reference coordinate system.

Rotational Kinematics Expressed by Quaternions

Time dependency of rigid body rotation (given by differential equation) can be derived from the product definition of two quaternions (5.17). If orientation $q(t)$ of the rigid body at time t is known then its orientation in time $q(t + \Delta t)$ can be written as

$$q(t + \Delta t) = q(t) \circ \Delta q(t) \tag{5.34}$$

where $\Delta q(t)$ defined the change of the rigid body orientation from $q(t)$ to $q(t + \Delta t)$. In other words $\Delta q(t)$ is the orientation of the body at time $t + \Delta t$ relative to its orientation at time t. The final orientation of the body $q(t + \Delta t)$ is therefore obtained by first rotating the body according to some reference frame for rotation $q(t)$, and then also for rotation $\Delta q(t)$ according to $q(t)$. $\Delta q(t)$ is defined by relation (5.10)

$$\Delta q(t) = \begin{bmatrix} \cos \Delta\varphi/2 \\ e_x \sin \Delta\varphi/2 \\ e_y \sin \Delta\varphi/2 \\ e_z \sin \Delta\varphi/2 \end{bmatrix} \tag{5.35}$$

where $e(t) = [e_x, e_y, e_z]^T$ is the rotation axis expressed in rigid body local coordinates at time t and $\Delta\varphi$ is the rotation angle during time period Δt. Assuming $e(t)$ and $\Delta\varphi$ are constant during the period Δt, the product of quaternions (5.34) can be reformulated using definition (5.17) as follows:

$$q(t + \Delta t) = \left(\cos\left(\frac{\Delta\varphi}{2}\right) I + \sin\left(\frac{\Delta\varphi}{2}\right) \begin{bmatrix} 0 & -e_x & -e_y & -e_z \\ e_x & 0 & e_z & -e_y \\ e_y & -e_z & 0 & e_x \\ e_z & e_y & -e_x & 0 \end{bmatrix} \right) q(t) \tag{5.36}$$

where I is a 4×4 identity matrix. For a short interval Δt we can approximate $\Delta\varphi \approx \sqrt{\omega_x^2 + \omega_y^2 + \omega_z^2} \Delta t$ where $\omega(t) = [\omega_x, \omega_y, \omega_z]^T$ is the vector of current angular rates, which can also be written in the form $\boldsymbol{\omega}(t) = \sqrt{\omega_x^2 + \omega_y^2 + \omega_z^2}\boldsymbol{e}$. For small angles, Eq. (5.36) can be approximated by

$$q(t + \Delta t) \approx \left(I + \frac{\Delta t \boldsymbol{\Omega}}{2} \right) q(t) \tag{5.37}$$

where

$$\boldsymbol{\Omega} = \begin{bmatrix} 0 & -\omega_x & -\omega_y & -\omega_z \\ \omega_x & 0 & \omega_z & -\omega_y \\ \omega_y & -\omega_z & 0 & \omega_x \\ \omega_z & \omega_y & -\omega_x & 0 \end{bmatrix} \tag{5.38}$$

A differential equation that describes rigid body orientation is obtained by limiting Δt toward zero:

$$\frac{dq}{dt} = \lim_{\Delta t \to 0} \frac{q(t + \Delta t) - q(t)}{\Delta t} = \frac{1}{2}\boldsymbol{\Omega} q \tag{5.39}$$

where angular rates in $\boldsymbol{\Omega}$ are given in the rigid body coordinates.

Rotational Kinematics Expressed by DCM

The differential equation for rigid body orientation presentation given by the DCM is derived in the following. Defining similarly as in Eq. (5.34),

$$R(t + \Delta t) = \Delta R(t) R(t) \tag{5.40}$$

where $R(t)$ is the orientation of the rigid body at time t, $R(t + \Delta t)$ is the orientation at time $t + \Delta t$, and $\Delta R(t)$ is the change of orientation (orientation of the body at time $t = t + \Delta t$) according to the orientation at time t.

Change of orientation $\Delta R(t)$ is defined as

$$\Delta R(t) = e^{\int_t^{t+\Delta t} \boldsymbol{\Omega}' dt} \tag{5.41}$$

where

$$\boldsymbol{\Omega}' = \begin{bmatrix} 0 & \omega_z & -\omega_y \\ -\omega_z & 0 & \omega_x \\ \omega_y & -\omega_x & 0 \end{bmatrix} \tag{5.42}$$

and $\boldsymbol{\omega}(t) = [\omega_x, \omega_y, \omega_z]^T$ is the vector of the current angular rates of the body.

Assuming $\boldsymbol{\Omega}'$ is a constant matrix in time period Δt we can approximate $\int_t^{t+\Delta t} \boldsymbol{\Omega}'(t) dt \approx \boldsymbol{\Omega}' \Delta t = B$. The exponent in relation (5.41) written in the Taylor series becomes

$$\Delta R(t) = e^B$$

$$= \left(I + B + \frac{B^2}{2!} + \frac{B^3}{3!} + \cdots \right)$$

$$= \left(I + B + \frac{B^2}{2!} - \frac{\sigma^2 B}{3!} - \frac{\sigma^2 B^2}{4!} + \frac{\sigma^4 B}{5!} + \cdots \right) \quad (5.43)$$

$$= \left(I + \frac{\sin \sigma}{\sigma} B + \frac{1 - \cos \sigma}{\sigma^2} B^2 \right)$$

where I is the 3×3 identity matrix and $\sigma = \Delta t \sqrt{\omega_x^2 + \omega_y^2 + \omega_z^2}$. For small angles σ relation (5.43) approximates to

$$\Delta R(t) = I + B = I + \Omega' \Delta t \quad (5.44)$$

which can be used to predict rotation matrix (5.40):

$$R(t + \Delta t) = \left(I + \Omega' \Delta t \right) R(t) \quad (5.45)$$

The final differential equation is obtained by limiting Δt to zero:

$$\frac{dR}{dt} = \lim_{\Delta t \to 0} \frac{R(t + \Delta t) - R(t)}{\Delta t} = \Omega' R \quad (5.46)$$

For the sake of completeness, also the equivalent presentation of rotation parameterization using Euler angles (notation 3–2–1) is given:

$$\dot{\varphi} = \omega_x + \omega_y \sin \varphi \tan \theta + \omega_z \cos \varphi \tan \theta$$

$$\dot{\theta} = \omega_y \cos \varphi - \omega_z \sin \varphi \quad (5.47)$$

$$\dot{\psi} = \frac{\omega_y \sin \varphi + \omega_z \cos \varphi}{\cos \theta}$$

where it can be seen that the notation becomes singular (the first end the third equation of Eq. (5.47) at $\theta = \pm \pi/2$).

5.2.4 Projective Geometry

Projection is the transformation of a space with $N > 0$ dimensions into a space with $M < N$ dimensions. Normally, under a general projective transformation some information is inevitably lost. However, if multiple projections of an object are available, it is in some cases possible to reconstruct the original object in N-dimensional space. Two of the most common projective transformations are *perspective projection* and *parallel projection* (linear transformation with a focal point in infinity).

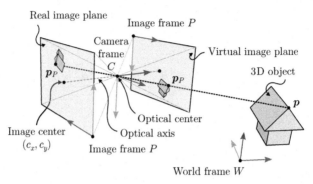

Fig. 5.1 Pinhole camera model.

According to the *pinhole camera model*, the 3D point $\boldsymbol{p}_C^T = [x_C \ y_C \ z_C]$ given in the camera frame C is projected to the 2D point $\boldsymbol{p}_P^T = [x_P \ y_P]$ in the image (picture) frame P as follows (see Fig. 5.1):

$$\underline{\boldsymbol{p}}_P = \frac{1}{z_C} \boldsymbol{S} \boldsymbol{p}_C \tag{5.48}$$

where $\underline{\boldsymbol{p}}_P$ is the representation of the point $\boldsymbol{p}_P$ in homogeneous coordinates, that is, $\underline{\boldsymbol{p}}_P^T = [x_P \ y_P \ 1]$. Matrix $\boldsymbol{S} \in \mathbb{R}^3 \times \mathbb{R}^3$ describes the *internal camera model*:

$$\boldsymbol{S} = \begin{bmatrix} \alpha_x f & \gamma & c_x \\ 0 & \alpha_y f & c_y \\ 0 & 0 & 1 \end{bmatrix} \tag{5.49}$$

The intrinsic camera parameters contained in $\boldsymbol{S}$ are the focal length f; the scaling factors α_x and α_y in horizontal and vertical directions, respectively; the optical axis image center (c_x, c_y); and the skewness γ. The camera model parameters are normally obtained or refined in the process of camera calibration. The perspective camera model (5.48) is nonlinear because of the term z_C^{-1} (the inverse of the object distance along the z-axis in the camera frame C). Although points, lines, and general conics are invariant under perspective transformation (i.e., the points transform into points, the lines into lines, and conics into conics), the projected image is a somewhat distorted representation of reality. In general, the angles between the lines and distance ratios are not preserved (e.g., the parallel lines do not transform into parallel lines). In some particular camera configurations the perspective projection can be approximated with an appropriate linear model [2]; this can enable, for example, simple camera calibration. The camera model (5.48) can also be written as

$$\underline{\boldsymbol{p}}_P \propto \boldsymbol{S}\boldsymbol{p}_C \tag{5.50}$$

The image of an object is formed on the screen behind the camera optical center (at the distance of focal length f along the negative z_C semiaxis), and it is rotated for 180 degree and scaled down. When creating a visual presentation of the camera model a virtual unrotated image can be assumed to be formed in front of the camera's optical center (on the positive z_C semiaxis at the same distance from the optical center as a real image) as it is shown in Fig. 5.1.

EXAMPLE 5.4

The internal camera model (5.49) has the following parameters: $\alpha_x f = \alpha_y f = 1000$, no skewness ($\gamma = 0$), and the image center is in the middle of the image with dimensions of 1024×768. Project the following set of 3D points given the camera frame to the image:

$$\boldsymbol{p}_C^T \in \{[-1\ 1\ 4], [1\ 1\ 5], [0\ -1\ 4], [-1\ 1\ -4], [4\ 1\ 5]\}$$

Solution

The projection of 3D points to an image according to Eq. (5.48) is given in Listing 5.1. The results are also shown graphically on the left-hand side of Fig. 5.2. Notice that the last point does not appear in the image since it is out of the image screen's boundaries. Moreover, the fourth point also should not appear in the image since the point is behind the camera. Remember that the mathematical model (5.48) does not take visibility constraints into account. Therefore additional checks need to be made that ensure that only visible points are projected to the image: that is, the projected points must be inside the image boundaries and the point to be projected must be in front of the camera. Therefore, only the first three points actually appear in the image (see right-hand side of Fig. 5.2).

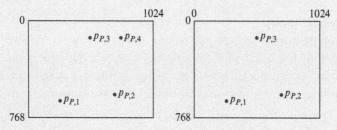

Fig. 5.2 Projected points from Example 5.4. On the *left* there are points projected to the image according to the camera model (5.48), and on the *right* there are points that are actually projected.

(Continued)

EXAMPLE 5.4—cont'd

Listing 5.1 Solution of Example 5.4

```
1  % Intrinsic camera parameters and image screen size
2  alphaF = 1000; % alpha*f, in px/m
3  s = [1024; 768]; % screen size, in px
4  c = s/2; % image centre, in px
5  S = [alphaF, 0, c(1); 0, alphaF, c(2); 0, 0, 1]; % Camera model
6
7  % Set of 3-Dn points in camera frame
8  pC = [-1 1 4; 1 1 5; 0 -1 4; -1 1 -4; 4 1 5].';
9
10 % Projection of points to image frame
11 pP = (S*pC)./repmat(pC(3,:), 3, 1)
```

```
pP =
         262        712        512        762       1312
         634        584        134        134        584
           1          1          1          1          1
```

From Eq. (5.48) it is clear that every point in the 3D space is transformed to a point in 2D space, but this does not hold for the inverse transformation. Since the perspective transformation causes loss of the scene depth, the point in the image space can only be back-projected to a ray in 3D space if no additional information is available. The scene can be reconstructed if scene depth is somehow recovered. There are various techniques that enable 3D reconstruction that can be based on depth cameras, structured light, visual cues, motion, and more. The position of the point in a 3D scene can also be recovered from corresponding images of the 3D point obtained from multiple (at least two) views. Three-dimensional reconstruction is therefore possible with a stereo camera.

Multiview Geometry

Multiview geometry is not important only because it enables scene reconstruction but some of the properties can be exploited in the design of machine vision algorithms (e.g., image point matching and image-based camera pose estimation). Consider that the rotation matrix $R_{C_2}^{C_1}$ and translation vector $t_{C_2}^{C_1}$ describe the relative pose between two cameras (Fig. 5.3). Under the assumption that camera centers do not coincide ($t_{C_2}^{C_1} \neq 0$), the relation (5.52) can be obtained after a short mathematical manipulation (cameras with identical internal parameters are assumed):

$$p_{C_1} = R_{C_2}^{C_1} p_{C_2} + t_{C_2}^{C_1}, \tag{5.51}$$

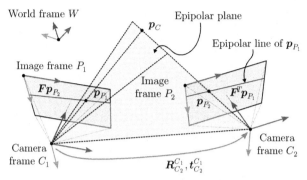

Fig. 5.3 Two-view perspective-camera geometry.

$$S^{-1}\underline{p}_{P_1} \propto R_{C_2}^{C_1} S^{-1} \underline{p}_{P_2} + t_{C_2}^{C_1}$$

$$\left[t_{C_2}^{C_1}\right]_\times S^{-1}\underline{p}_{P_1} \propto \left[t_{C_2}^{C_1}\right]_\times R_{C_2}^{C_1} S^{-1} \underline{p}_{P_2}$$

$$0 = \underline{p}_{P_1}^T S^{-T} \left[t_{C_2}^{C_1}\right]_\times R_{C_2}^{C_1} S^{-1} \underline{p}_{P_2}$$

$$0 = \underline{p}_{P_1}^T F \underline{p}_{P_2} \tag{5.52}$$

Notice that the cross product between the vectors $a^T = [a_1 \ a_2 \ a_3]$ and $b^T = [b_1 \ b_2 \ b_3]$ is written as $a \times b = [a]_\times b$, where $[a]_\times$ is a skew-symmetric matrix:

$$[a]_\times = \begin{bmatrix} 0 & -a_3 & a_2 \\ a_3 & 0 & -a_1 \\ -a_2 & a_1 & 0 \end{bmatrix} \tag{5.53}$$

The matrix F is known as a *fundamental matrix* that describes the *epipolar constraint* (5.52): the point $\underline{p}_{P_1}$ is on the line $F\underline{p}_{P_2}$ in the first image and the point $\underline{p}_{P_2}$ lies on the line $F^T\underline{p}_{P_1}$ in the second image. Another important relation in machine vision is $p_{C_1}^T E p_{C_2} = 0$, where $E = [t_{C_2}^{C_1}]_\times R_{C_2}^{C_1}$ is known as the *essential matrix*. The relation between the essential and fundamental matrix is $E = S^T F S$.

The epipolar constraint can be used to improve matching of image points in multiple images that belong to the same world point given the known pose between two calibrated cameras. Since the corresponding pair of the point $\underline{p}_{P_1}$ in the first image must lie on the line $F^T\underline{p}_{P_1}$ in the second image, the 2D searching over the entire image area is simplified to a 1D search along the epipolar line, and therefore the matching process can be sped up significantly and it can also be made more robust, since matches that do not satisfy the epipolar constraint are rejected.

EXAMPLE 5.5

A set of 3D points is observed from two cameras with relative displacement described by the rotation matrix $\boldsymbol{R}_{C_2}^{C_1} = \boldsymbol{R}_x(30°)\boldsymbol{R}_y(60°)$ and translation vector $\boldsymbol{t}_{C_2}^{C_1} = [4 \ -1 \ 2]^T$. Both cameras have the same internal camera model (5.49): $\alpha_x f = \alpha_y f = 1000$, no skewness ($\gamma = 0$), and image center is in the middle of the image with the dimensions of 1024×768. A set of projected points into the image of the first camera is

$$\boldsymbol{p}_{P_1}^T \in \{[262 \ 634], [762 \ 634], [512 \ 134], [443 \ 457], [412 \ 284]\}$$

and a set of point projected into the image of the second camera is

$$\boldsymbol{p}_{P_1}^T \in \{[259 \ 409], [397 \ 153], [488 \ 513], [730 \ 569], [115 \ 214]\}$$

The points in both images are shown in Fig. 5.4. Determine all possible image point correspondences taking into account the epipolar constraint. For all the matched point pairs reconstruct the position of the points in the 3D space with respect to both camera frames.

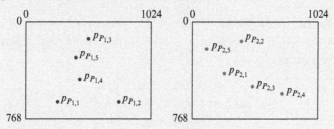

Fig. 5.4 Points in the two images from Example 5.5.

Solution

When matching points between the images from two cameras with a known relative pose between them, the points that belong to the same world point must satisfy the epipolar constraint (5.52). Eq. (5.52) has the form $\boldsymbol{l}^T \boldsymbol{p} = 0$, that is, the homogeneous point $\boldsymbol{p}^T = [x \ y \ 1]$ lies on the line $\boldsymbol{l}^T = [a \ b \ c]$, since $ax + by + c = 0$ is a line equation. Therefore, the epipolar constraint (5.52) can be utilized to find the epipolar lines $\boldsymbol{l}_{P_2}$ in the second image for every point $\boldsymbol{p}_{P_1}$ from the first image, and vice–versa, lines $\boldsymbol{l}_{P_1}$ for every point $\boldsymbol{p}_{P_2}$. The epipolar lines for the first and the second image are shown in Fig. 5.5. From Figs. 5.4 and 5.5 four possible point matches can be found graphically.

EXAMPLE 5.5—cont'd

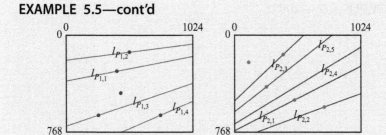

Fig. 5.5 Points and epipolar lines in the two images from Example 5.5.

The implementation of algorithm for point matching is given in Listing 5.2 (functions rotX and rotY are Matlab implementations of Eqs. 5.3, 5.4, respectively). The resulting point matches are collected in the index matrix pairs.

Listing 5.2 Point matching from Example 5.52

```
1  % Intrinsic camera parameters and image screen size
2  alphaF = 1000; % alpha*f, in px/m
3  s = [1024; 768]; % screen size, in px
4  c = s/2; % image centre, in px
5  S = [alphaF, 0, c(1); 0, alphaF, c(2); 0, 0, 1]; % Camera model
6
7  % Relative pose between the cameras
8  R = rotX(pi/6)*rotY(pi/3); t = [4; -1; 2];
9
10 % Set of points
11 pP1 = [262, 634; 762, 634; 512, 134; 443, 457; 412, 284].';
12 pP2 = [259, 409; 397, 153; 488, 513; 730, 569; 115, 214].';
13
14 N1 = size(pP1, 2); N2 = size(pP2, 2); % Number of points
15
16 % Fundamental matrix
17 tx = [0, -t(3), t(2); t(3), 0, -t(1); -t(2), t(1), 0];
18 F = S.'\tx*R/S;
19
20 epsilon = 1e-2; % Distance error tolerance
21 pP1 = [pP1; ones(1,N1)]; pP2 = [pP2; ones(1,N2)]; % Homogenous
22   points
23
24 % Epipolar lines in frame P1 associated with the points in frame P2
25 1P1 = F*pP2;
26 % Epipolar lines in frame P2 associated with the points in frame P1
27 1P2 = F.'*pP1;
28
29 % Find point pairs (evaluate fundamental constraint in frame P2)
30 pairs = [];
31 for i = 1:N1
32     d = abs(1P2(:,i).'*pP2);
33     k = find(d<epsilon);
```

(Continued)

EXAMPLE 5.5—cont'd

```
34      if ~isempty(k), pairs = [pairs, [i; k(1)]]; end
35 end

   pairs =
         1    2    3    5
         3    4    2    1
```

Singular Cases

The relation (5.52) becomes singular if there is no translation between the camera centers. In this case the following relation can be obtained from Eq. (5.51) if $t_{C_2}^{C_1}$ is set to zero:

$$\underline{p}_{P_1} \propto S R_{C_2}^{C_1} S^{-1} \underline{p}_{P_2} \tag{5.54}$$

A similar equation form can also be obtained in a case where all the points in the 3D space are confined to a single common plane. Without loss of generality, the plane $z_W = 0$ can be selected:

$$\underline{p}_{P_1} \propto S \begin{bmatrix} r_{W,1}^{C_1} & r_{W,2}^{C_1} & t_W^{C_1} \end{bmatrix} \begin{bmatrix} r_{W,1}^{C_2} & r_{W,2}^{C_2} & t_W^{C_2} \end{bmatrix}^{-1} S^{-1} \underline{p}_{P_2} \tag{5.55}$$

where $R_W^C = [r_{W,1}^C \ r_{W,2}^C \ r_{W,3}^C]$. The transformation from the world plane to the picture plane is

$$\underline{p}_P \propto S[r_{W,1}^C \ r_{W,2}^C \ t_W^C]\underline{p}_W \tag{5.56}$$

where in this particular case $\underline{p}_W^T = [x_W \ y_W \ 1]$. Eqs. (5.54)–(5.56) all have similar form: $\underline{p}' \propto H\underline{p}$. The matrix H is known as *homography*.

3D Reconstruction

In a stereo camera configuration the 3D position of a point can be reconstructed from both images of the point. This procedure requires finding corresponding image points among the views, which is one of the fundamental problems in machine vision. But once point correspondences have been established and the relative pose between the camera views is known (the translation between the cameras must be different from zero) and also the internal camera model is known, the 3D position of the point can be estimated. If both camera models are the same, the point depth in both camera frames (z_{C_1} and z_{C_2}) can be estimated by solving the following set of equations (e.g., using least squares method):

$$S^{-1}\underline{p}_{P_1} z_{C_1} - R_{C_2}^{C_1} S^{-1} \underline{p}_{P_2} z_{C_2} = t_{C_2}^{C_1} \tag{5.57}$$

EXAMPLE 5.6

For all the obtained point matches between two images from Example 5.52 reconstruct the 3D positions of the world points with respect to the first and the second camera frame.

Solution

Three-dimensional positions of the points can be obtained if the system of Eq. (5.57) is solved for every pair of matched image points from Example 5.5, and the solution is inserted into the inverse transformation of Eq. (5.48). The implementation of the solution in Matlab is given in Listing 5.3. The reconstructed 3D points in the first and the second camera frames are saved to variables $pC1$ and $pC2$, respectively.

Listing 5.3 3D point reconstruction from Example 5.57

```
1  M = size(pairs, 2);
2  pC1 = zeros(3,M);
3  for i = 1:M
4      a = pairs(1,i); b = pairs(2,i);
5      c1 = S\pP1(:,a);
6      c2 = -R*(S\pP2(:,b));
7      psi = [c1, c2]\t;
8      pC1(:,i) = psi(1)*c1;
9  end
10 pC2 = R.'*(pC1-repmat(t, 1, M));

   pC1 =
       -0.9989     0.9994          0    -0.2999
        0.9989     0.9994    -1.0005    -0.2999
        3.9956     3.9978     4.0018     2.9994
   pC2 =
       -0.1372     0.8638    -0.4988    -1.0973
        0.7333     0.7327    -1.0013     0.1066
        5.6930     3.9635     4.3308     4.3316
```

The reconstruction problem simplifies in a canonical stereo configuration of two cameras, where one camera frame is only translated along the x-axis for *baseline distance b* (as shown in Fig. 5.6), since the epipolar lines are

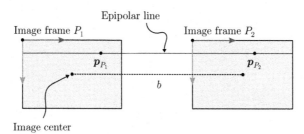

Fig. 5.6 Canonical stereo setup.

parallel and the epipolar line of the point $\boldsymbol{p}_{P_1}$ passes through that point (not only the point $\boldsymbol{p}_{P_2}$ in the other image). In the case of a digital image, this means that the matching point of a point in the first image is in the same image row in the second image. For the sake of simplicity, set $\alpha_x = \alpha_y = \alpha$ and $\gamma = 0$ in the camera model (5.49). The position of a 3D world point in the first camera frame can then be obtained from two images of the world point according to

$$\boldsymbol{p}_{C_1}^T = \frac{b}{d}\left[x_{P_1} - c_x \quad y_{P_1} - c_y \quad \alpha f\right] \tag{5.58}$$

where the *disparity* was introduced as $d = x_{P_1} - x_{P_2}$. The disparity contains information about scene depth as it can be seen from the last element in Eq. (5.58): $z_{C_1} = \alpha f b d^{-1}$. For all the points that are in front of the camera the disparity is positive, $d \geq 0$.

EXAMPLE 5.7

Assume a canonical stereo camera setup with baseline distance $b = 0.2$ and the following internal camera model parameters: $\alpha_x f = \alpha_y f = 1000$, $c_x = 512$, and $c_y = 384$. Reconstruct the position of a 3D point that projects to the points $\boldsymbol{p}_{P_1}^T = [351\ 522]$ and $\boldsymbol{p}_{P_2}^T = [236\ 522]$ in the first and second camera image, respectively.

Solution

Since this is a canonical stereo camera setup the solution can be obtained directly from Eq. (5.58) (see Listing 5.4).

Listing 5.4 3D point reconstruction from Example 5.58

```
1  % Intrinsic camera parameters and image screen size
2  alphaF = 1000; % alpha*f, in px/m
3  c = [512; 384]; % image centre, in px
4  S = [alphaF, 0, c(1); 0, alphaF, c(2); 0, 0, 1]; % Camera model
5
6  b = 0.2; % Stereo camera baseline, in m
7
8  % Matched points
9  pP1 = [351; 522];
10 pP2 = [236; 522];
11
12 % 3D reconstruction
13 d = pP1(1)-pP2(1);
14 pC1 = b/d*[pP1-c; alphaF]

   pC1 =
       -0.2800
        0.2400
        1.7391
```

5.3 POSE MEASUREMENT METHODS

In the following the main methodologies are introduced showing how sensors are used to estimate robot pose in the environment. Those approaches can measure relative pose advancement according to previously determined pose or absolute pose according to some reference-based coordinate system.

5.3.1 Dead Reckoning

Dead reckoning (also called deduced reckoning) approaches estimate the robot's (equipped with relative sensors) current pose from the known previous pose and relative measured displacements from the previous pose. Those displacements or motion increments (distance and orientation) are calculated from measured speeds over the elapsed time and heading. Common to those approaches is the use of path integration to estimate the current pose; therefore, the accumulation of different errors (error of integration method, measurement error, bias, noise, etc.) typically appears.

The two basic approaches used in mobile robotics are odometry and inertial navigation.

Odometry

Odometry is used to estimate the robot pose by integration of motion increments that can be measured or gathered from applied motion commands. Relative motion increments are in mobile robotics usually obtained from axis sensors (e.g., incremental encoder) that are attached to the robot's wheels. Using an internal kinematic model (see Eq. 2.1 for differential drive) these wheel rotation measurements are related to the position and the orientation changes of the mobile robot. The position and orientation changes given in the known time period between successive measurements can also be expressed by robot velocities. In some cases wheel angular velocities can be measured directly or inferred from known velocity commands (supposing that implemented speed controllers are accurate).

Let us assume there is a robot with differential drive and incremental encoder sensors mounted on the wheels. The measurements give relative rotation change of the left and right wheel, namely $\Delta\alpha_L(t)$ and $\Delta\alpha_R(t)$, from the previous orientation at time $t^- = t - \Delta t$. Supposing that there is pure rolling of the wheels the traveled distances are

$$\Delta d_L(t) = \Delta\alpha_L(t)R, \quad \Delta d_R(t) = \Delta\alpha_R(t)R$$

where R is the wheel's radius. Using the internal kinematic model (2.1) orientation displacement and traveled distance (position displacements) are

$$\Delta\varphi(t) = \frac{\Delta d_R(t) - \Delta d_L(t)}{L}$$

and

$$\Delta d(t) = \frac{\Delta d_R(t) + \Delta d_L(t)}{2}$$

where L is the distance between the wheels.

The robot pose estimate is obtained by integration of the external kinematic model (2.2) from measured robot velocities or as a summation of calculated robot position and orientation displacements. Applying trapezoidal numerical integration the estimated robot pose is

$$x(t) = x(t^-) + \Delta d(t) \cos\left(\varphi(t^-) + \frac{\Delta\varphi(t)}{2}\right)$$

$$y(t) = y(t^-) + \Delta d(t) \sin\left(\varphi(t^-) + \frac{\Delta\varphi(t)}{2}\right) \qquad (5.59)$$

$$\varphi(t) = \varphi(t^-) + \Delta\varphi(t)$$

However, due to the integral nature of odometry, cumulative error occurs. The main source of the error consists of the systematic and nondeterministic error sources. The former includes errors due to approximate kinematic models (e.g., wrong radius of the wheel), error due to accuracy of applied integration method, and measurement error (unknown bias), and the latter includes slippage of the wheels, noise, and the like. Therefore, independent use of odometry is applicable only for short-term pose estimation at a known initial pose. More commonly, odometry is used in combination with pose measurements from absolute sensors where odometry is used for prediction and filtration of absolute pose measurements to obtain better pose estimates.

In wheeled mobile systems sensors used in odometry usually are axis sensors (e.g., incremental optical or magnetic encoders, potentiometers) that are attached to the wheels' axes and measure rotation angle or velocity.

Inertial Navigation

An inertial navigation system (INS) is a self-contained technique to estimate vehicle position, orientation, and velocity by means of dead reckoning. INS includes motion sensors (accelerometers) and rotation sensors (gyroscopes) where position and orientation of the vehicle is estimated relatively to a known initial starting pose.

Output of accelerometer and gyroscope measurements are 3D vectors of accelerations and angular velocities in space. To estimate robot position and orientation double integration of acceleration measurement and single integration of angular velocity is required. Use of integration is the main cause for pose error of INSs because of accumulation of constant (systematic) errors (sensor drift, bad sensor calibration, etc.). Constant measurement error causes quadratic growth of the position estimate error with time while the error of orientation estimate grows linearly with time. Additionally the error of position estimate in INS is dependent on the error of orientation estimate because the accelerometer measures total acceleration including gravitation. Vehicle (INS unit) acceleration is obtained from acceleration measurement by subtracting gravity acceleration, which requires the accurate knowledge of the vehicle orientation. Any error in orientation therefore causes incorrect virtual acceleration due to inexact gravity canceling. This is especially problematic because vehicle accelerations are usually much smaller than gravity. A small signal-to-noise ratio (SNR) at lower accelerations causes additional troubles. These effects are clearly seen from the acceleration sensor model consisting of translational vehicle acceleration a, gravity $g = [0, 0, 0.981]^T$, and radial acceleration,

$$a_m = R_i^{w\,T}(a + R_e^w g + \omega \times v) + a_{\text{bias}} + a_{\text{noise}} \qquad (5.60)$$

where a_m is the acceleration measurement from a sensor in its local coordinates, R_e^w is the rotation matrix between Earth's coordinates and world coordinates where pose tracking is performed, R_i^w is the rotation matrix from the local INS coordinates to the world coordinates, ω is angular velocity in world coordinates, v is translational velocity, a_{bias} is acceleration bias, and a_{noise} is noise.

To estimate orientation a gyroscope measurement is used. The model of a gyroscope sensor is

$$\omega_m = \omega_i + \omega_{\text{bias}} + \omega_{\text{noise}} \qquad (5.61)$$

where ω_m is a vector of measured angular rates, ω_i is a true vector of angular rates of the body in local coordinates, ω_{bias} is sensor bias, and ω_{noise} is sensor noise.

To estimate INS unit orientation the unit angular rate estimates are obtained from Eq. (5.61) as follows: $\omega_i = \omega_m - \omega_{\text{bias}}$. The biased part is usually estimated online by means of some estimator (e.g., Kalman filter). Exceptionally, for shorter estimation periods, the bias can be supposed constant if quality gyroscope is used and initial calibration (estimation of ω_{bias}) is made. Note, however, that bias is changing with time. Therefore, the latter approach is sufficient for short-term use only. The simplest calibration may be obtained by averaging the gyroscope measurement (for N measurements) while holding the INS unit in a constant position ($\omega_{bias} = \frac{1}{N} \sum_{i=1}^{N} \omega_m$, when $\omega_i = 0$).

A relative estimate of INS unit orientation according to initial orientation is then obtained using rotational kinematics for quaternions (5.39) as follows:

$$\frac{d\boldsymbol{q}_w^i(t)}{dt} = \frac{1}{2}\boldsymbol{\Omega}(t)\boldsymbol{q}_w^i(t) \tag{5.62}$$

$$\boldsymbol{q}_w^i(t) = \int_0^t \frac{d\boldsymbol{q}_w^i(t)}{dt} dt + \boldsymbol{q}_w^i(0) \tag{5.63}$$

where $\boldsymbol{q}_w^i(t)$ is the quaternion that describes the rotation from world (w) coordinates to INS unit coordinates (i), $\boldsymbol{q}_w^i(0)$ is the initial orientation,

$$\boldsymbol{\Omega}(t) = \begin{bmatrix} 0 & -\omega_x & -\omega_y & -\omega_z \\ \omega_x & 0 & \omega_z & -\omega_y \\ \omega_y & -\omega_z & 0 & \omega_x \\ \omega_z & \omega_y & -\omega_x & 0 \end{bmatrix}$$

and $\boldsymbol{\omega}_i(t) = [\omega_x, \omega_y, \omega_z]^T$. To obtain the DCM $\boldsymbol{R}_w^i = \boldsymbol{R}_i^w{}^T$ use relation (5.18). In implementation of Eq. (5.63) a numerical integration (5.36) can be used.

To estimate INS unit position a translational acceleration in world coordinates is computed from Eq. (5.60):

$$\boldsymbol{a} = \boldsymbol{R}_i^w(\boldsymbol{a}_m - \boldsymbol{a}_{bias}) - \boldsymbol{R}_e^w\boldsymbol{g} - \boldsymbol{\omega} \times \boldsymbol{v} \tag{5.64}$$

where the angular velocity vector in world coordinates is $\boldsymbol{\omega} = \boldsymbol{R}_i^w\boldsymbol{\omega}_i$ and the bias term $\boldsymbol{a}_{bias}$ needs to be estimated online by means of some estimator (e.g., Kalman filter). Also, a proper acceleration sensor calibration is required. The velocity estimate $\boldsymbol{v}(t)$ and position $\boldsymbol{x}(t)$ are obtained by integration:

$$\boldsymbol{v}(t) = \int_0^t \boldsymbol{a}(t)dt + \boldsymbol{v}(0) \tag{5.65}$$

$$x(t) = \int_0^t v(t)dt + x(0) \tag{5.66}$$

EXAMPLE 5.8

Simulate measured acceleration and angular rates for a differential robot with an INS unit that drives on the curve $x(t) = \cos(t)$, $y(t) = \sin(2t)$, and $z(t) = 0$ (all defined in world coordinates). The time of simulation is 6 s and the sampling time is 1 ms. The INS unit is oriented so that its x-axis is tangent to trajectory, the y-axis is perpendicular to the x-axis, and the z-axis is aligned with the z-axis of the world coordinates.

From the measurements, estimate, the INS unit position and orientation in world coordinates. Additionally, consider sensor bias and noise and observe how this influences the estimated INS pose.

Solution

The INS unit angular rates in the simulation are obtained considering relation (5.46) to obtain the matrix of angular rates $\mathbf{\Omega}' = \frac{d\mathbf{R}}{dt}\mathbf{\Omega}'\mathbf{R}^T$. The simulated accelerometer and gyroscope measurements are modeled using Eqs. (5.60), (5.61) and are shown in Fig. 5.7.

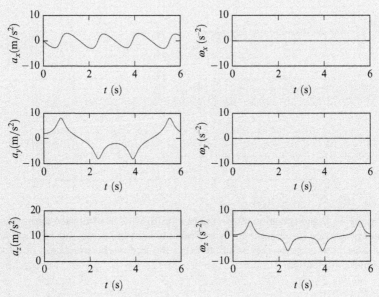

Fig. 5.7 Simulated acceleration (*left column*) and angular rate (*right column*) measurements of INS unit.

(Continued)

EXAMPLE 5.8—cont'd

The estimated position and orientation of INS in an ideal case with no noise and bias are shown in Fig. 5.8.

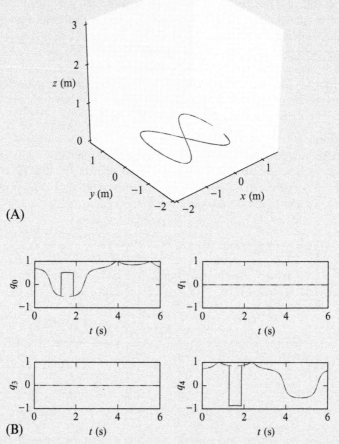

(A)

(B)

Fig. 5.8 Estimated position (A) and orientation (B) of INS unit from accelerometer and gyroscope measurements in an ideal situation (no noise and no bias). A smaller position error can be observed at the end of the simulation due to numeric integration.

The estimated pose in the case of noisy measurements and included bias are shown in Fig. 5.9, where fast growth of the position error can be observed.

EXAMPLE 5.8—cont'd

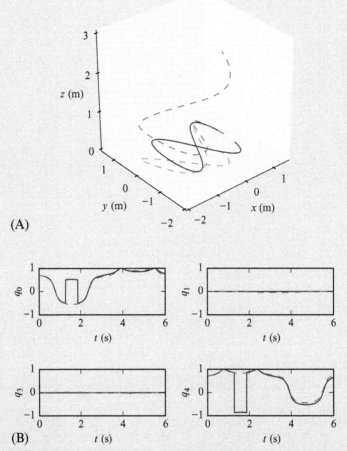

(A)

(B)

Fig. 5.9 Estimated position (A) and orientation (B) of the INS unit from the accelerometer and gyroscope measurements with included sensor noise and bias. Because bias is not compensated, a huge error in the position estimate and a smaller error in orientation appears.

A script to simulate the INS unit and estimate its pose is given in Listing 5.5 (functions `rotX`, `rotY`, and `rotZ` are Matlab implementations of Eqs. 5.3, 5.4, 5.5, respectively).

(*Continued*)

EXAMPLE 5.8—cont'd

Listing 5.5 Simulation of the INS unit

```
1  biasA = [1; 1; 1]*0.02; biasW = [1; 1; 1]*0.04; % Sensor bias
2  SigmaA = 0.1; SigmaW = 0.05; % Sensor noise
3
4  nSteps = 6000; dT = 0.001; t = 0; % Time samples, step size
5
6  for k = 1:nSteps
7      % INS motion simulation: get true robot pose, acceleration and rotation
8      x = [cos(t); sin(2*t); 0];     % Position
9      v = [-sin(t); 2*cos(2*t); 0]; % Velocity
10     a = [-cos(t);-4*sin(2*t); 0]; % Acceleration
11
12     fi = [0; 0; atan2(v(2), v(1))]; % Euler angles (from world to INS)
13     dfi = [0; 0; (v(1)*a(2) - v(2)*a(1))/(v(1)^2 + v(2)^2)]; % Derivative
14     Rx = rotX(fi(1)); Ry = rotY(fi(2)); Rz = rotZ(fi(3));
15     dRx = [0, 0, 0; 0,-sin(fi(1)), cos(fi(1)); 0,-cos(fi(1)),-sin(fi(1))];
16     dRy = [-sin(fi(2)), 0,-cos(fi(2)); 0, 0, 0; cos(fi(2)), 0,-sin(fi(2))];
17     dRz = [-sin(fi(3)), cos(fi(3)), 0;-cos(fi(3)),-sin(fi(3)), 0; 0, 0, 0];
18     R = Rx*Ry*Rz;
19     dR = dRx*Ry*Rz*dfi(1) + Rx*dRy*Rz*dfi(2) + Rx*Ry*dRz*dfi(3);
20     q = dcm2quat(R).'; % Quaterion from world to INS
21
22     % Gyro measurements
23     Omega = dR*R.'; % Skew-symmetric form of the angular rate vector wb
24     wb = -[Omega(3,2); Omega(1,3); -Omega(1,2)]; % Angular rates of the INS
25
26     % Accelerometer measurements
27     agDyn = a; % Dynamic acceleration in world coordinates
28     agGrav = [0; 0; 9.81]; % Gravitation
29     Rearth = eye(3); % Assume world and Earth frames are the same
30     wg = R.'*wb; % True angular rates in world frame
31     wgSkew = [0 -wg(3) wg(2); wg(3) 0 -wg(1); -wg(2) wg(1) 0];
32     vg = v; % Velocity in world frame
33
34     % Measured angular rates and accelerations
35     wbMea = wb + biasW + randn(3,1)*SigmaW;
36     abMea = R*(agDyn + Rearth*agGrav + wgSkew*vg) +biasA+randn(3,1)*SigmaA;
37
38     % Inertial navigation
39     if k==1 % Initialization
40         qEst = q; xEst = x; vEst = v; % Init true initial values
41     else % Update
42         % Gyro
43         wx = wbMea(1); wy = wbMea(2); wz = wbMea(3);
44         OMEGA = [ 0 -wx -wy -wz; ...
45                   wx   0  wz -wy; ...
46                   wy -wz   0  wx; ...
47                   wz  wy -wx   0];
48         dQest = 0.5*OMEGA*qEst;
49         qEst = qEst + dQest*dT; % Quaterninon integration
50         qEst = qEst/norm(qEst); % Quaternion normalization
51         % Acceleration
52         agGrav = [ 0; 0; 9.81]; % Gravity
53         R_ = quat2dcm(qEst.'); % Ratation from world to INS
54         Rearth = eye(3); % Earth frame is the same as world frame
55         wg_ = R_.'*[wx; wy; wz]; % Angular rates in world frame
56         wgSkew_ = [0 -wg_(3) wg_(2); wg_(3) 0 -wg_(1); -wg_(2) wg_(1) 0];
```

EXAMPLE 5.8—cont'd

```
57          Aest = R_.'*abMea - Rearth*agGrav - wgSkew_*vEst; % Estimated a
58          vEst = vEst + Aest*dT; % Estimated velocity
59          xEst = xEst + vEst*dT; % Estimated position
60      end
61      t = t + dT;
62 end
```

5.3.2 Heading Measurement

Heading measurement systems provide information about the vehicle orientation (or attitude) in space, the direction the vehicle is pointing. Several sensors can be used to estimate heading information such as a magnetometer, gyroscope, and accelerometer. Their information is integrated in sensor systems to estimate heading information (e.g., attitude, compass, or inclinometer).

The magnetometer and accelerometer provide absolute measurements of 3D direction vectors of Earth's magnetic field (strength and direction) and Earth's gravity direction vector (if the sensor unit is not accelerating), respectively. To lower heading estimation noise and improve accuracy also the relative measurements from the gyroscope are included in estimation filters.

To estimate orientation of the sensor unit according to some reference frame (e.g., Earth's fixed coordinates) at least two directional vectors are required. To illustrate the idea the first sensor can be a magnetometer whose measurement model reads

$$b_m = R_w^i R_e^w b_{\text{true}} + b_{\text{bias}} + b_{\text{noise}} \qquad (5.67)$$

where b_m is the measured magnetic field in sensor coordinates, b_{true} is Earth's magnetic field in Earth coordinates for some place on Earth, and rotation matrices are defined as in Eq. (5.60). b_{true} can be approximated as constant for some small area on Earth (e.g., 10×10 km^2). The second sensor is the accelerometer with a model given in Eq. (5.60), whose measurement in the case of nonaccelerating motion points in the direction of gravity and the defined z-axis of Earth's fixed coordinates.

Orientation of the sensor unit according to Earth's fixed coordinates is described by the DCM $R_e^i = R_i^{wT} R_e^w$. This can be obtained by creating a matrix of the Earth's coordinate axis vectors expressed in the local coordinates as follows. In static case accelerometer measures the direction vector from Earth's center to the INS position. This direction therefore defines the z-axis of Earth coordinates and is expressed in a local (sensor) frame as follows:

$$z_d = \frac{a_m}{\|a_m\|}$$

The direction of Earth axis x expressed in local coordinates is defined by the component of the magnetometer vector that is perpendicular to z_d:

$$x_d = \frac{b_m \times z_d}{\|b_m \times z_d\|}$$

where $\times$ denotes the cross product. Direction to the north defines Earth's y-axis expressed in local coordinates:

$$y_d = z_d \times x_d$$

The obtained rotation matrix is

$$R_e^i = [x_d, y_d, z_d] \tag{5.68}$$

from where the DCM between the world and Earth coordinates is $R_w^i = R_i^{wT} = \left(R_e^i R_e^{wT}\right)^T$. This orientation can also be described by quaternions using relations (5.19)–(5.22) or by Euler angles using relation (5.6).

Accurate orientation estimation of the vehicle is important also when performing odometry or inertial navigation to lower orientation error and consequentially the error in position estimates. Therefore, absolute heading measurements are often combined with dead reckoning as a correction measurement in the correction step of estimators (e.g., Kalman filter).

5.3.3 Active Markers and Global Position Measurement

Localization in the environment is possible also by observing markers located at known positions in the environment. These markers can be natural if they are already part of the environment (e.g., lights on the ceiling, Wi-Fi transmitters, etc.) or they can be artificial if they are placed in the environment for the purpose of localization (e.g., radiofrequency transmitters, ultrasonic or infrared transmitters (beacons), wires buried in the ground for robotic lawn mowers, GNSS satellites, etc.).

The main advantages of using active markers are simplicity, robustness, and fast localization. However, their operational cost, maintenance, and setup cost are relatively high.

To estimate the position or pose of the system usually the *triangulation* or *trilateration* approach is applied. Trilateration uses measured distances from more transmitters to estimate the position of the receiver (mounted on the vehicle). A very well-known trilateration approach is global positioning using a global navigation satellite system (GNSS) where active markers are GNSS satellites at known locations in space. Triangulation uses measured angles to three or more markers (e.g., light source) on known locations.

The basic idea of triangulation is illustrated in Fig. 5.10, where the robot measures relative angles α_i to the active markers. Suppose there are three markers as in Fig. 5.10, the current robot pose $q = [x, y, \varphi]^T$ and measured angles α_j ($j = 1, 2, 3$) are related by the following relations:

$$\tan(\alpha_1 + \varphi) = \frac{y_{m1} - y}{x_{m1} - x}$$

$$\tan(\alpha_2 + \varphi) = \frac{y_{m2} - y}{x_{m2} - x} \qquad (5.69)$$

$$\tan(\alpha_3 + \varphi) = \frac{y_{m3} - y}{x_{m3} - x}$$

To obtain the triangulation problem solution relations in Eq. (5.69) need to be solved for $q = [x, y, \varphi]^T$.

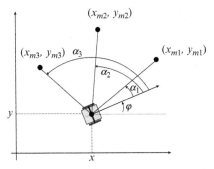

Fig. 5.10 Robot localization using triangulation where relative angles α_i to the markers are measured.

The basic idea of trilateration is illustrated in Fig. 5.14, where the current robot position, measured distances to markers, and their positions are related by the following set of equations:

$$d_1^2 = (x_{m1} - x)^2 + (y_{m1} - y)^2$$
$$d_2^2 = (x_{m2} - x)^2 + (y_{m2} - y)^2 \qquad (5.70)$$
$$d_3^2 = (x_{m3} - x)^2 + (y_{m3} - y)^2$$

In the following some examples of trilateration and triangulation are illustrated.

EXAMPLE 5.9

The robot is equipped with sensor which can measure direction to the active markers. There are three active markers placed at known locations $m_1 = [x_{m1}, y_{m1}]^T = [0, 0]^T$, $m_2 = [x_{m2}, y_{m2}]^T = [5, 3]^T$, and $m_3 = [x_{m3}, y_{m3}]^T = [1, 5]^T$. At a current robot pose $q = [x, y, \varphi]^T$ measured directions are given as a relative angles $\alpha_1 = -2.7691$, $\alpha_2 = -0.3585$, and $\alpha_3 = 1.4277$.

What is the current robot pose q?

There are many possible solutions to solve this triangulation problem, and here two possibilities are shown. In the first part of the solution, particle swarm optimization (PSO) is applied. Some more details about PSO can be found in Section 3.3.9. In the second part of the solution a popular geometric algorithm based on the intersection of circular arcs is used.

Solution A

The current robot pose $q = [x, y, \varphi]^T$ and measured angles α_j ($j = 1, 2, 3$) are related by the relations in Eq. (5.69). The task of the PSO algorithm is to find unknowns (x, y, and φ) so that the relations in Eq. (5.69) are valid. Each particle position presents one candidate solution (q_i) and during optimizations a swarm of particles updates toward more optimal solutions. The criteria for optimality of the ith particle solution is given as follows:

$$J_i = f(q_i) = \sum_{j=1}^{3} (\alpha_j - \hat{\alpha}_j)^2$$

EXAMPLE 5.9—cont'd

where $\hat{\alpha}_j$ is the simulated measurement of the ith particle that is obtained from Eq. (5.69):

$$\hat{\alpha}_j = \arctan \frac{y_{mj} - y}{x_{mj} - x} - \varphi$$

Note that the arctan function must return a correct angle in range $(-\pi, \pi]$ (in Matlab function `atan2` should be used).

The correct solution is $q = [2, 2.5, \pi/6]^T$, and the solution script is given in Listing 5.6 and the resulting situation is shown in Fig. 5.11.

Listing 5.6 Triangulation problem solved by PSO

```
1  m = [0, 0; 5, 3; 1, 5].'; % Markers
2  r0 = [2; 2.5; pi/6]; % True robot pose (x, y and fi), unknown.
3
4  % Measured angles
5  alpha = wrapToPi(atan2(m(2,:)-r0(2), m(1,:)-r0(1))-r0(3));
6
7  % Method using PSO
8  iterations = 50; % Number of iterations
9  omega = 0.5; % Inertia
10 c1 = 0.5; % Correction factor
11 c2 = 0.5; % Global correction factor
12 N = 25; % Swarm size
13
14 % Initial swarm position
15 swarm = zeros([3,N,4]);
16 swarm(1,:,1) = 3 + randn(1,N); % Init of x
17 swarm(2,:,1) = 3 + randn(1,N); % Init of y
18 swarm(3,:,1) = 0 + randn(1,N); % Init of fi
19 swarm(:,:,2) = 0; % Initial velocity
20 swarm(1,:,4) = 1000; % Best value so far
21
22 for iter = 1:iterations % PSO iteratively find best solution
23     % Evaluate particles parameters
24     for i = 1:N
25         % Compute predicted angle measurments using i-th particle
26         pEst = swarm(:,i,1); % Estimated particle parameters (x, y, fi)
27
28         % Compare obtained predicted angles to the true angle measurments
29         alphaEst = wrapToPi(atan2(m(2,:)-pEst(2), m(1,:)-pEst(1))-pEst(3));
30
31         % Compute cost function
32         cost = (alphaEst-alpha)*(alphaEst-alpha).';
33         if cost<swarm(1,i,4) % If new parameter is better, update:
34             swarm(:,i,3) = swarm(:,i,1); % param values (x, y and fi)
35             swarm(1,i,4) = cost;        % and best criteria value.
36         end
37     end
38     [~, gBest] = min(swarm(1,:,4)); % Global best particle parameters
39
40     % Updating parameters by velocity vectors
41     swarm(:,:,2) = omega*swarm(:,:,2) + ...
42         c1*rand(3,N).*(swarm(:,:,3)-swarm(:,:,1)) + ...
43         c2*rand(3,N).*(repmat(swarm(:,gBest,3), 1, N) - swarm(:,:,1));
```

(Continued)

EXAMPLE 5.9—cont'd

```
44      swarm(:,:,1) = swarm(:,:,1) + swarm(:,:,2);
45 end
46
47 r = swarm(:,gBest,1) % Solution, best pose estimate
```

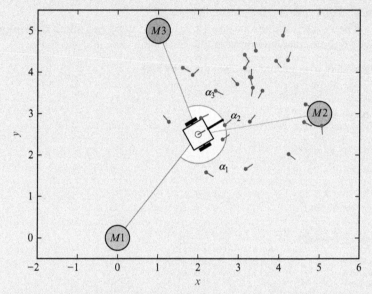

Fig. 5.11 Solution of triangulation problem from Example 5.9 obtained with particle swarm optimization (PSO). Initial poses of particles are shown with *dot-line markers*, and final particle poses are shown with *circle-line markers*.

Solution B

Among many analytical solutions for the triangulation problem, the most often used one computes the intersection of circular arcs. Here the basic idea is described and only the final analytical solution is used to compute robot pose. The complete derivation of the algorithm can be found in [3].

The algorithm is based on three circles where each circle is defined by three points: any pair of markers m_i, m_j, $(i, j = 1, 2, 3, i \neq j)$, and robot position x, y (see Fig. 5.12).

EXAMPLE 5.9—cont'd

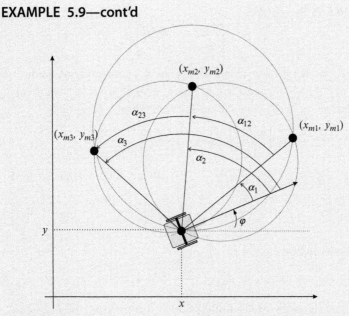

Fig. 5.12 Robot localization using the triangulation algorithm based on intersection of circles.

The robot position and the two markers are connected by two lines with the angle between the lines $\alpha_{ij} = \alpha_j - \alpha_i$. The centers and radius of those three circles are

$$c_{ij} = \begin{bmatrix} x_{cij} \\ y_{cij} \end{bmatrix} = \frac{1}{2} \left(m_i + m_j + \begin{bmatrix} (y_{mi} - y_{mj} \cot \alpha_{ij}) \\ (x_{mj} - x_{mi} \cot \alpha_{ij}) \end{bmatrix} \right) \tag{5.71}$$

$$r_{ij} = \frac{\| m_i - m_j \|}{2 \sin \alpha_{ij}} \tag{5.72}$$

Because $\alpha_{13} = \alpha_{12} + \alpha_{33}$ only two of those angles are independent and the two belonging circles (for α_{12} and α_{23}) are

$$(x - x_{12})^2 + (y - y_{12})^2 = r_{12}^2 \tag{5.73}$$

$$(x - x_{23})^2 + (y - y_{23})^2 = r_{23}^2 \tag{5.74}$$

The intersection of both circles (Eq. 5.73) is the solution for the robot position. For more details on how to obtain the analytical solution see [3]. Here, only the final solution is stated. A new temporary coordinate frame is defined so that m_2 is in its origin and m_3 lies on its x-axis. With these temporary coordinates the solution is easier to find. The rotation matrix from reference to temporary coordinates reads

(Continued)

EXAMPLE 5.9—cont'd

$$R = \begin{bmatrix} \cos\beta & -\sin\beta \\ \sin\beta & \cos\beta \end{bmatrix}$$

where $\beta = \tan\frac{y_{m3}-y_{m2}}{x_{m3}-x_{m2}}$. Markers expressed in temporary coordinates $(\bar{m}_i = [\bar{x}_{mi}, \bar{y}_{mi}]^T)$ are obtained by transformation $\bar{m}_i = R^{-1}(m_i - m_2)$. The intersection of the circles in temporal coordinates $(\bar{x}, \bar{y})$ reads

$$\begin{bmatrix} \bar{x} \\ \bar{y} \end{bmatrix} = \bar{x}_{m3}\frac{1 - \eta\cot\alpha_{12}}{1 - \eta^2}\begin{bmatrix} 1 \\ -\eta \end{bmatrix}$$

where $\eta = \frac{\bar{x}_{m3}-\bar{x}_{m1}-\bar{y}_{m1}\cot\alpha_{12}}{\bar{x}_{m3}\cot\alpha_{23}-\bar{y}_{m1}+\bar{x}_{m1}\cot\alpha_{12}}$. The solution in the reference coordinates is

$$\begin{bmatrix} x \\ y \end{bmatrix} = m_2 + R\begin{bmatrix} \bar{x} \\ \bar{y} \end{bmatrix}$$

and orientation of the robot is

$$\varphi = \arctan\frac{y_{m1} - y}{x_{m1} - x} - \alpha_1$$

A complete algorithm to obtain the solution is given in Matlab Listing 5.7. The solution is shown graphically in Fig. 5.13.

Listing 5.7 Triangulation problem solved using geometric approach

```
1  m = [0, 0; 5, 3; 1, 5].'; % Three markers
2  r0 = [2; 2.5; pi/6]; % True robot pose (x, y and fi), unknown.
3
4  % Measured angles
5  alpha = wrapToPi(atan2(m(2,:)-r0(2), m(1,:)-r0(1))-r0(3));
6
7  % Triangulation: compute robot pose from measured angles
8  f = atan2(m(2,3)-m(2,2), m(1,3)-m(1,2));
9  S = [cos(f) -sin(f); sin(f) cos(f)]; % Rotation for corrdinate frame in m2
10 m_ = S.'*(m - repmat(m(:,2),1,3)); % Transformed markers
11
12 cta = cot(alpha(2)-alpha(1));
13 ctb = cot(alpha(3)-alpha(2));
14 ni = (m_(1,3)-m_(1,1)-m_(2,1)*cta)/(m_(1,3)*ctb-m_(2,1)+m_(1,1)*cta);
15 p_ = m_(1,3)*(1-ni*ctb)/(1+ni^2)*[1; -ni];
16
17 % Solution
18 p = m(:,2) + S*p_ % Position
19 fi = wrapToPi(atan2(m(2,1)-p(2), m(1,1)-r0(1))-alpha(1)) % Orientation


p =
    2.0000
    2.5000
fi =
    0.5236
```

EXAMPLE 5.9—cont'd

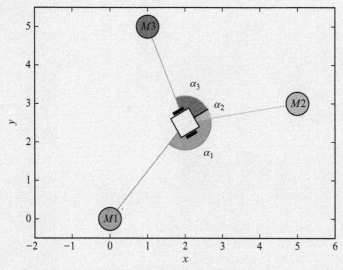

Fig. 5.13 Solution of the triangulation problem from Example 5.9 with a direct approach.

EXAMPLE 5.10

The robot is equipped with a sensor that can measure the distance to the active markers. This can be done by measuring the traveling time of the signal from the marker (transmitter) to the robot with the receiver.

There are three active markers placed at known locations: $m_1 = [x_{m1}, y_{m1}]^T = [0, 0]^T$, $m_2 = [x_{m2}, y_{m2}]^T = [5, 3]^T$, and $m_3 = [x_{m3}, y_{m3}]^T = [1, 5]^T$.

At a current robot position $r = [x, y]^T$ measured distances are $d_1 = 3.2016$ m, $d_2 = 3.0414$ m, and $d_3 = 2.6926$ m as illustrated in Fig. 5.14.

What is the current robot position r?

This can again be solved by some optimization algorithm similarly as in Example 5.9 with PSO or using some analytical solution.

Solution

The task of the trilateration algorithm is to find unknown position x, y so that the relations in Eq. (5.70) are valid. This can be done with PSO similarly as in Example 5.9, where the implementation needs to be modified to include relations (5.70) in computation of the criteria function.

(Continued)

EXAMPLE 5.10—cont'd

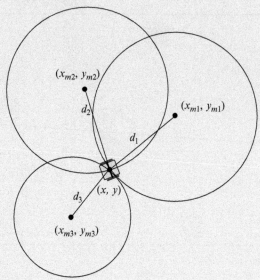

Fig. 5.14 Robot localization using trilateration where distances d_i to the markers are measured.

The solution of Eq. (5.70) can also be done analytically. Many different algorithms can be implemented; here, a simple solution is given. Subtracting the third equation in Eq. (5.70) from the first and the second one results in

$$d_1^2 - d_3^2 = (x_{m1} - x)^2 - (x_{m3} - x)^2 + (y_{m1} - y)^2 - (y_{m3} - y)^2$$
$$d_2^2 - d_3^2 = (x_{m2} - x)^2 - (x_{m3} - x)^2 + (y_{m2} - y)^2 - (y_{m3} - y)^2 \tag{5.75}$$

By rearranging

$$2(x_{m3} - x_{m1})x + 2(y_{m3} - y_{m1})y = d_1^2 - d_3^2 - x_{m1}^2 + x_{m3}^2 - y_{m1}^2 + y_{m3}^2$$
$$2(x_{m3} - x_{m2})x + 2(y_{m3} - y_{m2})y = d_2^2 - d_3^2 - x_{m2}^2 + x_{m3}^2 - y_{m2}^2 + y_{m3}^2 \tag{5.76}$$

we obtain linear equations in terms of x and y, which can be written in matrix $Ar = b$ where

$$A = \begin{bmatrix} 2(x_{m3} - x_{m1}) & 2(y_{m3} - y_{m1}) \\ 2(x_{m3} - x_{m2}) & 2(y_{m3} - y_{m2}) \end{bmatrix}$$

and

$$b = \begin{bmatrix} d_1^2 - d_3^2 - x_{m1}^2 + x_{m3}^2 - y_{m1}^2 + y_{m3}^2 \\ d_2^2 - d_3^2 - x_{m2}^2 + x_{m3}^2 - y_{m2}^2 + y_{m3}^2 \end{bmatrix}$$

EXAMPLE 5.10—cont'd

from where an unknown position is obtained:

$$r = A^{-1}b \tag{5.77}$$

The correct solution is $r = [2, 2.5]^T$, and Matlab script implementation is given in Listing 5.8.

Listing 5.8 Trilateration problem

```
1  m = [0, 0; 5, 3; 1, 5].'; % Markers
2  r0 = [2; 2.5; pi/6]; % True robot position (x and y), unknown.
3
4  % Measured distances to the markers
5  d = sqrt((m(1,:)-r0(1)).^2+(m(2,:)-r0(2)).^2);
6
7  % Trilateration: find the robot position from measured distances
8  N = size(m,2); A = zeros(N-1,2); b = zeros(N-1,1);
9  for i = 1:N-1
10     A(i,:) = 2*[m(1,N)-m(1,i), m(2,N)-m(2,i)];
11     b(i) = d(i)^2-d(N)^2-m(1,i)^2 + m(1,N)^2-m(2,i)^2 + m(2,N)^2;
12  end
13
14  r = A\b % Calculated position


r =
     2.0000
     2.5000
```

EXAMPLE 5.11

In Example 5.10 measured distances were exact, which is an unrealistic assumption. Measurements are normally corrupted with noise and other disturbances; therefore, an overdetermined system with more than three markers is needed to minimize error in the position estimate. To illustrate this use $n = 4$ markers at locations $m_1 = [x_{m1}, y_{m1}]^T = [0, 0]^T$, $m_2 = [x_{m2}, y_{m2}]^T = [5, 3]^T$, $m_3 = [x_{m3}, y_{m3}]^T = [1, 5]^T$, and $m_4 = [x_{m4}, y_{m4}]^T = [2, 4]^T$. And measured distances corrupted with noise are $d_1 = 3.2297$ m, $d_2 = 3.0697$ m, $d_3 = 2.7060$ m, and $d_4 = 1.4759$ m. Estimate the current robot position r.

Solution

The solution that minimize mean-square error $\|Ar - b\|$ of an overdetermined system with n active markers is obtained as follows. Matrix A becomes

$$A = \begin{bmatrix} 2(x_{mn} - x_{m1}) & 2(y_{mn} - y_{m1}) \\ 2(x_{mn} - x_{m2}) & 2(y_{mn} - y_{m2}) \\ \vdots & \vdots \\ 2(x_{mn} - x_{mn-1}) & 2(y_{mn} - y_{mn-1}) \end{bmatrix} \tag{5.78}$$

(Continued)

EXAMPLE 5.11—cont'd

and

$$b = \begin{bmatrix} d_1^2 - d_n^2 - x_{m1}^2 + x_{mn}^2 - y_{m1}^2 + y_{mn}^2 \\ d_2^2 - d_n^2 - x_{m2}^2 + x_{mn}^2 - y_{m2}^2 + y_{mn}^2 \\ \vdots \\ d_{n-1}^2 - d_n^2 - x_{mn-1}^2 + x_{mn}^2 - y_{mn-1}^2 + y_{mn}^2 \end{bmatrix} \tag{5.79}$$

Finally using pseudo inverse the solution is obtained:

$$r = (A^T A)^{-1} A^T b \tag{5.80}$$

For given distances the solution is $r = [1.9873, 2.5337]^T$ and the Matlab script implementation is given in Listing 5.9. The solution is also shown in Fig. 5.15.

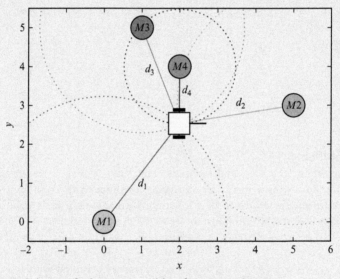

Fig. 5.15 Solution of trilateration problem from Example 5.11.

Listing 5.9 Trilateration problem of an overdetermined system

```
1  m = [0, 0; 5, 3; 1, 5; 2, 4].'; % Markers
2
3  % Measured distances to the markers
4  d = [3.2297, 3.0697, 2.7060, 1.4759];
5
6  % Trilateration: find the robot position from measured distances
7  N = size(m,2); A = zeros(N-1,2); b = zeros(N-1,1);
8  for i = 1:N-1
9      A(i,:) = 2*[m(1,N)-m(1,i), m(2,N)-m(2,i)];
```

EXAMPLE 5.11—cont'd

```
10      b(i) = d(i)^2-d(N)^2-m(1,i)^2 + m(1,N)^2-m(2,i)^2 + m(2,N)^2;
11  end
12
13  r = A\b % Calculated position


    r =
        1.9873
        2.5337
```

Global Navigation Satellite System

The most often used localization system using the trilateration principle is the Global Navigation Satellite System (GNSS). Satellites present active markers that transmit coded signals to the GNSS receiver station whose position needs to be estimated by trilateration. Satellites have very accurate atomic clocks and their positions are known (computed using Kepler elements and other TLE [two-line element set] parameters). There are several GNSS systems: Navstar GPS from ZDA, Glonass from Russia, and Galileo from Europe. The Navstar GPS system is designed to operate with at least 24 satellites that encircle Earth twice a day at a height of 20,200 km.

GNSS seems to be a very convenient sensor system for localization but it has some limitations that need to be considered when designing mobile systems. Its use is not possible where obstacles (e.g., trees, hills, buildings, indoors, etc.) can block the GNSS signal. Multiple signal reflections lead to multipath problems (wrong distance estimate). The expected accuracy of the system is approximately 5 m for single receiver system and approximately 1 cm for referenced systems (with additional receiver in reference station).

A simplified explanation of GNSS localization is as follows. GNSS receivers measure the traveling time of a GNSS signal from a certain satellite. Traveling time is the difference of the reception time t_r and transmission time t_t. Knowing that the signal travels at the speed of light c the distance between the receiver and satellite is computed. However, the receiver clock is not as accurate as the atomic clock on satellites. Therefore, some time bias or distance error appears, which is unknown but equal for all distances to satellites. The GNSS receiver therefore needs to estimate four parameters, namely its 3D position (x, y, and z) and time bias t_b.

If around each received satellite a sphere with measured distance is drawn, the following can be concluded. The intersection of two spheres

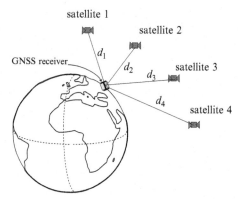

Fig. 5.16 GNSS localization requires reception of minimally four satellites to estimate the GNSS receiver position and the time drift.

is a circle and the intersection of three spheres are two points where the receiver can be located. Therefore, we need at least four spheres to reliably estimate the position of the receiver. If the receiver is on the Earth's surface then Earth can be considered as the fourth sphere to isolate a correct point obtained from the intersection of three satellite' spheres. Ideally only three satellite receptions would be enough. But as already mentioned the receiver clock is inaccurate and this causes unknown time bias t_b. The intersection of the four spheres (three from satellites and the fourth from Earth) is therefore not a point but an area. To estimate t_b and lower localization error also the reception of the fourth satellite is required. Therefore, a minimum of four satellite receptions is required for GNSS localization as shown in Fig. 5.16. GNSS localization therefore is needed to solve the following set of equations:

$$d_1 = c(t_{r1} - t_{t1} - t_d) = \sqrt{(x_1 - x)^2 + (y_1 - y)^2 + (z_1 - z)^2}$$

$$d_2 = c(t_{r2} - t_{t2} - t_d) = \sqrt{(x_2 - x)^2 + (y_2 - y)^2 + (z_2 - z)^2}$$

$$d_3 = c(t_{r3} - t_{t3} - t_d) = \sqrt{(x_3 - x)^2 + (y_3 - y)^2 + (z_3 - z)^2}$$

$$d_4 = c(t_{r4} - t_{t4} - t_d) = \sqrt{(x_4 - x)^2 + (y_4 - y)^2 + (z_4 - z)^2}$$

$$(5.81)$$

where unknowns are the receiver position x, y, z and the receiver time drift t_d. For ith satellite the known values are position (x_i, y_i, z_i), reception time t_{ri} and transmission time t_{ti}, and speed of light c.

5.3.4 Navigation Using Environmental Features

Features are a subset of patterns that can be robustly extracted from raw sensor measurements or other data. Some examples of features are lines, line segments, circles, blobs, edges, corners, and other patterns. Among other applications, features detected in the environment can be utilized for the purpose of mobile robot pose estimation (localization) and map building. For example, in the case of a 2D laser range scanner the obtained range (distance and angle) data can be processed to extract line segments (features). Line features in the local coordinate frame of the mobile robot can then be compared to the global map of the environment, which is also represented with a set of lines, in order to determine the mobile robot pose in the map. Nowadays, among the various sensors used in mobile robotics, the camera as a sensor is gaining more and more attention. Over the recent years many machine vision algorithms have been developed that enable detection of image features, which can also be used for measurement of a robot pose in the environment. Approaches that are based on features normally include the following steps: feature extraction, feature description, and feature matching. In the phase of feature extraction, raw data are processed in order to determine feature locations. The neighboring area around the detected feature location is normally used in order to describe the feature. Feature descriptors can then be used in order to find similar features in the process of feature matching.

Features are located at known locations. Therefore, their observation can improve knowledge about mobile robot location (lower location uncertainty). The list of features with their locations is called a *map*. This requires either an offline learning phase to construct a map of features or online localization and map building (simultaneous localization and mapping [SLAM]). The former approach is methodologically simpler but impractical in practice especially for larger environments. It requires the use of some reference localization system to map observed features, or this must be done manually. The latter approach builds a map simultaneously while localizing, and the main idea is to localize from observed features that are already in the map and storing newly observed features based on localized location. For reliable feature detection and robot localization usually as accurate as possible, dead reckoning is preferred. In case features are non or weakly separable (e.g., tree trunks in an orchard or detection of straight-line segments in buildings) the approximate information about the robot's location is needed to identify the observed features.

An approximate location of the robot in current time is obtained from the location at a previous time and dead reckoning prediction of relative motion from a previous location to the current one.

Features are said to be natural if they are already part of the environment, and they are artificial if they are made especially for localization purposes. Examples of natural features in structured environments (typically indoor) are walls, floor pates, lights, corners, and similar features, while in the nonstructured environments (usually outdoor) examples of features are three trunks, road signs, etc. Artificial features are made solely for the purpose of localization to be easily and robustly recognized (color patches, bar codes, lines on the floor, etc.).

Usually obtaining features requires some processing over raw sensor data in order to obtain more compact, more informative, and abstract presentation of the current sensor view (line presentation vs set of points). In some cases also a raw sensor measurement (LRF scan or camera image) can be used for the localization process by correlating the sensor view with a stored map.

Very often can visual features be detected with some image sensors. One of the simplest and most used features are straight lines. They can be detected in the environment by a camera or a laser range scanner (LRF).

Straight-Line Features

The use of straight lines as features in localization is a very common choice as they are the simplest geometric features. Straight-line features can be used to describe indoor or outdoor environments (walls, flat objects, road lines, etc.). By comparing currently observed feature parameters and the parameters of a priori known environment map the robot pose can be estimated.

A very popular sensor for this purpose is the LRF, which measures the cloud of reflection points in the environment. From this measured cloud of points (usually 2D) straight-line parameters are estimated using one of several line-extraction algorithms. The line fitting process usually requires two steps: the first one is identification of clusters that can be presented by a line, and the second is estimation of line fitting parameters for each cluster using the least squares method or something similar. Usually these two steps are performed iteratively.

Split-and-Merge Algorithm

A very popular algorithm for LRF data is the split-and-merge algorithm [4, 5], which is simple for realization and has low computational complexity and good performance. Its main operation is as follows. The algorithm requires batched data that is iteratively split to clusters where each cluster is described by a linear prototype (straight line for 2D data). The algorithm can only be applied to sorted data where consecutive data samples belong to the same straight line (data from LRF are sorted).

Initially all data samples belong to one cluster whose linear prototype (straight-line) parameters are identified. The initial cluster is then split in two new clusters at the sample which has the highest distance from the initial prototype if this distance is higher than the threshold d_{split}. The choice of d_{split} depends on data noise and must be selected higher than the expected measurement error due to the noise (e.g., three standard deviations). When clustering is finished collinear clusters are merged. This step is optional and is usually not needed in sorted data streams.

For each cluster j ($j = 1, \ldots, m$) the linear prototype is defined in normal form:

$$[z^T(k), 1]\theta_j = 0 \tag{5.82}$$

where $z(k) = [x(k), y(k)]^T$ is the kth sample ($k = 1, \ldots, n$ and n is the number of all samples) which lies on the straight line defined by parameters θ. For cluster j, which contains samples $k_j = 1, \ldots, n_j$ the parameter vector θ_j can be estimated using singular value decomposition as follows. The regression matrix,

$$\psi = \begin{bmatrix} z^T(1) & 1 \\ \vdots & \vdots \\ z^T(n_j) & 1 \end{bmatrix}$$

defines a set of homogenous equations in matrix form $\psi\theta_j = 0$, where the prototype parameters θ_j need to be estimated in the sense of a least-squares minimization. The solution presents the eigenvector ($p_r = [p_x, p_y, p_p]^T$) of the regression matrix $\psi^T\psi$ belonging to the minimum eigenvalue (calculated using singular value decomposition). The normal-form parameters of the prototype are obtained by normalization of p_r:

$$\theta_j = \pm\frac{p_r}{\sqrt{p_x^2 + p_y^2}}$$

where the sign is selected opposite to the third parameter of $\boldsymbol{p}_r$ (p_p). For data sample $\boldsymbol{z}(k)$ that is not on the prototype j the orthogonal distance is calculated by

$$d_j(k) = \left| [\boldsymbol{z}^T(k), 1]\boldsymbol{\theta}_j \right| \qquad (5.83)$$

In the case of 2D data the linear prototype can alternatively be estimated simply by connecting the first and the last data sample in the cluster. This, however, is not optimal in a least squares sense but it lowers the computational complexity and ensures that the sample that defines the split does not appear at the first or the last data sample.

An example of fitting raw LRF data to straight lines is given in Example 5.12.

EXAMPLE 5.12

For raw laser range scanner data consisting of 180 reflection points (see Fig. 5.17 and Listing 5.10) estimate straight–line clusters that best describe the scan. Clustering is done using distance threshold $d_{\mathrm{split}} = 0.06$ m.

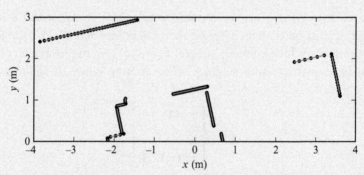

Fig. 5.17 Laser range scanner data and identified straight-line clusters using the split-and-merge algorithm.

Listing 5.10 Laser scan data
```
1  data = [-0 -2149   38 -2158   76 -2166  110 -2092  137 -1962  162 -1851 ...
2      185 -1761  222 -1809  255 -1817  289 -1822  324 -1835  358 -1840 ...
3      393 -1849  428 -1853  464 -1862  502 -1873  539 -1879  577 -1887 ...
4      617 -1898  656 -1905  697 -1915  738 -1921  780 -1929  824 -1942 ...
```

EXAMPLE 5.12—cont'd

```
 5    860 -1931   873 -1872   885 -1814   898 -1763   911 -1714   957 -1726 ...
 6    1001 -1734  1035 -1722  2407 -3852  2431 -3743  2452 -3635  2476 ...
 7    -3536  2497 -3437  2519 -3342  2539 -3250  2558 -3158  2576 -3070 ...
 8    2596 -2986  2614 -2903  2630 -2821  2649 -2744  2664 -2664  2683 ...
 9    -2591  2698 -2516  2715 -2445  2730 -2373  2742 -2301  2759 -2234 ...
10    2774 -2167  2790 -2102  2802 -2036  2817 -1973  2831 -1910  2845 ...
11    -1847  2857 -1785  2869 -1724  2885 -1666  2898 -1606  2908 -1546 ...
12    2924 -1490  2936 -1432  1146 -535   1149 -512   1155 -490   1160 -469 ...
13    1164 -447   1169 -425   1174 -404   1178 -383   1182 -361   1188 -341 ...
14    1190 -319   1195 -298   1199 -277   1205 -256   1210 -235   1214 -214 ...
15    1220 -193   1222 -172   1227 -151   1231 -129   1238 -108   1241 -87 ...
16    1248 -65  1253 -44   1254 -22   1259 0   1265 22   1268 44   1277 67 ...
17    1279 89  1287 113  1291 136   1295 159  1300 183  1306 207  1312 ...
18    231   1315 256  1320 280  1328 307  1178 294  1133 304  1092 313 ...
19    1050 321  1014 329   977 336   944 344   915 351   885 357   856 363 ...
20    829 369   805 375   778 380   756 385   735 391   713 395   694 401   674 ...
21    405   652 408   637 413   617 416   600 420   582 423   566 427   552 432 ...
22    537 435   520 436   507 441   493 444   478 446   463 447   452 452   439 ...
23    455   426 456   414 459   401 461   391 466   380 469   1920 2457   1944 ...
24    2580   1976 2720   2011 2872   2044 3030   2081 3204   2115 3385   2042 ...
25    3399   1971 3415   1903 3433   1832 3445   1764 3462   1694 3474   1629 ...
26    3493   1561 3506   1495 3522   1428 3533   1362 3549   1297 3564   1231 ...
27    3575   1167 3593   1102 3603   187 654   175 655   164 656   153 662 ...
28    141 662   130 667   118 668   106 671   94 670   83 677   71 678   60 681 ...
29    48 683   36 686   24 691   12 690].';
30 x = data(2:2:end)/1000;
31 y = data(1:2:end-1)/1000;
```

Solution

A simple algorithm implementation presented here computes line parameters of each cluster in a least squares sense. If a cluster needs to be split and the splitting sample appears as the first or the last point of the cluster then cluster cannot be split. If so the straight-line parameters are computed again to fit only the first and the last sample and the split is made.

Implementation of line estimation is given in Listing 5.11 (laser scan data is defined in Listing 5.10). Estimated line segments are shown in Fig. 5.17.

Listing 5.11 Line estimation using split-and-merge algorithm

```
 1  X = [x, y]; % Laser scan data
 2  [N, M] = size(X);
 3
 4  % Init
 5  C = 50; % Maximum number of clusters
 6  clusters = 1; % Last active cluster
 7  dMin = 0.06; % Threshold distance to split cluster
 8
 9  sizeOfCluster = zeros(C,1); % Number of points in clusters
10  clusterBounds = zeros(C,2); % Indexes of boundary points in clusters
11  clusterParams = zeros(C,M+1); % Cluster parameters
12  splitCluster = zeros(C,1); % Split flag
```

(Continued)

EXAMPLE 5.12—cont'd

```
13
14 % At the begining add all points to a single cluster
15 sizeOfCluster(clusters,1) = N;
16 clusterBounds(clusters,:)= [1, N]; % Points are ordered
17 splitCluster(clusters,1) = 1; % Initial cluster can be split
18
19 exit = false;
20 while ~exit
21     exit = true;
22     tmpLastCluster = clusters;
23     for i = 1:tmpLastCluster
24         if splitCluster(i)
25             p0 = clusterBounds(i,1); % First point in the cluster
26             p1 = clusterBounds(i,2); % Last point in the cluster
27
28             % Estimation of cluster params in the least-square sense (LSQ)
29             Psi = [X(p0:p1,:), ones(p1-p0+1,1)];
30             [~, ~, V] = svd(Psi);
31             thetaEst = V(:,3);
32             % Transform line ax+by+c=0 to normal form
33             s = -sign(thetaEst(3)); if s==0, s = 1; end
34             mi = 1/sqrt(thetaEst(1)^2+thetaEst(2)^2)*s;
35             Theta = thetaEst*mi;
36
37             % Estimation of simple line parameters (through the first and
38             % the last point). These params are used when the split point
39             % is at the cluster boundary.
40             if abs(X(p1,1)-X(p0,1))<100*eps % Vertical line
41                 a = 1; b = 0; c = -X(1,1);
42             else
43                 a = (X(p1,2)-X(p0,2))/(X(p1,1)-X(p0,1));
44                 b = -1;
45                 c = -a*X(p0,1) + X(p0,2);
46             end
47             % Transform line ax+by+c=0 to normal form
48             thetaEst = [a; b; c];
49             s = -sign(thetaEst(3)); if s==0, s = 1; end
50             mi = 1/sqrt(thetaEst(1)^2+thetaEst(2)^2)*s;
51             Theta0 = thetaEst*mi;
52
53             % Store LSQ params
54             clusterParams(i,:) = Theta.';
55             ind = p0:p1;
56             XX = X(ind,:);
57
58             % Calculate distance obtained from the first and the last
59             % cluster sample (simple line)
60             dik = [XX, ones(size(XX,1),1)]*Theta0;
61             [dd0, iii] = max(abs(dik)); ii0 = ind(iii);
62
63             % Calculate distance from line obtained in the LSQ sense
64             dik = [XX, ones(size(XX,1),1)]*Theta;
65             [dd, iii] = max(abs(dik)); ii = ind(iii);
66
67             % Cluster splitting
68             doSplit = 0;
69             if dd>dMin && (ii-p0)>=2 && (p1-ii)>=1 % Use LSQ line
```

EXAMPLE 5.12—cont'd

```
70                      if clusters<C
71                          iiFin = ii; % Split location
72                          doSplit = 1;
73                          clusterParams(i,:) = Theta.';
74                      end
75                  elseif dd0>dMin && (ii0-p0)>=2 && (p1-ii0)>=1 % Use simple line
76                      if clusters<C
77                          iiFin = ii0; % Split location
78                          doSplit = 1;
79                          clusterParams(i,:) = Theta0.';
80                      end
81                  else
82                      splitCluster(i) = 0;
83                  end
84
85
86                  if doSplit==1 && clusters<C
87                      % Split the cluster to cluster A and B
88                      clusters = clusters + 1; % New cluster
89                      % First and last point in cluster A
90                      clusterBounds(i,1);
91                      clusterBounds(i,2) = iiFin-1;
92                      splitCluster(i) = 1;
93                      % First and last point in cluster B
94                      clusterBounds(clusters,1) = iiFin+1;
95                      clusterBounds(clusters,2) = p1;
96                      splitCluster(clusters) = 1;
97
98                      exit = false;
99                  end
100         end
101     end
102 end
```

Evolving Straight-Line Clustering

Similarly as in split-and-merge clustering, straight lines can also be estimated in the case of streamed data. Clustering is done online and is iteratively updated when new data arrives. An example of a simple and computationally efficient algorithm is the evolving principal component clustering [6]. A brief description of its main steps are as follows.

The jth prototype that models the data $z(k_j)$ $(k_j = 1, \ldots, n_j)$ in the cluster j is as follows:

$$\left(z(k_j) - \mu_j\right)^T \cdot p_j = 0$$

where $\boldsymbol{\mu}_j(k_j)$ is the mean value of the data in cluster j that updates in each iteration (when new sample is available) by

$$\boldsymbol{\mu}_j(k_j) = \frac{k_j - 1}{k_j}\boldsymbol{\mu}_j(k_j - 1) + \frac{1}{k_j}\boldsymbol{z}(k_j) \qquad (5.84)$$

and $\boldsymbol{p}_j$ is the normal vector of the jth prototype. It is computed from the jth cluster covariance matrix (for 2D data),

$$\boldsymbol{\Sigma}_j(k_j) = \begin{bmatrix} \sigma_{11}^2 & \sigma_{12}^2 \\ \sigma_{21}^2 & \sigma_{22}^2 \end{bmatrix}$$

as follows:

$$\boldsymbol{p}_j = \frac{1}{\sqrt{\left(\frac{\sigma_{12}}{\sigma_{11}}\right)^2 + 1}}\begin{bmatrix} -\frac{\sigma_{12}}{\sigma_{11}} \\ 1 \end{bmatrix}$$

The covariance matrix is updated iteratively using

$$\boldsymbol{\Sigma}_j(k_j) = \frac{k_j - 2}{k_j - 1}\boldsymbol{\Sigma}_j(k_j - 1) + \frac{1}{k_j}\left(\boldsymbol{z}(k_j) - \boldsymbol{\mu}_j(k_j - 1)\right)\left(\boldsymbol{z}(k_j) - \boldsymbol{\mu}_j(k_j - 1)\right)^T$$
$$(5.85)$$

The current sample, $\boldsymbol{z}(k)$, needs to be classified in one of the existing prototypes j ($j \in \{1, \dots, m\}$). This is done by calculating the orthogonal distance $d_j(k)$ from each jth prototype:

$$d_j(k) = |(\boldsymbol{z}(k) - \boldsymbol{\mu}_j)^T \boldsymbol{p}_j| \qquad (5.86)$$

If $d_j(k) = 0$ the data sample lies on the linear prototype j. The sample belongs to the jth cluster if $d_j(k)$ for the jth cluster is the smallest and less than the predefined threshold distance d_{min} ($d_j(k) < d_{min}$). In [6] robust clustering is proposed, where d_{min} is estimated online from the jth cluster data.

The basic idea of the clustering algorithm is illustrated in Fig. 5.18. It can be applied online to sorted data arriving from some stream or to a batched data sample (as split-and-merge). Classification results of Example 5.12 would, at a similar clustering quality as the split-and-merge algorithm, require less computational effort.

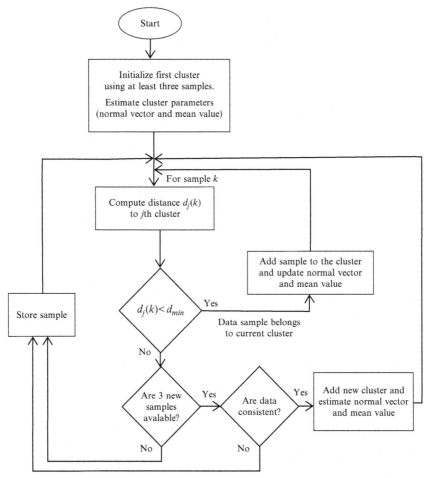

Fig. 5.18 Idea of evolving clustering for streamed data where clusters are defined by straight-line prototypes.

EXAMPLE 5.13

For raw laser range scanner data consisting of 180 reflection points (see Fig. 5.17 and the beginning of the solution for points coordinates) estimate straight-line clusters that best describes the scan. Clustering is done using the evolving straight-line clustering algorithm.

Solution

The current sample belongs to jth cluster if its distance $d_j(k)$ to the cluster's line is smaller than the threshold d_{min} ($d_j(k) < d_{min}$). d_{min} can be a

(Continued)

EXAMPLE 5.13—cont'd

constant or as in this example estimated online. The threshold distance parameter d_{min} is computed from the estimated cluster distance variance $\sigma_j(k_j)$ (distance variance of samples from the line). The variance recursive estimate reads $\sigma_j(k_j) = \sigma_j(k_j - 1)\frac{k_j-2}{k_j-1} + \frac{d_j^2(k)}{k_j}$ and the threshold is $d_{min} = \kappa_{max}\sqrt{\sigma_j}$, where $\kappa_{max} = 7$ is the tuning parameter.

Implementation of line estimation is given in Listing 5.12 (laser scan data is defined in Listing 5.10). Estimated line segments are shown in Fig. 5.19.

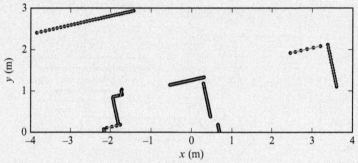

Fig. 5.19 Laser range scanner data and identified straight-line clusters using the evolving straight-line clustering algorithm.

Listing 5.12 Evolving straight-line clustering algorithm

```
1  X = [x, y]; % Laser scan data
2  Kapa_Max=7;
3
4  [n,m] = size(X); dimension = m;
5  %%%%%% init
6  Nr_cloud_max = 20; Current_clust = 1;
7
8  Nr_points_in_cloud = zeros(Nr_cloud_max,1); % vector specifying the number
9  % of data points in the cloud where rows correspond to different clouds
10 M_of_clouds = zeros(Nr_cloud_max, dimension); % matrix - rows correspond to
11 % different clouds and columns to elements of the input vector
12 V_of_clouds = zeros(Nr_cloud_max, dimension);
13 VarD_of_cloud = zeros(Nr_cloud_max,1); % distance variance of points
14 % from cluster
15 M_dist = zeros(Nr_cloud_max,1); % distances among points in cluster
16 Eig_of_clouds = zeros(Nr_cloud_max, dimension);
17 EigLat_of_cloud = zeros(Nr_cloud_max, dimension);
18 Points_in_buffer=zeros(6,dimension);  % stores up to 6 samples
19 StartEndX_points_in_cloud = zeros(Nr_cloud_max,2);
20
21 M_old=0;V_old=0;
22
```

EXAMPLE 5.13—cont'd

```
23 %%%%%%%%%%%%% init %%%%%%%%%%%%%%%%%%
24 % first cluster receives 3 points (the points must be collinear)
25 Nr_points_in_cloud(Current_clust,1)=3;
26 Nr_points_in_buffer=0;
27
28 N=3;
29 XX=X(1:N,:);
30 M = sum(XX)/N;
31 Mold = M;
32
33 V=[0, 0];
34 sumD=0;
35 for i=1:N
36     V=V+(XX(i,:)-M).*(XX(i,:)-M);     % variance of cluster points
37 end
38 V=V/(N-1);
39 Vold = V;
40
41 % covariance matrix of samples
42 Vmat=(XX-repmat(M,N,1))' * (XX-repmat(M,N,1)) / (N-1);
43
44 p = [sqrt(V(1,1)) sqrt(V(1,2))];
45 signC=sign(X(2,:)-X(1,:));
46 p = p.*signC / sqrt(p*p'); % 1. eigen vector, normalised
47 pL = [p(1,2) , -p(1,1)];   % 2. eigen vector, normalised
48
49 % distance from line
50 sumD=0;
51 for i=1:N
52     sumD=sumD + ( ((X(i,:) - M)*pL') /(pL*pL') )^2;
53 end
54 varD=sumD/(N-1);
55 varDold=varD;
56
57 % mean distance between points
58 d1= sqrt( (X(1,:)-X(2,:))*(X(1,:)-X(2,:))' );
59 d2= sqrt( (X(3,:)-X(2,:))*(X(3,:)-X(2,:))' );
60 d_med_toc=(d1+d2)/2;
61
62 % storing
63 M_of_clouds( Current_clust,:)=M;
64 V_of_clouds( Current_clust,:)=V;
65 Eig_of_clouds (Current_clust,:)=p;
66 EigLat_of_cloud( Current_clust,:)=pL;
67 VarD_of_cloud(Current_clust,1) = varD;
68
69 M_dist(Current_clust,1) =d_med_toc;
70 StartEndX_points_in_cloud(Current_clust,1)=1;
71 StartEndX_points_in_cloud(Current_clust,2)=3;
72
73
74 for k=4:n    % loop through all samples
75     % compute distance to current sample
76     pL=EigLat_of_cloud( Current_clust,:);
77     M=M_of_clouds( Current_clust,:);
78     V=V_of_clouds( Current_clust,:);
79     varD=VarD_of_cloud(Current_clust,1);
```

(Continued)

EXAMPLE 5.13—cont'd

```
80
81      d=abs( (X(k,:) - M)*pL');
82
83      if(Nr_points_in_cloud(Current_clust)>1)
84          d_med_toc=sqrt((X(k,:) - X(k-1,:))*(X(k,:) - X(k-1,:))');
85      end
86
87      % new point belongs to cluster if its distance to cluster is small
88      % enough and sample are close enough
89      if (d<Kapa_Max*sqrt(varD) & d_med_toc < 2*M_dist(Current_clust,1))
90          StartEndX_points_in_cloud(Current_clust,2)=k;
91          Nr_points_in_buffer=0;
92          j=Nr_points_in_cloud(Current_clust)+1;   % increase cluster samples
93          M=Mold+(X(k,:)-Mold)/j; % recursive mean
94          V=Vold*(j-2)/(j-1)+j*(M-Mold).^2;   % recursive variance
95
96          % covariance matrix of data
97          Vmat=Vmat*(j-2)/(j-1) + 1/j*(X(k,:)-Mold)'*(X(k,:)-Mold);
98
99          Mold=M;
100         Vold=V;
101
102         p = [sqrt(V(1,1)) sqrt(V(1,2))];
103         signC=sign(X(k,:)-X(StartEndX_points_in_cloud(Current_clust,1),:));
104
105         p = p.*signC / sqrt(p*p'); % 1. eigen vector, normalised
106         pL = [p(1,2) , -p(1,1)]; % 2. eigen vector, normalised
107
108         % recursive distance variance
109         d=abs( (X(k,:) - M)*pL') ;
110         varD=varDold*(j-2)/(j-1)+d^2/j;
111         varDold=varD;
112
113         % storing
114         M_of_clouds( Current_clust,:)=M;
115         V_of_clouds( Current_clust,:)=V;
116         Eig_of_clouds (Current_clust,:)=p;
117         EigLat_of_cloud( Current_clust,:)=pL;
118         Nr_points_in_cloud(Current_clust)=j;
119         VarD_of_cloud(Current_clust,1) = varD;
120
121     else    % new point does not belong to cluster, make new cluster
122
123         Nr_points_in_buffer= Nr_points_in_buffer+1;
124         Points_in_buffer(Nr_points_in_buffer, :)=X(k,:);
125
126         % new cluster need to have 3 consistent samples
127         if(Nr_points_in_buffer>=3)
128
129             XX=Points_in_buffer(1:Nr_points_in_buffer,:);
130             M = sum( XX )/Nr_points_in_buffer;
131
132             d_min_toc=0;
133             V=[0, 0];
134             for i=1:Nr_points_in_buffer
135                 d_tmp=(XX(i,:)-M).*(XX(i,:)-M);
136                 V=V+d_tmp;
```

EXAMPLE 5.13—cont'd

```
137              if (d_tmp>d_min_toc), d_min_toc=d_tmp; end
138         end
139         V=V/(Nr_points_in_buffer-1);
140
141         Mold = M;
142         Vold = V;
143
144         p = [sqrt(V(1,1)) sqrt(V(1,2))];
145         signC=sign(XX(Nr_points_in_buffer,:)-XX(1,:));
146         p = p.*signC / sqrt(p*p');
147         pL = [p(1,2) , -p(1,1)];
148
149         Vmat=(XX-repmat(M,N,1))' * (XX-repmat(M,N,1)) / (N-1);
150
151         % test consistency of samples
152         sumD=0;
153         for i=1:N
154            d=abs((XX(i,:) - M)*pL') ;
155            sumD=sumD+ d^2;
156         end
157         varD=sumD/(N-1);
158         varDold=varD;
159
160         d1= sqrt( (XX(1,:)-XX(2,:))*(XX(1,:)-XX(2,:))' );
161         d2= sqrt( (XX(3,:)-XX(2,:))*(XX(3,:)-XX(2,:))' );
162
163         if ( d<Kapa_Max/2*sqrt(varD) & abs(d2-d1)<min(d1,d2) )
164            % samples are consistent and correctly spaced
165            Current_clust = Current_clust+1;
166            M_of_clouds( Current_clust,:)=M;
167            V_of_clouds( Current_clust,:)=V;
168            Eig_of_clouds (Current_clust,:)=p;
169            EigLat_of_cloud( Current_clust,:)=pL;
170            Nr_points_in_cloud(Current_clust)=2;
171            VarD_of_cloud(Current_clust,1) = varD;
172
173            StartEndX_points_in_cloud(Current_clust,1)=k-2;
174            StartEndX_points_in_cloud(Current_clust,2)=k;
175
176            d_med_toc=(d1+d2)/2;
177            M_dist(Current_clust,1) = d_med_toc;
178         end
179         Nr_points_in_buffer=0;    % clear buffer
180
181      else    % wait until buffer has 3 samples
182         %
183      end
184   end
185 end        % end: loop through all samples
```

Hough Transform

The Hough transform [7] is a very powerful approach to estimating geometric primitives mostly used in image processing. The incoming data are transformed in the parameter space (e.g., straight-line parameters), and by locating the maxima the number and the parameters of the lines are obtained.

The algorithm requires quantization of parameter space. Fine quantization increases accuracy but requires quite some computational effort and storage. Several studies proposed methods to avoid quantization and increase accuracy of Hough transform, such as randomized Hough transform or adaptive implementations, as reported in [8, 9].

The Hough transform can reliably estimate the clusters in the presence of outliers. In its basic version the straight-line parameters α and d are defined by the linear prototype (5.82), where $\boldsymbol{\theta}_j = [\cos\alpha, \sin\alpha, -d]$. The normal line parameters' ranges $-\pi < \alpha \leq \pi$ and $d_{\min} < d \leq d_{\max}$ are presented in the accumulator by quantization to N discrete values for α and M discrete values for d.

For each data sample $x(k)$, $y(k)$ and all possible values for $\alpha(n) = -\pi + \frac{\pi n}{N}$ where $n \in \{1, \ldots, N\}$) solutions for parameter $d(n)$ are computed. Each pair $\alpha(n)$, $d(n)$ presents a possible straight line containing sample $x(k)$, $y(k)$. A corresponding location in the accumulator is increased by 1 for each computed parameter. When all data samples are processed the cells of the accumulator with the highest values are identified straight-line clusters. Some a priori knowledge is therefore required to properly select quantization of the parameter space and the threshold value to locate the maximum in the accumulator.

EXAMPLE 5.14

For raw laser range scanner data consisting of 180 reflection points (see Fig. 5.21 and the beginning of the solution for point coordinates) estimate straight-line clusters that best describe the scan. Clustering is done using the Hough transform where normal straight-line parameters α and d are quantized to 720 discrete values.

Solution

A possible solution in Matlab script is given in Listing 5.13 (laser scan data is defined in Listing 5.10) where the function houghpeaks is used to find maximums in the accumulator. The obtained accumulator is shown in Fig. 5.20 and identified clusters in Fig. 5.21.

EXAMPLE 5.14—cont'd

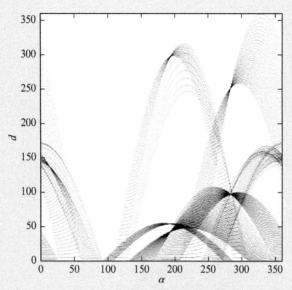

Fig. 5.20 Accumulator of the Hough transform where the ranges of $\alpha \in (-\pi, \pi]$ and $d \in [0, 4.5]$ are quantized to 360 discrete values.

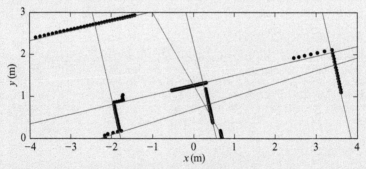

Fig. 5.21 Laser range scanner data and identified straight-line clusters using Hough transform.

Listing 5.13 Straight-line clusters using Hough transform

```
1 [x, y]; % Laser scan data
2 N = length(x);
3
4 dAlpha = pi/180; % Angle quant
5 nAlpha = round(2*pi/dAlpha);
```

(Continued)

EXAMPLE 5.14—cont'd

```
 6 nDist = nAlpha; % Use the same quantisation for distance
 7 lutDist = zeros(N,nAlpha); % Distance look-up table
 8 for i = 1:N % For each point compute lines for the range of alpha
 9     for j = 1:nAlpha
10         alpha = (j-1)*dAlpha-pi;
11         % Distance to the cordinate frame origin
12         d = x(i)*cos(alpha)+y(i)*sin(alpha);
13         if d<0
14             if alpha>pi, alpha = alpha - pi;
15             else          alpha = alpha + pi; end
16             jj = round((alpha+pi)/dAlpha);
17             lutDist(i,jj) = -d;
18         else
19             lutDist(i,j) = d;
20         end
21     end
22 end
23
24 % Find range for distance parameter
25 minLutDist = min(lutDist(:));
26 maxLutDist = max(lutDist(:));
27 dDist = (maxLutDist-minLutDist)/nDist; % Distance quant
28
29 % Realize accumulator of parameter space
30 A = zeros(nDist,nAlpha);
31 for i = 1:N % Go through all the points
32     for j = 1:nAlpha
33         k = round((lutDist(i,j)-minLutDist)/dDist)+1;
34         if k>nDist, k = nDist; end
35         A(k,j) = A(k,j)+1;
36     end
37 end
38 H = A(2:nAlpha,:); % Accumulators
39
40 % Locate maximums in the accumulator (the most probable lines)
41 nLines = 7; % Requested number of top most probable lines
42 peaks = houghpeaks(H, nLines, 'threshold', 3, 'NHoodSize', [31, 31]);
43 % Line parameters: distance from the origin and line angle
44 distAlpha = [peaks(:,1).'*dDist+minLutDist; peaks(:,2).'*dAlpha-pi];
```

Image Features

Several nice properties of the camera, computational capabilities of modern computers, and advances in the development of algorithms make the use of a camera an extremely appealing sensor for solving problems in robotics. A camera can be used to detect, identify, and track observed objects in the camera's field of view, since images are projections of the 3D objects in the environment (see Section 5.2.4). The digital image is a

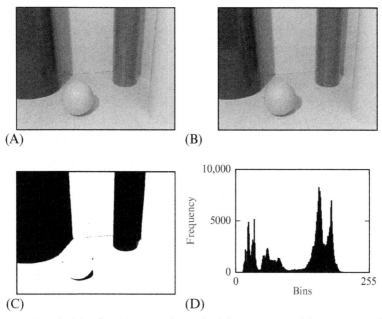

Fig. 5.22 Sample (A) color, (B) grayscale, and (C) binary image. (D) Histogram of the grayscale image.

2D discrete signal that is represented with a matrix of quantized numbers that represent either the presence or absence of light, light intensity, color, or some other quantity. The main types of images are color (Fig. 5.22A), grayscale (Fig. 5.22B), and binary (Fig. 5.22C). In machine vision grayscale images are normally sufficient when looking for particular patterns that are not color dependent. Binary images are seldom a result of image segmentation or are used for content masking. One of the simplest image segmentation methods is thresholding, where pixels with a grayscale level above a threshold are marked with a logical one, and all the other are set to logical zero. For example, thresholding grayscale image in Fig. 5.22B with a threshold of 70 results in a binary in Fig. 5.22C. The frequency of grayscale levels can be presented in a histogram that can be used in order to determine the most appropriate threshold level. In Fig. 5.22D a histogram with 256 bins that corresponds to 256 levels of the grayscale image in Fig. 5.22B is shown.

Over the recent years many machine vision algorithms have been developed that enable image-based object tracking. For this purpose image content is normally represented with image features. Features can be image regions with similar properties (e.g., similar color), patches of a particular pattern, or some other image cues (e.g., edges, corners, lines). In some situations *artificial markers* can be introduced into the environment that enable fast and reliable feature tracking (e.g., color markers or matrix barcodes to track mobile soccer robots). When this is not possible features need to be extracted from the image of a noncustomized scene. In some applications simple *color segmentation* can be used (e.g., detection of red apples in an orchard). In the recent years many approaches have been developed that enable extraction of *natural local image features* that are invariant to some image transformations and distortions.

Color Features

Detection of image regions with similar colors is not a trivial task due to inhomogeneous illumination, shadows, and reflections. Color-based feature tracking is therefore normally employed only in the environments where controlled illumination conditions can be established or the color of the tracked object is distinctive enough from all the other objects in the environment.

Color in a digital image is normally represented with three color components: red, green, and blue; this is known as the RGB color model. The color is restored from a combination of red, green, and blue color components that are filtered through red, green, and blue color filters, respectively. Two other color spaces that are seldom used in machine vision are HSL (hue-saturation-lightness) and HSV (hue-saturation-value). HSV color space is seldom used since it enables a more natural description of colors and better color segmentation than can be achieved in RGB space. The values from the RGB color space given in the range $[0, 255]$ can be converted to HSV color space according to the following:

$$
H = \begin{cases}
0 & \text{if } M - m = 0, \\
60\frac{G-B}{M-m} \mod 360 & \text{if } M = R, \\
60\frac{B-R}{M-m} & \text{if } M = G, \\
60\frac{R-G}{M-m} & \text{if } M = B,
\end{cases}
$$

$$
S = \begin{cases}
0 & \text{if } M = 0, \\
\frac{M-m}{M} & \text{otherwise,}
\end{cases}
$$

$$
V = \frac{M}{255},
$$

(5.87)

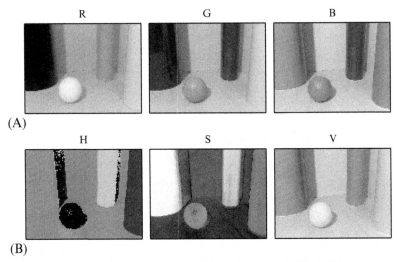

Fig. 5.23 RGB and HSV color components of the color image in Fig. 5.22A.

where $M = \max\{R, G, B\}$ and $m = \min\{R, G, B\}$. In Eq. (5.87) the saturation S and value V are in the range $[0, 1]$, and the hue H is in the range $[0, 360)$ that corresponds to angular degrees and wraps around. In Fig. 5.23 RGB and HSV components of the color image in Fig. 5.22A are shown.

Color histograms can be used to segment regions of a particular color. HSV histograms (with 32 bins) of several templates with similar color are shown in Fig. 5.24 (the color regions have been extracted from the image in Fig. 5.22A). Histograms of each template can be back-projected to the original image, that is, for every pixel in the image the frequency of the bin that belongs to the value of the pixel in the corresponding color channel in the image is set. The resulting back-projected grayscale images can be merged together as a linear combination of color channels. The results of blue, orange, red yellow, and wood color HSV histogram back-projection is shown in the first column of Fig. 5.25. In the resulting image the grayscale levels represent the measure of similarity of the pixels to the color template. Before the image is thresholded (see the third column in Fig. 5.25) some additional image filtering can be applied. The image can be smoothed with a 2D Gaussian filter (see the second column in Fig. 5.25) to remove some noisy peaks. The thresholded binary image can be filtered with, for example, some morphological or other binary image processing operations to remove or fill some connected regions.

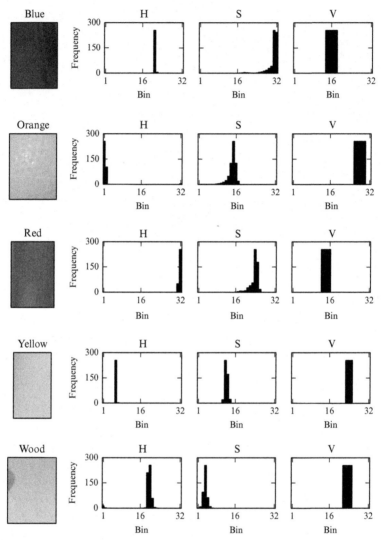

Fig. 5.24 Histograms in the HSV color space of several color templates.

The resulting image mask (see the fourth column in Fig. 5.25) can be used to filter out the detected regions in the original image (see the last column in Fig. 5.25). More on image processing algorithms can be found, among others, in [10]. In practical implementations of color-based image segmentation some filtering processes are skipped in order to speed-up algorithms' performance for the price of accuracy. Sometimes every

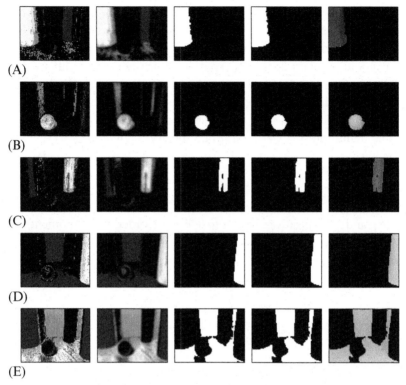

Fig. 5.25 Color-based feature extraction. *Each row* represents a different color. Steps from the *left* to the *right column*: color template histogram back-projection, blurring, thresholding, median filtering, and image masking.

pixel in the input image is considered part of the object if the color values of all components are within some lower and upper bounds that correspond to some color. In Fig. 5.24 it can be observed that hue and saturation values enable simple segmentation of colors. However, it should be pointed out that color-based segmentation is sensitive to changing illumination conditions in the environment. Although some approaches can compensate for inhomogeneous illumination conditions [11], the color-based image segmentation is the most commonly used only in situations where appropriate illumination conditions can be achieved.

The results of image segmentation is a binary image where image pixels that belong to the object are set to a nonzero value and all the other points are set to zero. There may be several or only a single connected region. If there are several connected regions an appropriate algorithm needs to be applied in order to find the positions and shapes of these objects [11]. To make searching of connected regions more robust some constraints can be introduced that reject areas with too small or too big areas, or some other region property. If the observed object is significantly different from the surrounding environment in a way that it can be robustly detected as a single region in a segmented image, the position and shape of this image region can be described with *image moments*.

The definition of a *raw moment* of a (binary) digital image $I(x, y)$ is the following:

$$m_{p,q} = \sum_x \sum_y x^p y^q I(x, y) \tag{5.88}$$

where p and q are positive integers ($p + q$ is the order of the moment). The moment $m_{0,0}$ represents object mass that is equal to the area of the object in the case of a binary image. Zero- and first-order raw moments can be used in order to find the center of the object (x_0, y_0) in the binary image:

$$x_0 = \frac{m_{10}}{m_{00}}, \quad y_0 = \frac{m_{01}}{m_{00}} \tag{5.89}$$

To describe object orientation and shape *central moments* can be used:

$$\mu_{p,q} = \sum_x \sum_y (x - x_0)^p (y - y_0)^q I(x, y) \tag{5.90}$$

which are invariant to in-image translation. Central moments can be used to fit an ellipsoid to the detected object in the image. The ellipse's semimajor axis a and semiminor axis b can be obtained from eigenvalues λ_a and λ_b ($\lambda_a \geq \lambda_b$) of the matrix $\mathbf{Q}$, which consists of second-order central moments:

$$\mathbf{Q} = \begin{bmatrix} \mu_{2,0} & \mu_{1,1} \\ \mu_{1,1} & \mu_{0,2} \end{bmatrix} \tag{5.91}$$

The ellipse axes are

$$a = \frac{2\sqrt{\lambda_a}}{m_{00}}, \quad b = \frac{2\sqrt{\lambda_b}}{m_{00}} \tag{5.92}$$

The orientation of the semimajor ellipse axis is given by the angle θ:

$$\theta = \frac{1}{2} \arctan \frac{2\mu_{1,1}}{\mu_{2,0} - \mu_{0,2}} \tag{5.93}$$

In Fig. 5.26 raw and central moments have been used in order to fit an ellipsoid over the binary patch (the result of orange color segmentation

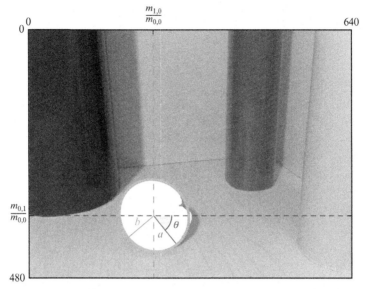

Fig. 5.26 Fitting of an ellipsoid over the mask that has been obtained as a result of an orange color segmentation.

from Fig. 5.25). The way ellipsoid parameters can be obtained from image moments is shown in Listing 5.14, where central moments were derived from raw moments. As it can be seen, image moments can be used to describe simple image features and determine locations of the features in an image. In [12] another set of moments is defined that are invariant to scale, translation, and rotation.

Listing 5.14 Ellipsoid parameters estimation

```
1  im = imread('colour_orange_mask.bmp')>128; % Ensure binary image
2
3  [x, y] = meshgrid(1:size(im,2), 1:size(im,1));
4
5  % Raw moments
6  m00 = sum(sum( (x.^0).*(y.^0).*double(im) ));
7  m10 = sum(sum( (x.^1).*(y.^0).*double(im) ));
8  m01 = sum(sum( (x.^0).*(y.^1).*double(im) ));
9  m11 = sum(sum( (x.^1).*(y.^1).*double(im) ));
10 m20 = sum(sum( (x.^2).*(y.^0).*double(im) ));
11 m02 = sum(sum( (x.^0).*(y.^2).*double(im) ));
12
13 % Area, x and y
14 area = m00
15 x0 = m10/m00
16 y0 = m01/m00
17
18 % Central moments
```

```
19  u00 = m00;
20  u11 = m11-x0*m01; % u11 = m11-y*m10;
21  u20 = m20-x0*m10;
22  u02 = m02-y0*m01;
23
24  % Ellipse
25  v = eig([u20, u11; u11, u02]); % Eigenvalues
26  a = 2*sqrt(v(2)/u00) % Semi-major axis
27  b = 2*sqrt(v(1)/u00) % Semi-minor axis
28  theta = atan2(2*u11, u20-u02)/2 % Orientation
```

```
area =
        14958
x0 =
  248.1951
y0 =
  359.7101
a =
   71.7684
b =
   66.6585
theta =
    0.8795
```

Artificial Pattern Markers

The introduction of artificial markers presents a minimal customization of the environment that can significantly simplify some vision-based tasks like, for example, object tracking, camera pose estimation, etc. An artificial pattern marker is a very distinctive pattern that can be robustly detected. A sample set of three artificial markers is shown in Fig. 5.27. Marker patterns are normally designed in a way that the markers can be reliably and precisely detected. Moreover, in the pattern of the marker, the marker ID is encoded; therefore, multiple markers can be tracked and identified through a sequence of images. One of the popular artificial marker detection algorithms is ArUco [13, 14]. Fig. 5.28 shows the detected artificial markers from Fig. 5.27 in two camera views.

Fig. 5.27 Artificial markers.

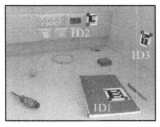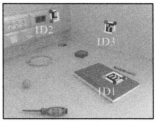

Fig. 5.28 Detected artificial markers in two camera views.

The algorithms for detection of artificial markers normally consist of a marker localization step, in which the position, orientation, and size of the marker is determined. This procedure needs to robustly filter out only true markers from the rest of the objects in the environment. Once the locations of markers in the image are known, the projection between the detected marker and marker's local coordinate frame can be established. Based on the pattern of the marker in the transformed local marker coordinate frame the ID of the marker is determined. Some artificial marker patterns are designed in a way that enable pattern correction that makes identification of patterns more reliable and also robust to noise and occlusions.

Natural Local Image Features

Images contain features that are inherently present in the environment. Over the years many machine vision algorithms have been developed that enable automatic detection of local features in images. Some of the important local feature algorithms are SIFT (Scale Invariant Feature Transform) [15], SURF (Speeded-Up Robust Features) [16], MSER (Maximally Stable Extremal Regions) [17], FAST (Features from Accelerated Segment Test) [18], and AGAST (Adaptive and Generic Accelerated Segment Test) [19]. Most of these algorithms are included in the open source computer vision library OpenCV [20, 21]. In robotic applications an important property of description of images with features is the algorithm efficiency to enable real-time performance. Different approaches have been developed for local feature extraction (localization of features), local feature description (representation of feature properties), and feature matching.

The goal of *feature extraction* is to detect and localize points of interest (local regions), normally in order to reduce the dimensionality of the raw image. Most commonly the features are extracted from grayscale images. Feature extractors can detect different types of image patterns: edges (e.g., Canny filter, Sobel filter, and Roberts cross), corners (e.g., Harris and Hessian operator, FAST), blobs (e.g., Laplacian of Gaussian, difference of Gaussians, MSER). To enable detection of features of different sizes, the raw image is seldom represented in scale-space [15]. Feature extraction should be invariant to some class of image transformations and distortions to enable repeatable detection of features across multiple views of the same scene. Normally it is desired that features are invariant to image translation, rotation, scaling, blur, illumination, and noise. Features should also be robust to some partial occlusions. The desirable property of the features is therefore locality. In mobile robotic applications it is normally required that features are also detected accurately and that the quantity of the features is sufficiently large, since this is a common prerequisite for accurate and robust implementation of algorithms that depend of extracted features (e.g., feature-based mobile robot pose estimation). An example of detected features in two camera views of the same scene is shown in Fig. 5.29. Every feature is given with position, orientation, and size.

The purpose of *feature description* is to describe each feature (based on the properties of the image pattern around the feature) in a way that enables identification of the same features across multiple images if the features reappear in the other images. The local pattern around the feature (e.g., the region marked with a square in Fig. 5.29) is used in order to determine the appropriate feature descriptor, which is normally represented as a feature vector. Local feature descriptors should be distinctive in a way that accurate feature identification is enabled regardless of different variations in the environment (e.g., illuminations changes, occlusions). Although local image features could be directly compared with, for example, convolution of local feature regions, this kind of approach does not have appropriate

Fig. 5.29 Detected features in two images of the same scene.

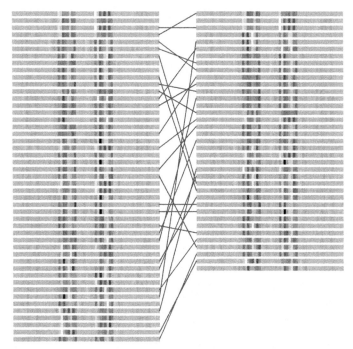

Fig. 5.30 Two-way matches between feature descriptors obtained from the features of the left and right image in Fig. 5.29. Each horizontal strip represents the values of a 64-element description vector encoded as different grayscale intensity levels.

distinctiveness, and it is also quite computationally demanding. Feature descriptor vectors tend to be a minimal representation of features that enable an adequate level of feature distinctiveness for reliable feature matching. Which feature extractor and feature is used depends on the particular application. An experimental comparison of various feature detectors can be found in, for example, [22]. An example of two sets of feature description vectors of features extracted from the left and right image in Fig. 5.29 is graphically represented in Fig. 5.30.

Feature matching is one of the basic problems in machine vision. Many machine vision algorithms—like scene depth estimation, image-based 3D scene reconstruction, and camera pose estimation depend on appropriate feature matches between multiple images. Features can be matched by comparing the distances between the feature descriptors (vectors). Depending on the type of feature descriptor different distance measures can be used. Normally Euclidean or Manhattan distance is used for real-valued feature descriptors, and Hamming distance is used for binary feature descriptors.

Due to imprecise feature localization, feature distortions, and repeated patterns, feature description matching does not always provide appropriate matches. Therefore an appropriate feature matching technique should be used that can eliminate spurious matches.

Given two sets of feature descriptors $\mathcal{A} = \{a_i\}_{i=1,2,...,N_A}$ and $\mathcal{B} = \{b_j\}_{j=1,2,...,N_B}$, the matching process should find the appropriate feature matches between the two sets. For every feature in the set $\mathcal{A}$ the closest feature in the set $\mathcal{B}$ (according to selected feature distance measure) can be selected as a possible match candidate. This is in general a surjective mapping, since some features in set $\mathcal{B}$ can have more than only one match from set $\mathcal{A}$. In normal conditions a feature in one set should have at most a single match in the other set. To overcome this issue, normally a two-way search is performed (i.e., for every feature in set $\mathcal{A}$ the closest feature in set $\mathcal{B}$ is found and vice versa) and only matches that are the same in both searches are retained. Additionally, if the distance between the best and the second best match is too small (below a selected threshold), the matched pair of features should be rejected, since there is a high probability of matching error. Also the pair of features should be rejected, if the distance between the matched feature descriptors is above a certain threshold, since this implies too high feature dissimilarity. In Fig. 5.30 the matches between two sets of feature descriptors are shown. Although the aforementioned filtering techniques were applied, not all incorrect matches were removed, as it is seen in Figs. 5.31 and 5.32 that there are still some incorrect matches. It should be noted that even if there are only a few outliers, these can have a significant influence on the results of the estimation algorithms that assume the features are matched correctly.

Some matched features may not satisfy system constraints: for example, if two matched features do not satisfy the epipolar constraint or if the reconstructed 3D point from the matched features would appear behind the camera, the match should be rejected. More constraints can be found in [23]. In wheeled mobile robotics additional constraints can be introduced based on the predicted robot states from known actions and the kinematic model. Incorporating a model of some geometric constraints into the feature matching process can be used to filter out the matched feature pairs that are not physically possible. For this purpose some robust model-fitting approach should be used. Commonly a RANSAC (Random Sample Consensus) method [24] is used that is able to fit the model even if there are many outliers.

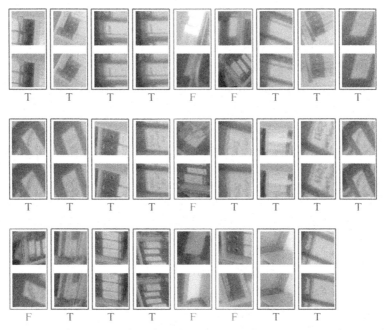

Fig. 5.31 Found feature pairs based on descriptor matching (see Fig. 5.30). Most of the feature pairs are true (*T*), but some are false (*F*).

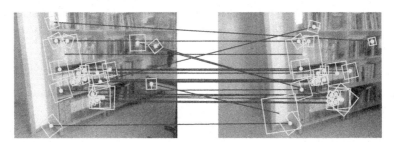

Fig. 5.32 Matched points.

The RANSAC algorithm can be explained on a line fitting problem. In the initial step the minimum number N_{min} of points that are required to fit the model are selected at random from the data. In the case of a line only two points are required ($N_{min} = 2$). The model (line) is then fitted to the selected (two) points. Around the fitted model (line) some confidence region is selected and the number of all data points in this region is counted. Then another N_{min} points are selected from the data, and the process is repeated. After some iterations the model with the largest number of points in the confidence region is selected as the best model. In the final step the model is refitted to all the points in the corresponding confidence region using a least squares method.

5.3.5 Matching of Environment Models: Maps

A map is a representation of the environment based on some feature parameters, which can be, for example, reflected points from obstacles obtained with a laser range scanner, a set of line segments that represent obstacle boundaries, a set of image features that represent objects in the environment, and the like.

The localization problem can be presented as optimal matching of two maps: local and global maps. A local map, which is obtained from the current sensor readings, presents the part of the environment that can be directly observed (measured) from the current robot pose (e.g., current reading of the LRF scanner). A global map is stored in the internal memory of the mobile system and presents the known or already visited area in the environment. Comparing both maps can lead to determination or improvement of the current mobile robot pose estimate.

When the mobile system moves around in the environment, the global map can be expanded and updated online with the local map, that is, the current sensor readings that represent previously undiscovered places are added to the global map. The approaches that enable this kind of online map building are known as *SLAM*. The basic steps of the SLAM algorithm are given in Algorithm 2.

ALGORITHM 2   Basic Steps of the SLAM Algorithm

Matching of local and global maps based on matching of feature pairs between the maps.

Estimation of current mobile robot pose based on the result of map matching.

Localization based on fusion of odometry data and comparison of maps.

Updating of the global map with a previously unobserved part of the environment (new features are added as new states in the map).

An example of the map that was obtained from a combination of LRF scanner measurements and odometry data using the SLAM algorithm is shown in Fig. 5.33.

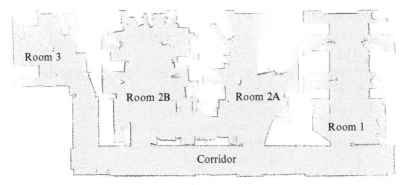

Fig. 5.33 Map of the indoor environment built from LRF scanner measurements and odometry data.

SLAM is the one of the basic approaches that is used in wheeled mobile robotics. Some of the common challenges in localization of mobile robots are the following:

- *Pose initialization* of the mobile robot at start-up cannot be determined with high probability. If the initial pose is not known, the global localization problem needs to be solved.
- The problem of *robot kidnapping* occurs when the real mobile robot pose in the environment changes instantly while the mobile robot is in operation (e.g., the mobile robot is transported to a new location or it is turned on at a different location as it has been turned off). Robust localization algorithms should be able to recover and estimate the real robot pose.
- Ensuring an appropriate pose estimate during the movement of the mobile robot. For this purpose, odometry data and measurements from absolute sensors are used.

The most commonly used algorithms for solving the SLAM problem are the extended Kalman filter, Bayesian filter, and particle filter (see Chapter 6).

5.4 SENSORS

This section provides a brief description of the main sensors used in wheeled mobile robotics, their characteristics, and classification.

5.4.1 Sensor Characteristics

Sensors' operation, their quality, and properties are characterized in different ways. The most common characteristics are as follows:

- *Range* defines the upper (y_{max}) and the lower limit (y_{max}) of the applied quantity that can be measured. The upper and lower limits are usually not symmetrical. Sensors should be used in the specified range, and exceeding this range may result in sensor damage.
- *Dynamic range* is the total range from the minimum to maximum range value. It can be given as a difference $R_{dyn} = y_{max} - y_{min}$ or more commonly as a ratio (in decibels) $R_{dyn} = A \log \frac{y_{max}}{y_{min}}$ where $A = 10$ for power-related measurements and $A = 20$ for the rest.
- *Resolution* is the smallest change of the measured quantity that can be detected by the sensor. If the sensor has an analog-to-digital converter then sensor resolution is usually the resolution of the converter (e.g., for 10 bit A/D and 5 V sensor range the resolution is $\frac{5\,V}{2^{10}}$).
- *Sensitivity* is the sensor output change per unit of the quantity being measured (e.g., distance sensor measures distance which is given as the voltage on the sensor output). Sensitivity can be constant over the whole sensor range (linear) or it can vary (nonlinearity).
- *Linearity* of the sensor is the sensor property where the sensor output is linearly dependent (proportional) from the measured quantity over the entire range. A linear sensor has constant sensitivity in the whole range.
- *Hysteresis* refers to the property that the sensor output trajectory (or its dependence on input) is different for situations when sensor input is increasing or decreasing.
- *Bandwidth* refers to the frequency by which the sensor can provide measurements (given Hz). It is the highest frequency (sampling rate) at which only 70.7% of the true value is measured.
- *Accuracy* is defined by the expected measurement error that is the difference between measured (m) and true value (v). Accuracy is computed from the relative measurement error using $accuracy = 1 - \frac{|m-v|}{v}$.
- *Precision* is degree of reproducibility of the sensor measurement at the same true value of quantity being measured. Real sensor output produces a range of values when the same true value is being measured many times. Precision is related to the variance of the measurement.
- *Systematic error* or deterministic error is caused by some factors that can be predicted or modeled (bias, temperature drift, sensor calibration, distortion caused by lens of the camera, etc.).

Table 5.1 Classification of sensors in mobile robotics according to their application, purpose (proprioception [PC]/exteroception [EC]), and emitted energy (active [A]/passive [P])

Classification	Applicability	Sensors	PC/EC	A/P
Tactile and haptic	Collision detection, security shutdown, closeness, wheel or motor rotation	Contact switches	EC	P
		Bumpers	EC	P
		Optical barriers	EC	A
		Proximity sensors	EC	P/A
		Contact arrays	EC	P
Axis	Wheel or motor rotation, joint orientation, localization by odometry	Incremental encoders	PC	A
		Absolute encoders	PC	A
		Potentiometer	PC	P
		Tachogenerators	PC	P
Heading	Orientation in reference frame, localization, inertial navigation	Gyroscope	PC	P
		Magnetometer	EC	P
		Compass	EC	P
		Inclinometer	EC	P
Speed	Inertial navigation	Accelerometer	EC	P
		Doppler radar	EC	A
		Camera	EC	P
Beacon	Object tracking, localization	IR beacons	EC	A
		WiFi transmitters	EC	A
		RF beacons	EC	A
		Ultrasound beacons	EC	A
		GNSS	EC	A/P

Table 5.1 Classification of sensors in mobile robotics according to their application, purpose (proprioception [PC]/exteroception [EC]), and emitted energy (active [A]/passive [P])—Cont'd

Classification	Applicability	Sensors	PC/EC	A/P
Ranging	Measure distance to obstacle, time-of-flight (TOF), localization	Ultrasound sensor	EC	A
		Laser range finder	EC	A
		Camera	EC	P/A
Machine-vision-based	Identification Object recognition Object tracking Localization Segmentation	Camera	EC	P/A
		TOF camera	EC	A
		Stereo camera	EC	P/A
		RFID	EC	A
		Radar	EC	A
		Optical triangulation	EC	A

- *Random error* or nondeterministic error is unpredictable and can only be described by a probability density function (e.g., normal distribution). This error is usually called noise and is usually characterized by the SNR.

5.4.2 Sensor Classifications

Wheeled mobile robots can measure the internal state using *proprioceptive* sensors or the external state of the environment using *exteroceptive* sensors. Examples of proprioceptive measurements include robot position and orientation, wheel or steering mechanism rotation, angular velocity of the wheels, state of the accumulators, temperature, and current on the motor. Examples of exteroceptive measurements, on the other hand, include distance to obstacles, image captured by camera, microphone, compass, GNSS, and others.

Sensors used for a environment detection are used for path planning purposes, obstacle detection, environment mapping, etc. Those sensors are *active* if they emit energy (electromagnetic radiation) in the environment and measures received response from environment (LRF, ultrasonic sensors, camera with integrated illumination, etc.). Sensors are *passive* if they receive energy that is already part of the environment. Passive sensors are therefore all sensors that are not active (camera without illumination, compass, gyroscope, accelerometer, etc.).

In Table 5.1 the most common sensors in mobile robotics are listed according to their usage. Additionally a brief description of their application, purpose (proprioception [PC] or exteroception [EC]), and emitted energy (active A or passive P) is included.

REFERENCES

[1] J.R. Wertz, Spacecraft Attitude Determination and Control, D. Reidel Publishing Company, Dordrecht, 1978.

[2] F. Chaumette, S. Hutchinson, Visual servo control, part I: basic approaches, IEEE Robot. Autom. Mag. 13 (2006) 82–90.

[3] J.M. Font-Llagunes, J.A. Batlle, Consistent triangulation for mobile robot localization using discontinuous angular measurements, Robot. Auton. Syst. 57 (9) (2009) 931–942.

[4] V. Nguyen, S. Gächter, A.M.N. Tomatis, R. Siegwart, A comparison of line extraction algorithms using 2D range data for indoor mobile robotics, Auton. Robot. 23 (2) (2007) 97–111.

[5] T. Pavlidis, S.L. Horowitz, Segmentation of plane curves, IEEE Trans. Comput. 23 (8) (1974) 860–870.

[6] G. Klančar, I. Škrjanc, Evolving principal component clustering with a low run-time complexity for LRF data mapping, Appl. Soft Comput. 35 (2015) 349–358.

[7] D.H. Ballard, Generalizing the Hough transform to detect arbitrary shapes, Pattern Recogn. 13 (2) (1981) 111–122.

[8] L. Xu, E. Oja, Randomized Hough transform (RHT): basic mechanisms, algorithms, and computational complexities, CVGIP: Image Und. 57 (2) (1993) 131–154.

[9] J. Basak, A. Das, Hough transform network: learning conoidal structures in a connectionist framework, IEEE Trans. Neural Netw. 13 (2) (2002) 381–392.

[10] D. Forsyth, J. Ponce, Computer Vision: A Modern Approach, Prentice Hall, Upper Saddle River, NJ, 2003.

[11] G. Klančar, M. Kristan, S. Kovačič, O. Orqueda, Robust and efficient vision system for group of cooperating mobile robots with application to soccer robots, ISA Trans. 43 (2004) 329–342.

[12] P.I. Corke, Visual Control of Robots: High-Performance Visual Servoing, Wiley, New York, 1996.

[13] S. Garrido-Jurado, R. Munoz-Salinas, F.J. Madrid-Cuevas, M.J. Marín-Jiménez, Automatic generation and detection of highly reliable fiducial markers under occlusion, Pattern Recogn. 47 (6) (2014) 2280–2292.

[14] S. Garrido-Jurado, R. Munoz-Salinas, F.J. Madrid-Cuevas, R. Medina-Carnicer, Generation of fiducial marker dictionaries using mixed integer linear programming, Pattern Recogn. 51 (2016) 481–491.

[15] D.G. Lowe, Distinctive image features from scale-invariant keypoints, Int. J. Comput. Vis. 60 (2004) 91–110.

[16] H. Bay, A. Ess, T. Tuytelaars, L. Van Gool, Speeded-up robust features (SURF), Comput. Vis. Image Underst. 110 (2008) 346–359.

[17] J. Matas, O. Chum, M. Urban, T. Pajdla, Robust wide-baseline stereo from maximally stable extremal regions, Image Vis. Comput. 22 (2004) 761–767.

[18] E. Rosten, T. Drummond, Fusing points and lines for high performance tracking, in: Tenth IEEE International Conference on Computer Vision (ICCV'05), vol. 2, 2005, pp. 1508–1515.

[19] E. Mair, G.D. Hager, D. Burschka, M. Suppa, G. Hirzinger, Adaptive and generic corner detection based on the accelerated segment test, in: Computer Vision-ECCV, 2010, Springer, Berlin, Heidelberg, 2010, pp. 183–196.

[20] The OpenCV reference manual: release 2.4.3, 2012.

[21] G. Bradski, A. Kaehler, Learning OpenCV, first ed., O'Reilly, Sebastopol, CA, 2008.

[22] K. Mikolajczyk, C. Schmid, A performance evaluation of local descriptors, IEEE Trans. Pattern Anal. Mach. Intell. 27 (2005) 1615–1630.

[23] B. Cyganek, J.P. Siebert, An Introduction to 3D Computer Vision Techniques and Algorithms, John Wiley & Sons, Chichester, UK, 2009.

[24] M.A. Fischler, R.C. Bolles, Random sample consensus: a paradigm for model fitting with applications to image analysis and automated cartography, Commun. ACM 24 (6) (1981) 381–395.

CHAPTER 6

Nondeterministic Events in Mobile Systems

6.1 INTRODUCTION

Deterministic events in real world situations are rare. Let us take a look at some examples.

Consider a case where we are measuring the speed of a vehicle. The true speed of the vehicle can never be determined with absolute precision, since the speed sensors have inherent uncertainty (sensitivity, finite resolution, bounded measurement range, physical constraints, etc.). Sensor measurements are susceptible to environmental noise and disturbances, the performance (characteristics) of the sensor normally changes over time, and the sensors sometimes fail or break down. Therefore, the sensors may not provide relevant information about the quantity we are measuring.

Like for sensors, similar conclusions can be drawn for actuators. For example, consider a case of a motor drive. Due to noise, friction, wear, unknown workload, or mechanical failure, the true angular speed cannot be determined based on input excitation.

Furthermore, some algorithms are intrinsically uncertain. Many algorithms are normally designed in a way that produce a result with limited precision (that is satisfactory for a given problem), since the result is required to be obtained in a limited time frame. In mobile robotics, as in many other systems that need to run in real time, the speed of computation is normally more crucial than the absolute precision. Given limited computational resources, real-time implementations of many algorithms with desired response times are not feasible without reduction of algorithm precision.

Moreover, the working environment of mobile agents (wheeled mobile robots) is also inherently uncertain. The uncertainties are normally lower in structured environments (e.g., human-made buildings with planar floors and walls, etc.) and higher in dynamic time-varying environments (e.g., domestic environments, road traffic, human public spaces, forests, etc.).

Wheeled Mobile Robotics
http://dx.doi.org/10.1016/B978-0-12-804204-5.00006-8

The development of the autonomous wheeled mobile system must address the problem of uncertainties that are due to various sources.

6.2 BASICS OF PROBABILITY

Let X be a random variable and x be an outcome of the random variable X. If the sample (outcome) space of the random variable X is a set with finite (countable) number of outcomes (e.g., coin tossing has only two possible outcomes, head or tail), X is a *discrete random variable*. In the case that the sample space is a set of real numbers (e.g., the weight of a coin), X is a *continuous random variable*. In this section a brief review of basic concepts from probability theory is given. A more involved overview of the topic can be found in many textbooks on statistics (e.g., [1]).

6.2.1 Discrete Random Variable

A discrete random variable X has a finite or countable sample space, which contains all the possible outcomes of the random variable X. The probability that the outcome of a discrete random variable X is equal to x is denoted as

$$P(X = x)$$

or in a shorter notation $P(x) = P(X = x)$. The range of the probability function $P(x)$ is inside the bounded interval between 0 and 1 for every value in the sample space of X, that is, $P(x) \in [0, 1] \; \forall \; x \in X$. The sum of the probabilities of all the possible outcomes of the random variable X is equal to 1:

$$\sum_{x \in X} P(x) = 1 \tag{6.1}$$

The probability of two (or more) events occurring together (e.g., random variable X takes value x and random variable Y takes value y) is described by *joint probability*:

$$P(x, y) = P(X = x, Y = y) \tag{6.2}$$

If random variables X and Y are independent, the joint probability (6.2) is simply a product of marginal probabilities:

$$P(x, y) = P(x)P(y)$$

Conditional probability, denoted by $P(x|y)$, gives the probability of random variable X taking the value x if it is a priori known that the random variable Y has value y. If $P(y) > 0$, then the conditional probability can be obtained from

$$P(x|y) = \frac{P(x, y)}{P(y)} \tag{6.3}$$

In the case of independent random variables X and Y the relation (6.3) becomes trivial:

$$P(x|y) = \frac{P(x, y)}{P(y)} = \frac{P(x)P(y)}{P(y)} = P(x)$$

One of the fundamental results in probability theory is the *theorem of total probability*:

$$P(x) = \sum_{y \in Y} P(x, y) \tag{6.4}$$

In the case that conditional probability is available, the theorem (6.4) can be given in another form:

$$P(x) = \sum_{y \in Y} P(x|y)P(y) = \boldsymbol{p}^T(x|Y)\boldsymbol{p}(Y) \tag{6.5}$$

In (6.5) the *discrete probability distribution* was introduced, defined as column vector $\boldsymbol{p}(Y)$:

$$\boldsymbol{p}(Y) = [P(y_1), P(y_2), \dots, P(y_N)]^T$$

where $N \in \mathbb{N}$ is the number of all the possible states the random variable Y can take. Similarly, the column vector $\boldsymbol{p}(x|Y)$ is defined as follows:

$$\boldsymbol{p}(x|Y) = [P(x|y_1), P(x|y_2), \dots, P(x|y_N)]^T$$

Discrete probability distributions can be visualized with histograms (Fig. 6.1). In accordance with Eq. (6.1), the height of all the columns in a histogram sums to one. When all the states of the random variable are equally plausible, the random variable can be described with a uniform distribution (Fig. 6.1B).

6.2.2 Continuous Random Variable

The range of a continuous random variable belongs to an interval (either finite or infinite) of real numbers. In the continuous case, it holds

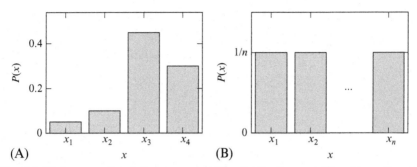

Fig. 6.1 (A) An example of discrete probability distribution and (B) uniform discrete probability distribution (the height of all the columns sums to one).

$P(X = x) = 0$, since the continuous random variable X has an infinite sample space. Therefore a *probability density function* $p(x)$ is introduced that has a bounded range between 0 and 1, that is, $p(x) \in [0, 1]$. The probability that the outcome of a continuous random variable X is less than a is

$$P(X < a) = \int_{-\infty}^{a} p(x)\, dx$$

Examples of a probability density function are shown in Fig. 6.2. Similarly as Relation (6.1) holds for a discrete random variable, the integral of probability density function over the entire sample space of a continuous random variable X is also equal to one:

$$P(-\infty < X < +\infty) = \int_{-\infty}^{+\infty} p(x)\, dx = 1$$

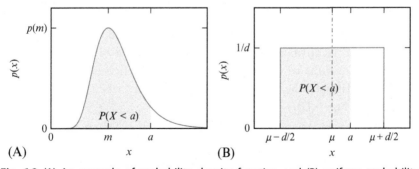

Fig. 6.2 (A) An example of probability density function and (B) uniform probability density function (the area under the entire curve integrates to one).

Table 6.1 Selected equations from probability theory for discrete and continuous random variables

Description	Discrete random variable	Continuous random variable
Total probability	$\sum_{x \in X} P(x) = 1$	$\int_{-\infty}^{+\infty} p(x)\, dx = 1$
Theorem of total probability	$P(x) = \sum_{y \in Y} P(x, y)$	$p(x) = \int_{-\infty}^{+\infty} p(x, y)\, dy$
	$P(x) = \sum_{y \in Y} P(x|y)P(y)$	$p(x) = \int_{-\infty}^{+\infty} p(x|y)p(y)\, dy$
Mean	$\mu_X = \sum_{x \in X} x P(x)$	$\mu_X = \int_{-\infty}^{+\infty} x p(x)\, dx$
Variance	$\sigma_X^2 = \sum_{x \in X} (x - \mu_X)^2 P(x)$	$\sigma_X^2 = \int_{-\infty}^{+\infty} (x - \mu_X)^2 p(x)\, dx$

Similar relations that hold for a discrete random variable (introduced in Section 6.2.1) can also be extended to a probability density function. Some important relations for discrete and continuous random variables are gathered side-by-side in Table 6.1.

The probability distributions of the random variable are seldom described by various statistics. The mean value of a continuous random variable X is defined as a mathematical expectation:

$$\mu_X = E\{X\} = \int_{-\infty}^{+\infty} x p(x)\, dx \tag{6.6}$$

One of the basic parameters that describe the shape of distribution is variance that is defined as follows:

$$\sigma_X^2 = \text{var}\{X\} = E\{(X - E\{X\})^2\} = \int_{-\infty}^{+\infty} (x - \mu_X)^2 p(x)\, dx \tag{6.7}$$

These properties can also be defined for discrete random variables, and they are given in Table 6.1.

The mean value μ and the variance σ^2 are the only two parameters that are required to uniquely describe one of the most important probability density functions, a normal distribution (Fig. 6.3), which is presented with a Gaussian function:

$$p(x) = \frac{1}{\sqrt{2\pi\sigma^2}} e^{-\frac{1}{2}\frac{(x-\mu)^2}{\sigma^2}} \tag{6.8}$$

In a multidimensional case, when a random variable is a vector $\boldsymbol{x}$, the normal distribution takes the following form:

$$p(\boldsymbol{x}) = \det(2\pi\, \boldsymbol{\Sigma})^{-\frac{1}{2}} e^{-\frac{1}{2}(\boldsymbol{x}-\boldsymbol{\mu})^T \boldsymbol{\Sigma}^{-1}(\boldsymbol{x}-\boldsymbol{\mu})}$$

where $\boldsymbol{\Sigma}$ is the covariance matrix. An example of a 2D Gaussian function is shown in Fig. 6.4. Covariance is a symmetric matrix where the element

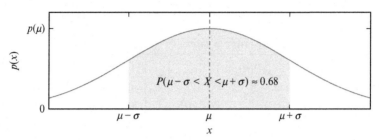

Fig. 6.3 Normal distribution (probability density function).

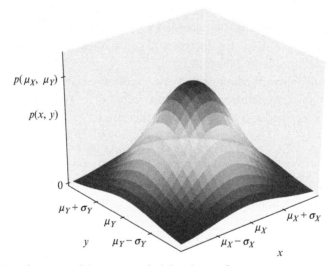

Fig. 6.4 Two-dimensional Gaussian probability density function.

in the row i and column j is a covariance $\text{cov}\{X_i, X_j\}$ between the random variables X_i and X_j.

The covariance $\text{cov}\{X, Y\}$ is a measure of the linear relationship between random variables X and Y:

$$\sigma_{XY}^2 = \text{cov}\{X, Y\} = \text{E}\{(X - \mu_X)(Y - \mu_Y)\}$$
$$= \int_{-\infty}^{+\infty} \int_{-\infty}^{+\infty} (X - \mu_X)(Y - \mu_Y)p(x, y)\,dx\,dy \quad (6.9)$$

where $p(x, y)$ is the joint probability density function of X and Y. The relation (6.9) can be simplified to

$$\sigma_{XY}^2 = \text{E}\{XY\} - \mu_X\mu_Y \quad (6.10)$$

If random variables X and Y are *independent*, then $E\{XY\} = E\{X\}E\{Y\}$ and the covariance (6.10) is zero: $\sigma^2_{XY} = 0$. However, null covariance does not imply that random variables are independent. Covariance of two identical random variables is the following variance: $\text{cov}\{X, X\} = \text{var}\{X\} = \sigma^2(X)$.

6.2.3 Bayes' Rule

Bayes' rule is one of the fundamental pillars in probabilistic robotics. The definition of *Bayes' rule* is

$$p(x|y) = \frac{p(y|x)p(x)}{p(y)} \tag{6.11}$$

for a continuous space and

$$P(x|y) = \frac{P(y|x)P(x)}{P(y)} \tag{6.12}$$

for a discrete space. *Bayes' rule* normally allows us to calculate the probability that is hard to determine, based on the probabilities that are presumably easier to obtain.

EXAMPLE 6.1

This is an introductory example of using Bayes' rule (6.11) in mobile systems. Let a random variable X represent the state of a mobile system that we want to estimate (e.g., a distance of the mobile system from the obstacle) based on the measurement Y that is stochastically dependent on random state X. Since the result of a measurement is uncertain, we would like to know the probability density of the estimated state $X = x$ based on the measurement $Y = y$.

Solution

We would like to know the probability $p(x|y)$, and this can be calculated from Eq. (6.11). The probability $p(x)$ describes the knowledge about the random variable X before the measurement y is made, and it is therefore known as a priori estimate. The conditional probability $p(x|y)$ gives the knowledge about the random variable X after the measurement is made, and it is therefore also known as a posteriori estimate. The probability $p(y|x)$ contains information about the influence of the state X on the measurement Y, and therefore it represents the model of a sensor (e.g., probability distribution of distance measurement to the obstacle if the mobile system is at specific distance from the obstacle). The probability $p(y)$ contains distribution of the measurement y, that is, the confidence

(Continued)

EXAMPLE 6.1—cont'd

in the measurement. The probability $p(y)$ can be determined using the total probability theorem $p(y) = \int p(y|x)p(x) \, dx$. Therefore, the current estimate $p(x|y)$ can be obtained based on a known statistical model of the sensor (i.e., $p(y|x)$) and a priori estimate $p(x)$.

EXAMPLE 6.2

There are three different paths that all lead to the same goal location. The mobile system picks the first path in 7 out of 10 cases, the second path only in 1 of 10 cases, and the third path in 1 of 5 cases. There is a 5%, 10%, and 8% chance of an obstacle on the first, second, and third paths, respectively.

1. Determine the probability that the mobile system encounters an obstacle when going toward the goal.
2. The mobile system encountered on an obstacle. What is the probability that this happened on the first path?

Solution

Let us denote the first, second, and third path with symbols A_1, A_2, and A_3, respectively, and the obstacle on any path with symbol B. The probabilities of selecting a specific path are the following: $P(A_1) = 0.7$, $P(A_2) = 0.1$, and $P(A_3) = 0.2$. The probability of encountering an obstacle on the first path is $P(B|A_1) = 0.05$, on the second path $P(B|A_2) = 0.1$, and on the third path $P(B|A_3) = 0.08$.

1. The probability that the mobile system encounters an obstacle is

$$P(B) = P(B|A_1)P(A_1) + P(B|A_2)P(A_2) + P(B|A_3)P(A_3)$$
$$= 0.05 \cdot 0.7 + 0.1 \cdot 0.1 + 0.08 \cdot 0.2$$
$$= 0.061$$

2. The probability that the mobile system gets stuck on the first path can be calculated using Bayes' rule (6.12):

$$P(A_1|B) = \frac{P(B|A_1)P(A_1)}{P(B)}$$
$$= \frac{P(B|A_1)P(A_1)}{P(B|A_1)P(A_1) + P(B|A_2)P(A_2) + P(B|A_3)P(A_3)}$$
$$= \frac{0.05 \cdot 0.7}{0.05 \cdot 0.7 + 0.1 \cdot 0.1 + 0.08 \cdot 0.2}$$
$$= 0.5738$$

EXAMPLE 6.2—cont'd

Using Matlab script shown in Listing 6.1 the probabilities that the mobile robot gets stuck on either of the paths are also calculated.

Listing 6.1 Implementation of the solution of Example 6.2

```
1  % Probabilities of selecting the first, second or third path:
2  % p(A) = [P(A1), P(A2), P(A3)]
3  p_A = [0.7 0.1 0.2]
4  % Probabilities of encountering on an obstacle on the first, second and
5  % third path: p(B|A) = [P(B|A1), P(B|A2), P(B|A3)]
6  p_BA = [0.05 0.1 0.08]
7
8  % Probability of an obstacle: P(B)
9  P_B = p_BA*p_A.'
10
11 % Probability of mobile robot getting stuck on the first, second and third
12 % path: p(A|B) = [P(A1|B), P(A2|B), P(A3|B)]
13 p_AB = (p_BA.*p_A)./P_B


p_A =
    0.7000    0.1000    0.2000
p_BA =
    0.0500    0.1000    0.0800
P_B =
    0.0610
p_AB =
    0.5738    0.1639    0.2623
```

EXAMPLE 6.3

A mobile robot is equipped with a dirt detection sensor that can detect the cleanliness of the floor underneath the mobile robot (Fig. 6.5). Based on the sensor readings we would like to determine whether the floor below the mobile robot is clean or not; therefore, the discrete random state has two possible values. The probability that the floor is clean is 40%. The following statements can be made about the dirt detection sensor: if the floor is clean, the sensor correctly determines the state of the floor as clean in 80%; and if the floor is dirty, the sensor correctly determines the state of the floor as dirty in 90%. The probability of incorrect measurement is small: if the floor is clean, the sensor will incorrectly determine the state of the floor in 1 out of 5 cases, but if the floor is dirty the sensor performance is even better since it makes an incorrect measurement only in 1 out of 10 cases. What is the probability that the floor is clean if the sensor detects the floor to be clean?

(Continued)

EXAMPLE 6.3—cont'd

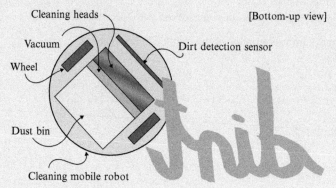

Fig. 6.5 Cleaning mobile robot.

Solution

Let us denote the state of the floor with discrete random variable $X \in \{clean, dirty\}$ and the measurement of the sensor with random variable $Z \in \{clean, dirty\}$. The probability that the floor is clean can therefore be written as $P(X = clean) = 0.4$, and the sensor measurement model can be mathematically represented as follows:

$$P(Z = clean | X = clean) = 0.8$$
$$P(Z = dirty | X = dirty) = 0.9$$

Let us introduce a shorter notation:

$$P(x) = P(X = clean) = 0.4$$
$$P(\bar{x}) = P(X = dirty) = 1 - P(x) = 0.6$$
$$P(z|x) = P(Z = clean | X = clean) = 0.8$$
$$P(\bar{z}|x) = P(Z = dirty | X = clean) = 1 - P(z|x) = 0.2$$
$$P(\bar{z}|\bar{x}) = P(Z = dirty | X = dirty) = 0.9$$
$$P(z|\bar{x}) = P(Z = clean | X = dirty) = 1 - P(\bar{z}|\bar{x}) = 0.1$$

We would like to determine $P(x|z)$ that can be calculated from Bayes' rule (6.12):

$$P(x|z) = \frac{P(z|x)P(x)}{P(z)}$$

EXAMPLE 6.3—cont'd

Using the total probability theorem, the probability that the sensor detects floor as clean can be calculated as follows:

$$P(z) = P(z|x)P(x) + P(z|\bar{x})P(\bar{x})$$
$$= 0.8 \cdot 0.4 + 0.1 \cdot 0.6$$
$$= 0.38$$

The probability that we are looking for is then

$$P(x|z) = \frac{0.8 \cdot 0.4}{0.38} = 0.8421$$

The obtained probability of a clean floor in the case the sensors detect the floor as clean is high. However, there is a more than 15% chance that the floor is actually dirty. Therefore, the mobile system can incorrectly assume that no cleaning is required; and if the mobile robot does not take any cleaning actions, the floor remains dirty.

6.3 STATE ESTIMATION

In this section an introduction into state estimation is given. State estimation is a process in which the true states of the system are estimated based on the measured data and knowledge about the system. Even if the system states are directly accessible for measurement, the measured data is normally corrupted by noise and other disturbances. This normally makes the raw measured data unsuitable for further use without appropriate signal filtering (e.g., direct use in calculation of control error). In many cases the states of the system may be estimated even if the states are not directly measurable. This can be achieved when the system is said to be observable. When implementing a state estimation algorithm, the system observability should be checked. The most important properties of state estimating algorithms are estimate convergence and estimation bias. This section therefore gives an overview of some practical considerations that should be taken into account before a particular state estimation algorithm is implemented.

6.3.1 Disturbances and Noise

All the system dynamics that are not taken into account in the model of the system, all unmeasurable signals, and modeling errors can be thought of as system disturbances. Under linearity assumption all the disturbances can be presented in a single term $n(t)$ that is added to the true signal $y_0(t)$:

$$y(t) = y_0(t) + n(t) \tag{6.13}$$

Disturbances can be classified into several classes: high-frequency, quasi-stationary stochastic signal (e.g., measurement noise), low-frequency, non-stationary signal (e.g., drift), periodic signal, or some other type of signal (e.g., spikes, data outliers). One of the most important stochastic signals is *white noise*.

Frequency spectrum and signal distribution are the most important properties that describe a signal. Signal distribution gives the probability that amplitude takes a particular value. Two of the most common signal distributions are uniform distribution and Gaussian (normal) distribution (see Section 6.2). The frequency spectrum of a signal represents the interdependence of a signal in every time moment, which is related to distribution of frequency components composing the signal. In the case of white noise, the distribution of all frequency components is uniform. Therefore the signal value in every time step is independent of the previous signal's values.

6.3.2 Estimate Convergence and Bias

As already mentioned, the state estimation provides the estimates of internal states based on the measured input-output signals, the relations among the variables (model of the system), and some statistical properties of signals, for example, variances, and other information about the system. All these pieces of information should be somehow fused in order to obtain accurate and precise estimation of internal states. Unfortunately, the measurements, the model, and the a priori known properties of the signals are inherently uncertain due to noises, disturbances, parasitic dynamics, wrong assumptions about the system model, and other sources of error. As a result, the state estimates are in general different from the actual states.

All the above-mentioned problems cause a certain level of uncertainty in signals. Mathematically, the problem can be coped within a stochastic environment where signals are treated as random variables. The signals are then represented by their probability density functions or sometimes simplified by the mean value and the variance. An important issue in a stochastic environment is the quality of a certain state estimate. Especially, the convergence of a state estimate toward the true value has to be analyzed. In particular two important questions should be asked:

1. Is the mathematical expectation of the estimate the same as the true value? If so, the estimate is *bias-free*. The estimate is *consistent* if it improves with time (larger observation interval) and it converges to the true value as the interval of observation becomes infinite.

2. Does the variance of the estimation error converge to zero as the observation time approaches infinity? If so and the estimate is consistent, the estimate is *consistent in the mean square*. This means that with growing observation time the accuracy and the precision of the estimates become excellent (all the estimates are in the vicinity of the true value).

The answers to the above questions are relatively easy to obtain when the assumptions about the system are simple (perfect model of the system, Gaussian noises, etc.). When a more demanding problem is treated, the answers become extremely difficult and analytical solutions are no longer possible. However, what needs to be remembered is that state estimators provide satisfactory results if some important assumptions are not violated. This is why it is extremely important to read the fine print when using a certain state estimation algorithm. Otherwise, these algorithms can provide the estimations that are far from the actual states. And the problem is that we are usually not even aware of this.

6.3.3 Observability

The states that need to be estimated are normally hidden, since information about the states is usually not directly accessible. The states can be estimated based on measurement of the systems' outputs that are directly or indirectly dependent on system states. Before the estimation algorithm is implemented the following question should be answered: Can the states be estimated uniquely in a finite time based on observation of the system outputs? The answer to this question is related to the *observability* of the system. If the system is observable, the states can be estimated based on observing the system outputs. The system may only be partially observable; if this is the case, only a subset of system states can be estimated, and the remaining system states may not be estimated. If the system is fully observable, all the states are output connected.

Before we proceed to the definition of observability, the concept of *indistinguishable* states should be introduced. Consider a general nonlinear system in the following form ($x \in \mathbb{R}^n$, $u \in \mathbb{R}^m$, and $y \in \mathbb{R}^l$):

$$\dot{x}(t) = f(x(t), u(t))$$
$$y(t) = h(x(t))$$

(6.14)

Two states x_0 and x_1 are indistinguishable if for every admissible input $u(t)$ on a finite time interval $t \in [t_0, t_1]$, identical outputs are obtained [2]:

$$y(t, x_0) \equiv y(t, x_1) \quad \forall \, t \in [t_0, t_1] \tag{6.15}$$

A *set of all indistinguishable states* from the state x_0 is denoted as $\mathcal{I}(x_0)$. The definitions of observability can now be given as follows. The system is *observable at* x_0 if a set of indistinguishable states $\mathcal{I}(x_0)$ contains only the state x_0, that is, $\mathcal{I}(x_0) = \{x_0\}$. The system is *observable* if a set of indistinguishable states $\mathcal{I}(x)$ contains only the state x for every state x in the domain of definition, that is, $\mathcal{I}(x) = \{x\} \; \forall \; x$. Notice that observability does not imply that estimation of x from the observed output may be possible for every input $u(t)$, $t \in [t_0, t_1]$. Also note that a long time of observation may be required to distinguish between the states. Various forms of observability have been defined [2]: local observability (stronger concept of observability), weak observability (weakened concept of observability), and local weak observability. In the case of autonomous linear systems these various forms of observability are all equivalent.

Checking the observability of a general nonlinear systems (6.14) requires advanced mathematical analysis. The *local weak observability* of the system can be checked with a simple algebraic test. For this purpose a *Lie derivative* operator is introduced as the time derivative of h along the system trajectory x:

$$L_f[h(x)] = \frac{\partial h(x)}{\partial x} f(x) \tag{6.16}$$

Consider the system (6.14) is autonomous ($u(t) = 0$ for every t). To check the local weak observability the system output y needs to be differentiated several times (until the rank of the matrix $\mathbf{Q}$, which is defined later on, increases):

$$y = h(x) = L_f^0[h(x)]$$

$$\dot{y} = \frac{\partial h(x)}{\partial x} \frac{dx}{dt} = \frac{\partial h(x)}{\partial x} f(x) = L_f[h(x)] = L_f^1[h(x)]$$

$$\ddot{y} = \frac{\partial}{\partial x} \left(\frac{\partial h(x)}{\partial x} f(x) \right) f(x) = L_f[L_f[h(x)]] = L_f^2[h(x)] \tag{6.17}$$

$$\vdots$$

$$\frac{d^i y}{dt^i} = L_f^i[h(x)]$$

The time derivatives of the system output (6.17) can be arranged into a matrix $L(x)$:

$$L(x) = \begin{bmatrix} L_f^0[h(x)] \\ L_f^1[h(x)] \\ L_f^2[h(x)] \\ \vdots \\ L_f^i[h(x)] \end{bmatrix} \tag{6.18}$$

The rank of matrix $Q(x_0) = \frac{\partial}{\partial x}L(x)|_{x_0}$ determines the local weak observability of the system at x_0. If the rank of the matrix $Q(x_0)$ is equal to the number of states, that is, $\text{rank}(Q(x_0)) = n$, the system is said to satisfy the *observability rank condition at* x_0, and this is a sufficient, but not necessary, condition for local weak observability of the system at x_0. If the observability rank condition is satisfied for every x from the domain of definition, the system is locally and weakly observable. A more involved study on the subject of observability can be found in, for example, [2–4].

The observability rank condition simplifies for time-invariant linear systems. For a system with n states in the form $\dot{x}(t) = Ax(t) + Bu(t)$, $y(t) = Cx(t)$ the Lie derivatives in Eq. (6.18) are

$$L_f^0[h(x)] = Cx(t)$$
$$L_f^1[h(x)] = C(Ax(t) + Bu(t))$$
$$L_f^2[h(x)] = CA(Ax(t) + Bu(t)) \tag{6.19}$$
$$\vdots$$

Partial differentiation of Lie derivatives (Eq. 6.19) leads to the *Kalman observability matrix* Q:

$$Q^T = \begin{bmatrix} C^T & A^T C^T & \cdots & (A^T)^{n-1} C^T \end{bmatrix} \tag{6.20}$$

The system is observable if the rank of the observability matrix has n independent rows, that is, the rank of the observability matrix equals the number of states:

$$rank(Q) = n \tag{6.21}$$

6.4 BAYESIAN FILTER

6.4.1 Markov Chains

Let us focus on systems for which a proposition about a *complete state* can be assumed. This means that all the information about the system in any given time can be determined from the system states. The system can be described based on the current states, which is a property of the Markov process. In Fig. 6.6, a hidden Markov process (chain) is shown where the states are not accessible directly and can only be estimated from measurements that are stochastically dependent on the current values of the states. Under these assumptions (Markov process), the current state of the system is dependent only on the previous state and not the entire history of the states:

$$p(x_k|x_0, \ldots, x_{k-1}) = p(x_k|x_{k-1})$$

Similarly, the measurement is also assumed to be independent of the entire history of system states if only the current state is known:

$$p(z_k|x_0, \ldots, x_k) = p(z_k|x_k)$$

The structure of a hidden Markov process where the current state x_k is dependent not only on the previous state x_{k-1} but also on the system input u_{k-1} is shown in Fig. 6.7. The input u_{k-1} is the most recent external action that influences the internal states from time step $k - 1$ to the current time step k.

6.4.2 State Estimation From Observations

A Bayesian filter is the most general form of an algorithm for calculation of probability distribution. A Bayesian filter is a powerful statistical tool that can be used for location (system state) estimation in the presence of system and measurement uncertainties [5].

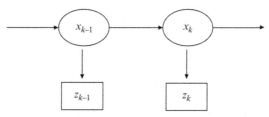

Fig. 6.6 Hidden Markov process. Measurement z_k is stochastically dependent on current state x_k, which is not accessible directly, and it is dependent on the previous state x_{k-1}.

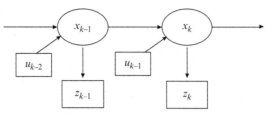

Fig. 6.7 Hidden Markov process with external inputs. Measurement z_k is stochastically dependent on the current state x_k. State x_k is not accessible directly, and it is dependent on the previous state x_{k-1} and the most recent external input u_{k-1}.

The probability distribution of the state after measurement ($p(x|z)$) can be estimated if the statistical model of the sensor $p(z|x)$ (probability distribution of the measurement given a known state) and the probability distribution of the measurement $p(z)$ are known. This was shown in Example 6.3 for a discrete random variable.

Let us now take a look at the case of estimating the state x when multiple measurements z are obtained in a sequence. We would like to estimate $p(x_k|z_1, \dots, z_k)$, that is, the probability distribution of the state x in time step k while considering the sequence of all the measurements until the current time step. The Bayesian formula can be written in a recursive form:

$$p(x_k|z_1, \dots, z_k) = \frac{p(z_k|x_k, z_1, \dots, z_{k-1})p(x_k|z_1, \dots, z_{k-1})}{p(z_k|z_1, \dots, z_{k-1})}$$

that can be rewritten into a shorter form:

$$p(x_k|z_{1:k}) = \frac{p(z_k|x_k, z_{1:k-1})p(x_k|z_{1:k-1})}{p(z_k|z_{1:k-1})} \tag{6.22}$$

The meanings of the terms in Eq. (6.22) are as follows:

- $p(x_k|z_{1:k})$ is the estimated state probability distribution in time step k, updated with measurement data.
- $p(z_k|x_k, z_{1:k-1})$ is the measurement probability distribution in time step k if the current state x_k and previous measurements up to time $k-1$ are known.
- $p(x_k|z_{1:k-1})$ is the predicted state probability distribution based on previous measurements.
- $p(z_k|z_{1:k-1})$ is the measurement probability (confidence in measurement) in time step k.

Normally it holds that the current measurement z_k in Eq. (6.22) is independent on the previous measurements $z_{1:k-1}$ (complete state assumption, Markov process) if the system state x_k is known:

$$p(z_k|x_k, z_{1:k-1}) = p(z_k|x_k)$$

Therefore, Eq. (6.22) simplifies into the following:

$$p(x_k|z_{1:k}) = \frac{p(z_k|x_k)p(x_k|z_{1:k-1})}{p(z_k|z_{1:k-1})} \tag{6.23}$$

The derivation of Eq. (6.23) is given in Example 6.4.

EXAMPLE 6.4

As an exercise, derive Eq. (6.23) from Eq. (6.22) if the complete state assumption is taken into account $(p(z_k|x_k, z_{1:k-1}) = p(z_k|x_k))$.

Solution

The derivation of Eq. (6.23) is the following:

$$\begin{aligned}
p(x_k|z_{1:k}) &= \frac{p(z_{1:k}|x_k)p(x_k)}{p(z_{1:k})} \\
&= \frac{p(z_k, z_{1:k-1}|x_k)p(x_k)}{p(z_k, z_{1:k-1})} \\
&= \frac{p(z_k|z_{1:k-1}, x_k)p(z_{1:k-1}|x_k)p(x_k)}{p(z_k|z_{1:k-1})p(z_{1:k-1})} \\
&= \frac{p(z_k|z_{1:k-1}, x_k)p(x_k|z_{1:k-1})p(z_{1:k-1})p(x_k)}{p(z_k|z_{1:k-1})p(z_{1:k-1})p(x_k)} \\
&= \frac{p(z_k|z_{1:k-1}, x_k)p(x_k|z_{1:k-1})}{p(z_k|z_{1:k-1})} \\
&= \frac{p(z_k|x_k)p(x_k|z_{1:k-1})}{p(z_k|z_{1:k-1})}
\end{aligned}$$

The recursive formula (6.23), which updates the state based on previous measurements, contains a prediction term $p(x_k|z_{1:k-1})$. The state estimation can be separated into two steps: a prediction step and a correction step.

Prediction Step

The prediction $p(x_k|z_{1:k-1})$ is made as follows:

$$p(x_k|z_{1:k-1}) = \int p(x_k|x_{k-1}, z_{1:k-1})p(x_{k-1}|z_{1:k-1}) \, dx_{k-1}$$

The complete state assumption yields a simpler form:

$$p(x_k|z_{1:k-1}) = \int p(x_k|x_{k-1})p(x_{k-1}|z_{1:k-1})\,dx_{k-1} \qquad (6.24)$$

where $p(x_k|x_{k-1})$ represents probability distribution of state transitions and $p(x_{k-1}|z_{1:k-1})$ is the corrected probability distribution of the estimated state from the previous time step.

Correction Step

The estimated state probability distribution after the measurement in time step k and predicted state probability distribution in the prediction step are known as the following:

$$p(x_k|z_{1:k}) = \frac{p(z_k|x_k)p(x_k|z_{1:k-1})}{p(z_k|z_{1:k-1})} \qquad (6.25)$$

The probability $p(z_k|z_{1:k-1})$ that describes the measurement confidence can be determined from the relation

$$p(z_k|z_{1:k-1}) = \int p(z_k|x_k)p(x_k|z_{1:k-1})\,dx_k$$

The Most Probable State Estimate

How can the estimated probability density distribution $p(x_k|z_{1:k})$ be used in calculation of the most probable state x_k? The best and most probable state estimate (mathematical expectation) $E\{x_k\}$ is defined as the value that minimizes the average square error of the measurement:

$$E\{x_k\} = \int x_k p(x_k|z_{1:k})\,dx_k$$

Moreover, the value of $x_{k_{max}}$, which maximizes the posterior probability $p(x_k|z_{1:k})$, can also be estimated:

$$x_{k_{max}} = \max_{x_k} p(x_k|z_{1:k})$$

EXAMPLE 6.5

Consider Example 6.3 again and determine the probability that the floor is clean if the sensor in time step $k = 2$ again detects the floor to be clean.

(*Continued*)

EXAMPLE 6.5—cont'd
Solution
In time step $k = 1$ of Example 6.3 the following situation was considered: the floor was clean and the sensor also detected the floor to be clean ($z_1 = clean$). The following conditional probability was calculated:

$$P(x_1|z_1) = \frac{0.8 \cdot 0.4}{0.38} = 0.8421$$

In the next time step $k = 2$ the sensor returns a reading $z_2 = clean$ with probability $P(z_2|x_2) = 0.8$ of a correct sensor reading and with probability $P(z_2|\bar{x}_2) = 0.1$ of an incorrect sensor reading (the sensor characteristic was assumed to be time invariant).

Let us first evaluate the predicted probability $P(x_2|z_1)$, that is, the floor is dirty based on the previous measurement. Considering Eq. (6.24), where the operation of integration is substituted for the operation of summation, yields

$$P(x_k|z_{1:k-1}) = \sum_{x_{k-1} \in X} P(x_k|x_{k-1})P(x_{k-1}|z_{1:k-1})$$

For now assume that the mobile system only detects the state of the floor, but it cannot influence the state of the floor (i.e., the mobile system does not perform any floor cleaning and also cannot make the floor dirty). Therefore, the state transition probability is simply $P(x_2|x_1) = 1$ and $P(x_2|\bar{x}_1) = 0$. Hence,

$$P(x_2|z_{1:1}) = P(x_2|x_1)P(x_1|z_1) + P(x_2|\bar{x}_1)P(\bar{x}_1|z_1)$$
$$= 1 \cdot 0.8421 + 0 \cdot 0.1579$$
$$= 0.8421$$

which is a logical result since if the measurement z_2 is not considered, no new information about the system has been obtained. Therefore, the state of the floor has the same probability in the time step $k = 2$ as it was the state probability in the previous time step $k = 1$.

The measurement can be fused with the current estimate in the correction step, using relation (6.25):

$$P(x_2|z_{1:2}) = \frac{P(z_2|x_2)P(x_2|z_{1:1})}{P(z_2|z_{1:1})}$$
$$= \frac{0.8 \cdot 0.8421}{P(z_2|z_{1:1})}$$

where the value of the normalization factor needs to be calculated. The missing factor is the probability of a clean floor in time step $k = 2$, and it can

EXAMPLE 6.5—cont'd

be determined with the summation of all the possible state combinations that result in the current measurement z_k, considering the outcomes of the previous measurements $z_{1:k-1}$:

$$P(z_k|z_{1:k-1}) = \sum_{x_k \in X} P(z_k|x_k)P(x_k|z_{1:k-1})$$

In our case,

$$P(z_2|z_1) = P(z_2|x_2)P(x_2|z_1) + P(z_2|\bar{x}_2)P(\bar{x}_2|z_1)$$
$$= 0.8 \cdot 0.8421 + 0.1 \cdot 0.1579$$
$$= 0.6895$$

The final result, the probability that the floor is clean if the sensor has detected the floor to be clean twice in a row, is

$$P(x_2|z_{1:2}) = 0.9771$$

EXAMPLE 6.6

Again consider Example 6.3 and determine how the state probability changes if the floor is clean and the sensor makes three measurements $z_{1:3} = (clean, clean, dirty)$.

Solution

The first two conditional probabilities were already calculated in Examples 6.3 and 6.5, the third is

$$P(x_3|z_{1:3}) = \frac{P(z_3|x_3)P(x_3|z_{1:2})}{P(z_3|z_{1:2})}$$
$$= \frac{0.2 \cdot 0.9771}{0.2 \cdot 0.9771 + 0.9 \cdot (1 - 0.9771)}$$
$$= 0.9046$$

where $P(Z = dirty|X = clean) = 1 - P(Z = clean|X = clean)$. The state probabilities in three consequential state measurements taken at time steps $i = 1, 2, 3$ are therefore

$$P(x_k|z_{1:i}) = (0.8421, 0.9771, 0.9046)$$

The posterior state probability distribution for time steps $k = 1, 2, 3$ is also shown in Fig. 6.8. Implementation of the solution in Matlab is shown in Listing 6.2.

(Continued)

EXAMPLE 6.6—cont'd

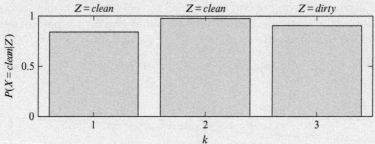

Fig. 6.8 Posterior state probability distributions in three time steps from Example 6.6.

Listing 6.2 Implementation of the solution of Example 6.6

```
1  % Probabilities of clean and dirty floor
2  P_Xc = 0.4    % P(X=clean)
3  P_Xd = 1-P_Xc % P(X=dirty)
4  % Conditional probabilities of the dirt sensor measurements
5  P_ZcXc = 0.8    % P(Z=clean|X=clean)
6  P_ZdXc = 1-P_ZcXc % P(Z=dirty|X=clean)
7  P_ZdXd = 0.9    % P(Z=dirty|X=dirty)
8  P_ZcXd = 1-P_ZdXd % P(Z=clean|X=dirty)
9
10 disp('Time step k = 1: Z=clean')
11 % Measurement probability in the case clean floor is detected
12 P_Zc_k1 = P_ZcXc*P_Xc + P_ZcXd*P_Xd
13 % Probability of clean floor after the measurement is made (Bayes' rule)
14 P_XcZc_k1 = P_ZcXc*P_Xc/P_Zc_k1
15 P_XdZc_k1 = 1-P_XcZc_k1;
16
17 disp('Time step k = 2: Z=clean')
18 % Measurement probability in the case clean floor is detected
19 P_Zc_k2 = P_ZcXc*P_XcZc_k1 + P_ZcXd*P_XdZc_k1
20 % Probability of clean floor after the measurement is made (Bayes' rule)
21 P_XcZc_k2 = P_ZcXc*P_XcZc_k1/P_Zc_k2
22 P_XdZc_k2 = 1-P_XcZc_k2;
23
24 disp('Time step k = 3: Z=dirty')
25 % Measurement probability in the case dirty floor is detected
26 P_Zd_k3 = P_ZdXc*P_XcZc_k2 + P_ZdXd*P_XdZc_k2
27 % Probability of clean floor after the measurement is made (Bayes' rule)
28 P_XcZd_k3 = P_ZdXc*P_XcZc_k2/P_Zd_k3
29 P_XdZd_k3 = 1-P_XcZd_k3;
30 P_Xc =
31     0.4000
32 P_Xd =
33     0.6000
```

EXAMPLE 6.6—cont'd

```
P_ZcXc =
    0.8000
P_ZdXc =
    0.2000
P_ZdXd =
    0.9000
P_ZcXd =
    0.1000
Time step k = 1: Z=clean
P_Zc_k1 =
    0.3800
P_XcZc_k1 =
    0.8421
Time step k = 2: Z=clean
P_Zc_k2 =
    0.6895
P_XcZc_k2 =
    0.9771
Time step k = 3: Z=dirty
P_Zd_k3 =
    0.2160
P_XcZd_k3 =
    0.9046
```

6.4.3 State Estimation From Observations and Actions

In Section 6.4.2 the state of the system was estimated based only on observing the environment. Normally the actions of the mobile system can have an effect on the environment; therefore, the system states can change due the mobile system actions (e.g., the mobile system moves, performs cleaning actions, etc.). Every action that the mobile system makes has an inherent uncertainty. Therefore, the outcome of the action is not deterministic, but it has an inherent uncertainty (probability). The probability $p(x_k|x_{k-1}, u_{k-1})$ describes the probability of transition from the previous state into the next state given a known action. The action u_{k-1} takes place in time step $k - 1$ and is active until time step k. Therefore, this action is also known as the most recent or current action. In a general case, the actions that take place in the environment increase the level of uncertainty about our knowledge of the environment, and the measurements made in the environment normally decrease the level of uncertainty.

Let us take a look at what effects the actions and measurements have on our knowledge about the states. We would like to determine the probability $p(x_k|z_{1:k}, u_{0:k-1})$, where z indicates the measurements and u indicates the actions.

As in Section 6.4.2, the Bayesian theorem can be written as follows:

$$p(x_k|z_{1:k}, u_{0:k-1}) = \frac{p(z_k|x_k, z_{1:k-1}, u_{0:k-1})p(x_k|z_{1:k-1}, u_{0:k-1})}{p(z_k|z_{1:k-1}, u_{0:k-1})} \quad (6.26)$$

where the following applies:

- $p(x_k|z_{1:k}, u_{0:k-1})$ is the estimated state probability distribution in time step k, updated with measurement data and known actions.
- $p(z_k|x_k, z_{1:k-1}, u_{0:k-1})$ is the measurement probability distribution in time step k, if the current state x_k, previous actions and measurements up to time step $k-1$ are known.
- $p(x_k|z_{1:k-1}, u_{0:k-1})$ is the predicted state probability distribution based on previous measurements and actions made up to time step k.
- $p(z_k|z_{1:k-1}, u_{0:k-1})$ is the measurement probability (confidence in measurement) in time step k.

Note that action indices are in the range from 0 to $k-1$, since the actions in the previous time steps influence the behavior of the system states in step k.

Furthermore, the current measurement z_k in Eq. (6.26) can be described only with system state x_k, since the previous measurements and actions do not give any additional information about the system (complete state assumption, Markov process):

$$p(z_k|x_k, z_{1:k-1}, u_{0:k-1}) = p(z_k|x_k)$$

Hence, the relation (6.26) can be simplified:

$$p(x_k|z_{1:k}, u_{0:k-1}) = \frac{p(z_k|x_k)p(x_k|z_{1:k-1}, u_{0:k-1})}{p(z_k|z_{1:k-1}, u_{0:k-1})} \quad (6.27)$$

The recursive rule (6.27), update of the state-probability estimate based on the previous measurements and actions, also includes prediction $p(x_k|z_{1:k-1}, u_{0:k-1})$, where the probability distribution for state estimate is made based on previous measurements $z_{1:k-1}$ and all the actions $u_{0:k-1}$. Therefore, the state estimation procedure can be split into a prediction step and a correction step. In the prediction step the latest measurement is not known yet, but the most recent action is known. Therefore, the state probability distribution can be estimated based on the system model. When the new measurement is available, the correction step can be made.

Prediction Step

The prediction $p(x_k|z_{1:k-1}, u_{0:k-1})$ can be evaluated using the total probability theorem:

$$p(x_k|z_{1:k-1}, u_{0:k-1}) = \int p(x_k|x_{k-1}, z_{1:k-1}, u_{0:k-1}) p(x_{k-1}|z_{1:k-1}, u_{0:k-1}) \, dx_{k-1}$$

where due to the complete state the following assumption holds:

$$p(x_k|x_{k-1}, z_{1:k-1}, u_{0:k-1}) = p(x_k|x_{k-1}, u_{k-1})$$

Moreover, the most recent action u_{k-1} is not required for a state estimate in the previous time step:

$$p(x_{k-1}|z_{1:k-1}, u_{0:k-1}) = p(x_{k-1}|z_{1:k-1}, u_{0:k-2})$$

This yields the final form of the prediction step:

$$p(x_k|z_{1:k-1}, u_{0:k-1}) = \int p(x_k|x_{k-1}, u_{k-1}) p(x_{k-1}|z_{1:k-1}, u_{0:k-2}) \, dx_{k-1}$$

$$(6.28)$$

where $p(x_k|x_{k-1}, u_{k-1})$ is the state transition probability distribution and $p(x_{k-1}|z_{1:k-1}, u_{0:k-2})$ is the corrected state probability distribution from the previous time step.

Correction Step

The state estimate when the measurement in time step k is available and previously calculated prediction is then as follows:

$$p(x_k|z_{1:k}, u_{0:k-1}) = \frac{p(z_k|x_k) p(x_k|z_{1:k-1}, u_{0:k-1})}{p(z_k|z_{1:k-1}, u_{0:k-1})} \qquad (6.29)$$

The probability $p(z_k|z_{1:k-1}, u_{0:k-1})$ that represents the confidence in the measurement is the following:

$$p(z_k|z_{1:k-1}, u_{0:k-1}) = \int p(z_k|x_k) p(x_k|z_{1:k-1}, u_{0:k-1}) \, dx_k$$

General Algorithm for a Bayesian Filter

The general form of the Bayesian algorithm is given as a pseudo code in Algorithm 3. The conditional probability of the correction step $p(x_k|z_{1:k}, u_{0:k-1})$, which gives the state probability distribution estimate based on known actions and measurements, is also known as *belief*, and it is normally denoted as

$$bel(x_k) = p(x_k|z_{1:k}, u_{0:k-1})$$

The conditional probability of the prediction $p(x_k|z_{1:k-1}, u_{0:k-1})$ is therefore normally denoted as

$$bel_p(x_k) = p(x_k|z_{1:k-1}, u_{0:k-1})$$

The Bayesian filter is estimating the probability distribution of the state estimate. In a moment, when the information about the current action u_{k-1} is known, the prediction step can be made, and when the new measurement is available, the correction step can be made. Let us introduce the normalization factor $\eta = \frac{1}{\alpha} = \frac{1}{p(z_k|z_{1:k-1}, u_{0:k-1})}$. The Bayesian filter algorithm (Algorithm 3) first calculates the prediction and then the correction that is properly normed.

ALGORITHM 3   Bayesian Filter

function BAYESIAN_FILTER($bel(x_{k-1})$, u_{k-1}, z_k)
    $\alpha \leftarrow 0$
    for all x_k **do**
        $bel_p(x_k) \leftarrow \int p(x_k|x_{k-1}, u_{k-1})bel(x_{k-1})\, dx_{k-1}$
        $bel'(x_k) \leftarrow p(z_k|x_k)bel_p(x_k)$
        $\alpha \leftarrow \alpha + bel'(x_k)$
    end for
    for all x_k **do**
        $bel(x_k) \leftarrow \frac{1}{\alpha}bel'(x_k)$
    end for
    return $bel(x_k)$
end function

As it can be observed from Algorithm 3, an integral must be evaluated to determine the probability distribution in the prediction step. Hence the implementation of the Bayesian algorithm is limited to simple continuous cases, where an explicit integral solution can be determined, and discrete cases with countable number of states, where the operation of integration can be substituted for the operation of summation.

EXAMPLE 6.7

A mobile robot is equipped with a sensor that can detect if the floor is clean or not ($Z \in \{clean, dirty\}$). The mobile system is also equipped with a cleaning system (a set of brushes, a vacuum pump, and a dust bin) that can clean the floor, but the cleaning system is engaged only if the robot believes the floor needs cleaning ($U \in \{clean, null\}$). Again we would like to determine if the floor is clean or not ($X \in \{clean, dirty\}$).

The initial probability (belief) that the floor is clean is

$$bel(X_0 = clean) = 0.5$$

The correctness of the sensor measurements is given with a statistical model of the sensor:

$$P(Z_k = clean|X_k = clean) = 0.8, \qquad P(Z_k = dirty|X_k = clean) = 0.2$$
$$P(Z_k = dirty|X_k = dirty) = 0.9, \qquad P(Z_k = clean|X_k = dirty) = 0.1$$

The outcome probabilities if the robot decides to perform floor cleaning are the following:

$$P(X_k = clean|X_{k-1} = clean, U_{k-1} = clean) = 1$$
$$P(X_k = dirty|X_{k-1} = clean, U_{k-1} = clean) = 0$$
$$P(X_k = clean|X_{k-1} = dirty, U_{k-1} = clean) = 0.8$$
$$P(X_k = dirty|X_{k-1} = dirty, U_{k-1} = clean) = 0.2$$

If the cleaning system is not activated, the following outcome probabilities can be assumed:

$$P(X_k = clean|X_{k-1} = clean, U_{k-1} = null) = 1$$
$$P(X_k = dirty|X_{k-1} = clean, U_{k-1} = null) = 0$$
$$P(X_k = clean|X_{k-1} = dirty, U_{k-1} = null) = 0$$
$$P(X_k = dirty|X_{k-1} = dirty, U_{k-1} = null) = 1$$

Assume that the mobile system first makes an action and then receives the measurement data. Determine the belief $bel_p(x_k)$ based on the action taken (prediction) and the belief $bel(x_k)$ based on measurement $bel(x_k)$ (correction) for the following sequence of actions and measurements:

k	U_{k-1}	Z_k
1	null	dirty
2	clean	clean
3	clean	clean

(*Continued*)

EXAMPLE 6.7—cont'd
Solution
Without loss of generality let us introduce the following notation: $X_k \in \{clean, dirty\}$ as $X_k \in \{x_k, \bar{x}_k\}$, $Z_k \in \{clean, dirty\}$ as $Z_k \in \{z_k, \bar{z}_k\}$, and $U_k \in (clean, null)$ as $U_k \in \{u_k, \bar{u}_k\}$.

Let us use Algorithm 3. In time step $k = 1$, when the action $\bar{u}_0 = null$ is active we can determine the predicted belief that the floor is clean is the following:

$$bel_p(x_1) = \sum_{x_0 \in X} P(x_1|x_0, \bar{u}_0)bel(x_0)$$
$$= P(x_1|\bar{x}_0, \bar{u}_0)bel(\bar{x}_0) + P(x_1|x_0, \bar{u}_0)bel(x_0)$$
$$= 0 \cdot 0.5 + 1 \cdot 0.5$$
$$= 0.5$$

and the predicted belief that the floor is dirty:

$$bel_p(\bar{x}_1) = \sum_{x_0 \in X} P(\bar{x}_1|x_0, \bar{u}_0)bel(x_0)$$
$$= P(\bar{x}_1|\bar{x}_0, \bar{u}_0)bel(\bar{x}_0) + P(\bar{x}_1|x_0, \bar{u}_0)bel(x_0)$$
$$= 1 \cdot 0.5 + 0 \cdot 0.5$$
$$= 0.5$$

Since no action has been taken, the state probabilities are unchanged. Based on the measurement $\bar{z}_1 = dirty$ the belief can be corrected:

$$bel(x_1) = \eta p(\bar{z}_1|x_1)bel_p(x_1) = \eta 0.2 \cdot 0.5 = \eta 0.1$$

and

$$bel(\bar{x}_1) = \eta p(\bar{z}_1|\bar{x}_1)bel_p(\bar{x}_1) = \eta 0.9 \cdot 0.5 = \eta 0.45$$

After the normalization factor η is evaluated:

$$\eta = \frac{1}{0.1 + 0.45} = 1.82$$

The final values of the belief can be determined:

$$bel(x_1) = 0.182, \qquad bel(\bar{x_1}) = 0.818$$

The procedure can be repeated for time step $k = 2$, where $u_1 = clean$ and $z_2 = clean$:

EXAMPLE 6.7—cont'd

$$bel_p(x_2) = 0.8364, \qquad bel_p(\bar{x}_2) = 0.1636$$
$$bel(x_2) = 0.9761, \qquad bel(\bar{x}_2) = 0.0239$$

And it can be continued for time step $k = 3$, where $u_2 = clean$ and $z_3 = clean$:

$$bel_p(x_3) = 0.9952, \qquad bel_p(\bar{x}_3) = 0.0048$$
$$bel(x_3) = 0.9994, \qquad bel(\bar{x}_3) = 0.0006$$

The solution of Example 6.7 in Matlab is given in Listing 6.3.

Listing 6.3 Implementation of the solution of Example 6.7

```
1  % Notation: X === X(k), X' === X(k-1)
2  disp('Initial belief of clean and dirty floor')
3  bel_Xc = 0.5; % bel(X=clean)
4  bel_X = [bel_Xc 1-bel_Xc] % bel(X=clean), bel(X=dirty)
5
6  disp('Conditional probabilities of the dirt sensor measurements')
7  P_ZcXc = 0.8;      % P(Z=clean|X=clean)
8  P_ZdXc = 1-P_ZcXc; % P(Z=dirty|X=clean)
9  P_ZdXd = 0.9;      % P(Z=dirty|X=dirty)
10 P_ZcXd = 1-P_ZdXd; % P(Z=clean|X=dirty)
11 p_ZX = [P_ZcXc, P_ZcXd; ...
12        P_ZdXc, P_ZdXd]
13
14 disp('Outcome probabilities in the case of cleaning')
15 P_XcXcUc = 1;         % P(X=clean|X'=clean,U'=clean)
16 P_XdXcUc = 1-P_XcXcUc; % P(X=dirty|X'=clean,U'=clean)
17 P_XcXdUc = 0.8;       % P(X=clean|X'=dirty,U'=clean)
18 P_XdXdUc = 1-P_XcXdUc; % P(X=dirty|X'=dirty,U'=clean)
19 p_ZXUc = [P_XcXcUc, P_XdXcUc; ...
20         P_XcXdUc, P_XdXdUc]
21
22 disp('Outcome probabilities in the case of no cleaning action is taken')
23 P_XcXcUn = 1;         % P(X=clean|X'=clean,U'=null)
24 P_XdXcUn = 1-P_XcXcUn; % P(X=dirty|X'=clean,U'=null)
25 P_XcXdUn = 0;         % P(X=clean|X'=dirty,U'=null)
26 P_XdXdUn = 1-P_XcXdUn; % P(X=dirty|X'=dirty,U'=null)
27 p_ZXUn = [P_XcXcUn, P_XdXcUn; ...
28         P_XcXdUn, P_XdXdUn]
29
30 U = {'null', 'clean', 'clean'};
31 Z = {'dirty', 'clean', 'clean'};
32 for k=1:length(U)
33     fprintf('Prediction step: U(%d)=%s\n', k-1, U{k})
34     if strcmp(U(k), 'clean')
35         belp_X = bel_X*p_ZXUc
36     else
37         belp_X = bel_X*p_ZXUn
38     end
39
40     fprintf('Correction step: Z(%d)=%s\n', k, Z{k})
41     if strcmp(Z(k), 'clean')
42         bel_X = p_ZX(1,:).*belp_X;
```

(Continued)

EXAMPLE 6.7—cont'd

```
43    else
44        bel_X = p_ZX(2,:).*belp_X;
45    end
46    bel_X = bel_X/sum(bel_X)
47 end

   Initial belief of clean and dirty floor
   bel_X =
       0.5000    0.5000
   Conditional probabilities of the dirt sensor measurements
   p_ZX =
       0.8000    0.1000
       0.2000    0.9000
   Outcome probabilities in the case of cleaning
   p_ZXUc =
       1.0000         0
       0.8000    0.2000
   Outcome probabilities in the case of no cleaning action is taken
   p_ZXUn =
       1    0
       0    1
   Prediction step: U(0)=null
   belp_X =
       0.5000    0.5000
   Correction step: Z(1)=dirty
   bel_X =
       0.1818    0.8182
   Prediction step: U(1)=clean
   belp_X =
       0.8364    0.1636
   Correction step: Z(2)=clean
   bel_X =
       0.9761    0.0239
   Prediction step: U(2)=clean
   belp_X =
       0.9952    0.0048
   Correction step: Z(3)=clean
   bel_X =
       0.9994    0.0006
```

6.4.4 Localization Example

A simple example will be used for presentation of the localization principle that presents the basic idea behind the Monte Carlo localization algorithm. Consider a mobile system that moves in an environment and has a capability of environment sensing. Based on the information from sensor readings and applied movement actions, the localization algorithm should determine the pose of the mobile system in the environment. To implement the localization algorithm the total probability theorem and Bayes' rule are used, and they represent the fundamental part in the Bayesian filter. In the following sections, the process of measurement in the presence of sensor uncertainties is presented, followed by the process of action in the presence of actuator uncertainties. The mobile system makes actions to change the state of the environment (the mobile system pose changes as the mobile system moves).

EXAMPLE 6.8

Consider a round path on which the mobile system can move in a forward or backward direction. The path is made out of a countable number of bright and dark tiles in some random order, where the width of a tile matches the width of the path. The mobile system can be positioned on any of the tiles that are marked with numbers. Without loss of generality, consider that the path consists of five tiles as shown in Fig. 6.9. Every tile presents a cell in which the mobile system can be positioned; therefore, this is a discrete representation of the environment. The mobile system knows the map of the environment (the sequence of bright and dark tiles is known), but it does not know which cell it is occupying. The initial belief of mobile system position is therefore given with a uniform

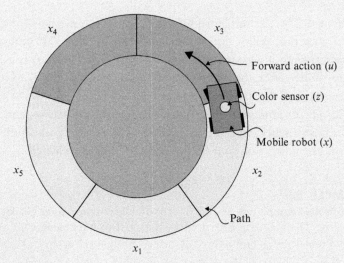

Fig. 6.9 The path in the environment is made up of five black and white tiles. The mobile system can move between the tiles and detects the color of the current tile.

(*Continued*)

EXAMPLE 6.8—cont'd

distribution, since every cell has equal probability. The mobile system has a sensor for detection of bright and dark tiles, but the sensor measurements are uncertain. The mobile system has an implemented capability for moving for desired number of tiles in a forward or backward direction, but the movement is inherently uncertain (the mobile system can move too little or too much). This sample case will be used in the explanation of the basic concept of the process of environment detection and the process of movement in the environment.

Solution

The configuration presented in this example is considered in Examples 6.9 to 6.14, where solutions to particular problems, presented later on, are given.

6.4.5 Environment Sensing

The measurement made in the environment can be used to improve the estimation of the state X (e.g., location) in this environment. Imagine that we are sleepwalking around the house in the middle of the night. When we wake up, we can figure out where we are by using our senses (sight, touch, etc.).

Mathematically, the initial knowledge about the environment can be described with probability distribution $p(x)$ (prior). This distribution can be improved when a new measurement z (with belief $p(z|x)$) is available if the probability after the measurement is determined $p(x|z)$. This can be achieved using the Bayesian rule $p(x|z) = bel(x) = \frac{p(z|x)p(x)}{p(z)}$. The probability $p(z|x)$ represents the statistical model of the sensor, and $bel(x)$ is state-estimate belief after the measurement is made. In the process of perception, the correction step of the Bayesian filter is evaluated.

EXAMPLE 6.9

Considering Example 6.8, assume that the mobile system detects a dark cell $Z = dark$. The mobile system detects a dark cell with probability 0.6 and makes a mistake with probability 0.2 (detects a bright cell as dark). Hence,

$$p(Z = dark|X = x_d) = 0.6, \quad d \in \{3, 4\}$$
$$p(Z = dark|X = x_b) = 0.2, \quad b \in \{1, 2, 5\}$$

EXAMPLE 6.9—cont'd

where index b indicated bright cells and index d indicated dark cells. In the beginning, the mobile system does not know its position. This can be described with a uniform probability distribution $P(X = x_i) = bel(x_i) = 0.2$, $i \in \{1, \ldots, 5\}$. Calculate the location probability distribution after a single measurement.

Solution

We would like to determine the distribution of conditional probability $p(X_1|Z = dark)$, the state-estimate belief after the measurement is made. The desired probability can be determined using the correction step of the Bayesian filter:

$$p(X_1|Z = dark) = \frac{p(Z = dark|X_1) * p(X_1)}{P(Z = dark)}$$

$$= \frac{[0.2, 0.2, 0.6, 0.6, 0.2]^T * [0.2, 0.2, 0.2, 0.2, 0.2]^T}{P(Z = dark)}$$

$$= \frac{[0.04, 0.04, 0.12, 0.12, 0.04]^T}{P(Z = dark)}$$

where the operator $*$ represents the operation of element-wise multiplication of vector elements.

We need to calculate the probability of detecting a dark cell $P(Z = dark)$. Therefore the total probability must be evaluated, that is, the probability of detecting a dark cell considering all the cells:

$$P(Z = dark) = \sum_i P(Z = dark|X_1 = x_i)P(X_1 = x_i)$$

$$= p^T(Z = dark|X_1)p(X_1)$$

$$= [0.2, 0.2, 0.6, 0.6, 0.2][0.2, 0.2, 0.2, 0.2, 0.2]^T$$

$$= 0.36$$

The posterior probability distribution is therefore the following:

$$p(X_1|Z = dark) = [0.11, 0.11, 0.33, 0.33, 0.11]^T$$

Hence, we can conclude that the position of the mobile system is three times more likely to be in cells 3 or 4 than in the remaining three cells. The probability distributions are also shown graphically in Fig. 6.10. The solution of this example is also given in Listing 6.4.

(Continued)

EXAMPLE 6.9—cont'd

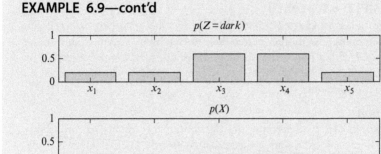

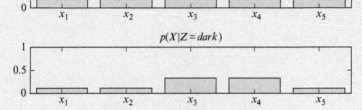

Fig. 6.10 Probability distributions from Example 6.9.

Listing 6.4 Implementation of the solution of Example 6.9

```
1  disp('Sensor measurement distribution p(Z=dark|X)')
2  p_ZdX = [0.2 0.2 0.6 0.6 0.2]
3  disp('Sensor measurement distribution p(Z=bright|X)')
4  p_ZbX = 1-p_ZdX
5  disp('Initial distribution p(X)')
6  p_X = ones(1,5)/5
7
8  disp('Probability of detecting a dark cell P(Z=dark)')
9  P_Zd = p_ZdX*p_X.'
10
11 disp('Posterior distribution p(X|Z=dark)')
12 p_XZd = p_ZdX.*p_X/P_Zd

   Sensor measurement distribution p(Z=dark|X)
   p_ZdX =
       0.2000    0.2000    0.6000    0.6000    0.2000
   Sensor measurement distribution p(Z=bright|X)
   p_ZbX =
       0.8000    0.8000    0.4000    0.4000    0.8000
   Initial distribution p(X)
   p_X =
       0.2000    0.2000    0.2000    0.2000    0.2000
   Probability of detecting a dark cell P(Z=dark)
   P_Zd =
       0.3600
   Posterior distribution p(X|Z=dark)
   p_XZd =
       0.1111    0.1111    0.3333    0.3333    0.1111
```

EXAMPLE 6.10

Answer the following questions about Example 6.9:

1. Can multiple measurements improve the estimated mobile system position (the mobile system does not move between the measurements)?
2. What is the probability distribution of the mobile system position if the mobile system detects a tile as dark twice in a row?
3. What is the probability distribution of the mobile system position if the mobile system detects a tile as dark and then as bright?
4. What is the probability distribution of the mobile system position if the mobile system detects a tile as dark, then as bright, and then again as dark?

Solution

1. Multiple measurements can improve the estimate about the mobile system position if the probability of the correct measurement is higher than the probability of the measurement mistake.
2. The probability distribution if the sensor detects cell as dark twice in a row is the following:

$$\boldsymbol{p}(X_2|Z_1 = dark, Z_2 = dark) = p(X_2|z_1, z_2)$$
$$= \frac{\boldsymbol{p}(z_2|X_2) * \boldsymbol{p}(X_2|z_1)}{P(z_2|z_1)}$$
$$= \frac{[0.2, 0.2, 0.6, 0.6, 0.2]^T * [0.11, 0.11, 0.33, 0.33, 0.11]^T}{P(z_2|z_1)}$$

where jth element in the probability distribution $\boldsymbol{p}(X_2|z_1)$ is given by $P(X_2 = x_j|z_1) = \boldsymbol{p}^T(X_2 = x_j|X_1)\boldsymbol{p}(X_1|z_1) = P(X_1 = x_j|z_1)$, since we have no influence on the states (see Example 6.5); we only observe the states with measurements. The conditional probability in the denominator is

$$P(z_2|z_1) = \sum_{x_i} P(z_2|X_2 = x_i)P(X_2 = x_i|z_1)$$
$$= \boldsymbol{p}^T(z_2|X_2)\boldsymbol{p}(X_2|z_1)$$
$$= [0.2, 0.2, 0.6, 0.6, 0.2][0.11, 0.11, 0.33, 0.33, 0.11]^T$$
$$= 0.4667$$

(*Continued*)

EXAMPLE 6.10—cont'd

Finally, the solution is

$$p(X_2|z_1, z_2) = \frac{[0.2, 0.2, 0.6, 0.6, 0.2]^T * [0.11, 0.11, 0.33, 0.33, 0.11]^T}{[0.2, 0.2, 0.6, 0.6, 0.2][0.11, 0.11, 0.33, 0.33, 0.11]^T}$$

$$= [0.0476, 0.0476, 0.4286, 0.4286, 0.0476]^T$$

3. The bright cell is detected correctly with probability $p(Z = bright|X = bright) = 1 - p(Z = dark|X = bright) = 0.8$ and incorrectly with probability $p(Z = bright|X = dark) = 1 - p(Z = dark|X = dark) = 0.4$. The second measurement can be made based on the probability distribution $p(X_2|Z_1 = dark)$:

$$p(X_2|Z_1 = dark, Z_2 = bright) = p(X_2|z_1, z_2)$$

$$= \frac{[0.8, 0.8, 0.4, 0.4, 0.8]^T * [0.11, 0.11, 0.33, 0.33, 0.11]^T}{P(Z_2 = bright|Z_1 = dark)}$$

$$P(Z_2 = bright|Z_1 = dark)$$

$$= \sum_{x_i} P(Z_2 = bright|X_2 = x_i)P(X_2 = x_i|Z_1 = dark)$$

$$= p^T(Z_2 = bright|X_2)p(X_2|Z_1 = dark)$$

$$= [0.8, 0.8, 0.4, 0.4, 0.8][0.11, 0.11, 0.33, 0.33, 0.11]^T = 0.533$$

$$p(X_2|Z_1 = dark, Z_2 = bright) = [0.167, 0.167, 0.25, 0.25, 0.167]^T$$

4. The state probability distribution after three measurements were made is

$$p(X_3|Z_1 = dark, Z_2 = bright, Z_3 = dark)$$

$$= [0.083, 0.083, 0.375, 0.375, 0.083]^T$$

The implementation of the solution in Matlab is shown in Listing 6.5. The posterior state probability distributions for three time steps are presented graphically in Fig. 6.11.

Listing 6.5 Implementation of the solution of Example 6.10

```
1 p_ZdX = [0.2 0.2 0.6 0.6 0.2];
2 p_ZbX = 1-p_ZdX;
3 p_X = ones(1,5)/5;
4
5 disp('Probability of detecting a dark tile P(Z1=dark)')
6 P_z1 = p_ZdX*p_X.'
7 disp('Posterior distribution p(X1|Z1=dark)')
8 p_Xz1 = p_ZdX.*p_X/P_z1
9
10 disp('Probability of detecting a bright tile P(Z2=bright|Z1=dark)')
```

EXAMPLE 6.10—cont'd

```
11 P_z2 = p_ZbX*p_Xz1.'
12 disp('Posterior distribution p(X2|Z1=dark,Z2=bright)')
13 p_Xz2 = p_ZbX.*p_Xz1/P_z2
14
15 disp('Probability of detecting a dark tile P(Z3=dark|Z1=dark,Z2=bright)')
16 P_z3 = p_ZdX*p_Xz2.'
17 disp('Posterior distribution p(X3|Z1=dark,Z2=bright,Z3=dark)')
18 p_Xz3 = p_ZdX.*p_Xz2/P_z3
```

```
Probability of detecting a dark tile P(Z1=dark)
P_z1 =
    0.3600
Posterior distribution p(X1|Z1=dark)
p_Xz1 =
    0.1111    0.1111    0.3333    0.3333    0.1111
Probability of detecting a bright tile P(Z2=bright|Z1=dark)
P_z2 =
    0.5333
Posterior distribution p(X2|Z1=dark,Z2=bright)
p_Xz2 =
    0.1667    0.1667    0.2500    0.2500    0.1667
Probability of detecting a dark tile P(Z3=dark|Z1=dark,Z2=bright)
P_z3 =
    0.4000
Posterior distribution p(X3|Z1=dark,Z2=bright,Z3=dark)
p_Xz3 =
    0.0833    0.0833    0.3750    0.3750    0.0833
```

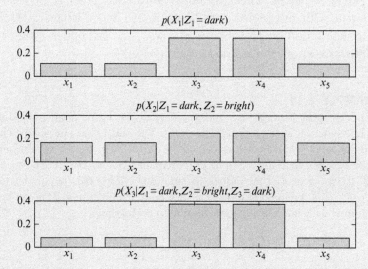

Fig. 6.11 Posterior state probability distributions in three time steps from the fourth case in Example 6.9.

6.4.6 Motion in the Environment

Mobile systems can move in the environment using some actuators (e.g., motorized wheels) and a control system. Every movement has small or large uncertainty; therefore, the movement of the mobile system through the environment increases the uncertainty about the mobile system state (pose) in the environment.

Consider that we are standing in well-known environment. We close our eyes and make several steps. After a few steps we know approximately where we are, since we know how big are our steps and we know in which direction the steps were made. Therefore, we can imagine where we are. However, the lengths of our steps are not known precisely, and also the directions of the steps are hard to estimate; therefore, our knowledge about our pose in space decreases with time as more and more steps are made.

In the case of movement without observing the states through measurement, the relation (6.28) can be reformulated:

$$p(x_k|u_{0:k-1}) = \int_{-\infty}^{+\infty} p(x_k|x_{k-1}, u_{k-1})p(x_{k-1}|u_{1:k-2}) \, dx_{k-1}$$

The belief in the new state $p(x_k|u_{0:k-1})$ depends on the belief in the previous time step $p(x_{k-1}|u_{0:k-2})$ and conditional state transition probability $p(x_k|x_{k-1}, u_{k-1})$. The probability distribution $p(x_k|, u_{0:k-1})$ can be determined with integration (or summation in the discrete case) of all possible state transition probabilities $p(x_k|x_{k-1}, u_{k-1})$ from previous states x_{k-1} into state x_k, given the known action u_{k-1}.

EXAMPLE 6.11

Consider again Example 6.8 and assume that the initial position of the mobile system is in the first cell ($X_0 = x_1$). The initial state can be written with probability distribution $p(X_0) = [1, 0, 0, 0, 0]$. The mobile system can move between cells, the outcome of movement action is correct in 80%, in 10% the movement of the mobile system is one cell less than required, and in 10% the mobile system moves one cell more than required. This can be described with the following state transition probabilities:

$$P(X_k = x_i|X_{k-1} = x_j, U_{k-1} = u) = 0.8; \quad \text{for } i = j + u$$
$$P(X_k = x_i|X_{k-1} = x_j, U_{k-1} = u) = 0.1; \quad \text{for } i = j + u - 1$$
$$P(X_k = x_i|X_{k-1} = x_j, U_{k-1} = u) = 0.1; \quad \text{for } i = j + u + 1$$

EXAMPLE 6.11—cont'd

The mobile system has to make a movement for two cells in the counter-clockwise direction ($U_0 = 2$). Determine the mobile system position belief after the movement is made.

Solution

The probability distribution (belief) after the movement is made can be determined if the probabilities of the mobile system positions in every cell are calculated (total probability). The mobile system can arrive into the first cell only from cell 3 (too long move), cell 4 (correct move), and cell 5 (too short move). This yields the probability distribution of the transition into the first cell $p(X_1 = x_1|X_0, U_0 = 2) = [0, 0, 0.1, 0.8, 0.1]^T$. After the movement the mobile system is in the first cell with probability

$$P(X_1 = x_1|U_0 = 2) = \sum_{x_i} P(X_1 = x_1|X_0 = x_i, U_0 = 2)P(X_0 = x_i)$$

$$= p^T(X_1 = x_1|X_0, U_0 = 2)p(X_0)$$

$$= [0, 0, 0.1, 0.8, 0.1][1, 0, 0, 0, 0]^T = 0$$

The probability that the mobile system after the movement is in the second cell is

$$P(X_1 = x_2|U_0 = 2) = \sum_{x_i} P(X_1 = x_2|X_0 = x_i, U_0 = 2)P(X_0 = x_i)$$

$$= p^T(X_1 = x_2|X_0, U_0 = 2)p(X_0)$$

$$= [0.1, 0, 0, 0.1, 0.8][1, 0, 0, 0, 0]^T = 0.1$$

Similarly, the probabilities of all the other cells can be calculated:

$$P(X_1 = x_3|U_0 = 2) = [0.8, 0.1, 0, 0, 0.1][1, 0, 0, 0, 0]^T = 0.8$$

$$P(X_1 = x_4|U_0 = 2) = [0.1, 0.8, 0.1, 0, 0][1, 0, 0, 0, 0]^T = 0.1$$

$$P(X_1 = x_5|U_0 = 2) = [0, 0.1, 0.8, 0.1, 0][1, 0, 0, 0, 0]^T = 0$$

The mobile position belief after the movement is therefore

$$p(X_1|U_0 = 2) = [0, 0.1, 0.8, 0.1, 0]^T$$

The posterior state probability distributions are presented graphically in Fig. 6.12. The implementation of the solution in Matlab is shown in Listing 6.6.

(Continued)

EXAMPLE 6.11—cont'd

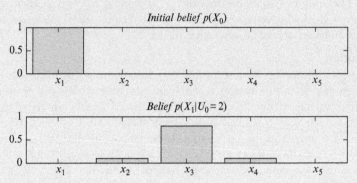

Fig. 6.12 Posterior state probability distributions in two time steps from Example 6.11.

Listing 6.6 Implementation of the solution of Example 6.11

```
1  disp('Initial belief p(X0)')
2  p_X0 = [1 0 0 0 0]
3
4  P_xxu_null = 0.8; % P(X=i|X'=j,U'=u), i=j+u
5  P_xxu_less = 0.1; % P(X=i|X'=j,U'=u), i=j+u-1
6  P_xxu_more = 0.1; % P(X=i|X'=j,U'=u), i=j+u+1
7
8  disp('Belief p(X1|U0=2)');
9  p_xXu = [0 0 P_xxu_more P_xxu_null P_xxu_less]; % for U=2
10 p_Xu = zeros(1,5);
11 for i=1:5
12     p_Xu(i) = p_xXu*p_X0.';
13     p_xXu = p_xXu([end 1:end-1]);
14 end
15 p_X1 = p_Xu

   Initial belief p(X0)
   p_X0 =
        1      0      0      0      0
   Belief p(X1|U0=2)
   p_X1 =
        0    0.1000    0.8000    0.1000         0
```

EXAMPLE 6.12

What is the belief of mobile system position if after the movement made in Example 6.11 the mobile system makes another move in the counter-clockwise direction, but this time only for a single cell ($U_1 = 1$)?

EXAMPLE 6.12—cont'd
Solution
The probability distribution (belief) after the movement can again be determined if the probability that the mobile system is in a particular cell is calculated for every cell (total probability). In the case of a movement for a single cell, the first cell can be reached from cell 1 (too short move), cell 4 (too long move), and cell 5 (correct move). The mobile system can arrive to the second cell from cells 1, 2, and 5, and so on. After the movement is made, the following probabilities can be calculated:

$$P(X_2 = x_1 | U_0 = 2, U_1 = 1) = [0.1, 0, 0, 0.1, 0.8][0, 0.1, 0.8, 0.1, 0]^T = 0.01$$

$$P(X_2 = x_2 | U_0 = 2, U_1 = 1) = [0.8, 0.1, 0, 0, 0.1][0, 0.1, 0.8, 0.1, 0]^T = 0.01$$

$$P(X_2 = x_3 | U_0 = 2, U_1 = 1) = [0.1, 0.8, 0.1, 0, 0][0, 0.1, 0.8, 0.1, 0]^T = 0.16$$

$$P(X_2 = x_4 | U_0 = 2, U_1 = 1) = [0, 0.1, 0.8, 0.1, 0][0, 0.1, 0.8, 0.1, 0]^T = 0.66$$

$$P(X_2 = x_5 | U_0 = 2, U_1 = 1) = [0, 0, 0.1, 0.8, 0.1][0, 0.1, 0.8, 0.1, 0]^T = 0.16$$

The position belief after the second movement is

$$p(X_2 | U_0 = 2, U_1 = 1) = [0.01, 0.01, 0.16, 0.66, 0.16]^T$$

and it is also shown in the bottom of Fig. 6.13. Note that the mobile system is most probably in cell 4. However, the probability distribution does not have as significant peak as it was before the second movement action was made (compare the middle with the bottom probability distribution in Fig. 6.13; the peak dropped from 80% to 66%). This observation is in accordance with the statement that every movement (action) increases the level of uncertainty about the states of the environment.

The implementation of the solution in Matlab is shown in Listing 6.7.

Listing 6.7 Implementation of the solution of Example 6.12

```
1 disp('Initial belief p(X0)')
2 p_X0 = [1 0 0 0 0]
3
4 P_xxu_null = 0.8; % P(X=i|X'=j,U'=u), i=j+u
5 P_xxu_less = 0.1; % P(X=i|X'=j,U'=u), i=j+u-1
6 P_xxu_more = 0.1; % P(X=i|X'=j,U'=u), i=j+u+1
7
8 disp('Belief p(X1|U0=2)');
9 p_xXu = [0 0 P_xxu_more P_xxu_null P_xxu_less]; % for U=2
10 p_Xu = zeros(1,5);
11 for i=1:5
12     p_Xu(i) = p_xXu*p_X0.';
13     p_xXu = p_xXu([end 1:end-1]);
14 end
15 p_X1 = p_Xu
16
17 disp('Belief p(X2|U1=1)');
18 p_xXu = [P_xxu_less 0 0 P_xxu_more P_xxu_null]; % for U=1
```

(*Continued*)

EXAMPLE 6.12—cont'd

```
19 p_Xu = zeros(1,5);
20 for i=1:5
21     p_Xu(i) = p_xXu*p_X1.';
22     p_xXu = p_xXu([end 1:end-1]);
23 end
24 p_X2 = p_Xu

Initial belief p(X0)
p_X0 =
     1     0     0     0     0
Belief p(X1|U0=2)
p_X1 =
     0   0.1000   0.8000   0.1000     0
Belief p(X2|U1=1)
p_X2 =
   0.0100   0.0100   0.1600   0.6600   0.1600
```

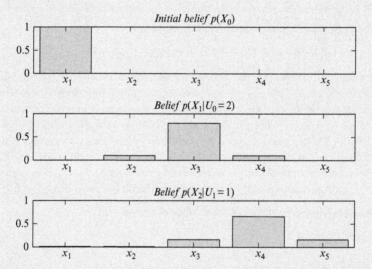

Fig. 6.13 Posterior state probability distributions in three time steps from Example 6.12.

EXAMPLE 6.13

Consider that the mobile system in Example 6.11 is initially in the first cell, $p(X_0) = [1, 0, 0, 0, 0]$. In every time step the mobile system makes one step in counter-clockwise direction.

EXAMPLE 6.13—cont'd

1. What is the belief into mobile system position after 10 time steps?
2. To which value does the belief converge after an infinite number of time steps?

Solution

1. The state belief after 10 time steps is

$$p(X_{10}|U_{0:9}) = [0.29, 0.22, 0.13, 0.13, 0.22]^T$$

2. After an infinite number of time steps a uniform distribution is obtained, since all the cells become equally possible to be occupied by the mobile system:

$$p(X_\infty|U_{0:\infty}) = [0.2, 0.2, 0.2, 0.2, 0.2]^T$$

These results were also validated in Matlab (Listing 6.8) and are shown graphically in Fig. 6.14.

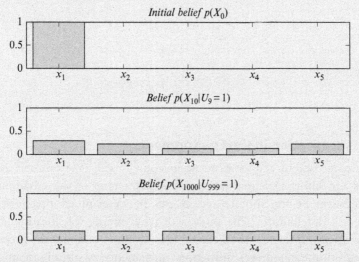

Fig. 6.14 Posterior state probability distributions in three time steps from Example 6.13.

Listing 6.8 Implementation of the solution of Example 6.13

```
1 disp('Initial belief p(X0)')
2 p_X0 = [1 0 0 0 0]
3
4 P_xxu_null = 0.8; % P(X=i|X'=j,U'=u), i=j+u
5 P_xxu_less = 0.1; % P(X=i|X'=j,U'=u), i=j+u-1
```

(*Continued*)

EXAMPLE 6.13—cont'd

```
6 P_xxu_more = 0.1; % P(X=i|X'=j,U'=u), i=j+u+1
7
8 p_X = p_X0;
9 for k=1:1000
10     p_xXu = [P_xxu_less 0 0 P_xxu_more P_xxu_null]; % for U=1
11     p_Xu = zeros(1,5);
12     for i=1:5
13         p_Xu(i) = p_xXu*p_X.';
14         p_xXu = p_xXu([end 1:end-1]);
15     end
16     p_X = p_Xu;
17     if k==10
18         disp('Belief p(X10|U9=1)');
19         p_X10 = p_X
20     elseif k==1000
21         disp('Belief p(X1000|U999=1)');
22         p_X1000 = p_X
23     end
24 end

Initial belief p(X0)
p_X0 =
     1     0     0     0     0
Belief p(X10|U9=1)
p_X10 =
     0.2949    0.2243    0.1283    0.1283    0.2243
Belief p(X1000|U999=1)
p_X1000 =
     0.2000    0.2000    0.2000    0.2000    0.2000
```

6.4.7 Localization in the Environment

The mobile system can estimate its location in the environment even if it does not know its initial location but has a map of the environment. The location of the mobile system can be determined precisely with probability distribution. The process of determining the location in the environment is known as localization. Localization combines the process of observation (measurement) and action (movement). As already mentioned, measurements made in the environment increase the knowledge about the location, but the movement of the mobile system through the environment decreases this information.

Localization is a process in which the mobile system repeatedly updates the probability distribution that represents the knowledge about the mobile system's location in the environment. The peak in the probability distribution (if it exists) represents the most probable mobile system location.

The localization process is a realization of the Bayesian filter (Algorithm 3), which combines the processes of movement and perception.

EXAMPLE 6.14

Consider the mobile system that moves around the environment as shown in Example 6.8. The mobile system first makes a move and then observes the environment. The initial pose of the mobile system is not known. This can be described with uniform probability distribution $p(X_0) = bel(X_0) = [0.2, 0.2, 0.2, 0.2, 0.2]$.

The movement action for u_k cells in the counter-clockwise direction is accurate in 80%; in 10% the movement is either a cell shorter or longer than required:

$$p(X_k = x_i | X_{k-1} = x_j, U_{k-1} = u_{k-1}) = 0.8; \quad \text{for } i = j + u_{k-1}$$
$$p(X_k = x_i | X_{k-1} = x_j, U_{k-1} = u_{k-1}) = 0.1; \quad \text{for } i = j + u_{k-1} - 1$$
$$p(X_k = x_i | X_{k-1} = x_j, U_{k-1} = u_{k-1}) = 0.1; \quad \text{for } i = j + u_{k-1} + 1$$

The mobile system detects a dark cell correctly with probability 0.6, and the probability of detecting a bright cell correctly is 0.8. This can be written down in mathematical form as

$$P(Z = dark | X = dark) = 0.6, \qquad P(Z = bright | X = dark) = 0.4$$
$$P(Z = bright | X = bright) = 0.8, \qquad P(Z = dark | X = bright) = 0.2$$

In every time step, the mobile system receives a command of moving for a single cell in the counter-clockwise direction ($u_{k-1} = 1$). The sequence of the first three measurements is $z_{1:3} = [bright, dark, dark]$.
1. What is the belief in the first time step $k = 1$?
2. What is the belief in the second time step $k = 2$?
3. What is the belief in the third time step $k = 3$?
4. In which cell is the mobile system most likely after the third step?

Solution

After every movement is made a prediction step of the Bayesian filter (Algorithm 3) is evaluated, and a correction step after the measurement is made.
1. The prediction step is evaluated based on the known movement action. The small symbol x_i, $i \in \{1, \ldots, 5\}$ denotes that the location (state) of the mobile system is in cell i, and big symbol X_k denotes the vector of all possible states in time step k.

(Continued)

EXAMPLE 6.14—cont'd

$$bel_p(X_1 = x_1) = \sum_{x_i} P(X_1 = x_1|X_0 = x_i, u_0)bel(X_0 = x_i)$$

$$= \boldsymbol{p}^T(X_1 = x_1|X_0, u_0)\boldsymbol{bel}(X_0)$$

$$= [0.1, 0, 0, 0.1, 0.8][0.2, 0.2, 0.2, 0.2, 0.2]^T = 0.2$$

$$bel_p(X_1 = x_2) = [0.8, 0.1, 0, 0, 0.1][0.2, 0.2, 0.2, 0.2, 0.2]^T = 0.2$$

$$bel_p(X_1 = x_3) = [0.1, 0.8, 0.1, 0, 0][0.2, 0.2, 0.2, 0.2, 0.2]^T = 0.2$$

$$bel_p(X_1 = x_4) = [0, 0.1, 0.8, 0.1, 0][0.2, 0.2, 0.2, 0.2, 0.2]^T = 0.2$$

$$bel_p(X_1 = x_5) = [0, 0, 0.1, 0.8, 0.1][0.2, 0.2, 0.2, 0.2, 0.2]^T = 0.2$$

Therefore, the complete probability distribution (belief) of the prediction step is

$$\boldsymbol{bel}_p(X_1) = [0.2, 0.2, 0.2, 0.2, 0.2]^T$$

After the measurement is obtained the correction step of the Bayesian filter is evaluated:

$$bel(X_1 = x_1) = \eta\, p(Z_1 = bright|x_1)\, bel_p(X_1 = x_1) = \eta\, 0.8 \cdot 0.2 = \eta\, 0.16$$

$$bel(X_1 = x_2) = \eta\, p(Z_1 = bright|x_2)\, bel_p(X_1 = x_2) = \eta\, 0.8 \cdot 0.2 = \eta\, 0.16$$

$$bel(X_1 = x_3) = \eta\, p(Z_1 = bright|x_3)\, bel_p(X_1 = x_3) = \eta\, 0.4 \cdot 0.2 = \eta\, 0.08$$

$$bel(X_1 = x_4) = \eta\, p(Z_1 = bright|x_4)\, bel_p(X_1 = x_4) = \eta\, 0.4 \cdot 0.2 = \eta\, 0.08$$

$$bel(X_1 = x_5) = \eta\, p(Z_1 = bright|x_5)\, bel_p(X_1 = x_5) = \eta\, 0.8 \cdot 0.2 = \eta\, 0.16$$

After considering the normalization factor,

$$\eta = \frac{1}{0.16 + 0.16 + 0.08 + 0.08 + 0.16} = 1.56$$

the updated probability distribution (belief) is obtained:

$$\boldsymbol{bel}(X_1) = [0.25, 0.25, 0.125, 0.125, 0.25]^T$$

The same result can be obtained from the following:

$$\boldsymbol{bel}(X_1) = \frac{\boldsymbol{p}^T(Z_1 = bright|X_1) * \boldsymbol{bel}_p^T(X_1)}{\boldsymbol{p}^T(Z_1 = bright|X_1)\boldsymbol{bel}_p(X_1)}$$

$$= \frac{[0.8, 0.8, 0.4, 0.4, 0.8] * [0.2, 0.2, 0.2, 0.2, 0.2]}{[0.8, 0.8, 0.4, 0.4, 0.8][0.2, 0.2, 0.2, 0.2, 0.2]^T}$$

$$= [0.25, 0.25, 0.125, 0.125, 0.25]^T$$

2. The procedure from the first case can be repeated again on the last result to obtain state belief in time step $k = 1$. First, a prediction step is repeated:

EXAMPLE 6.14—cont'd

$$bel_p(X_2 = x_1) = [0.1, 0, 0, 0.1, 0.8][0.25, 0.25, 0.125, 0.125, 0.25]^T = 0.237$$

$$bel_p(X_2 = x_2) = [0.8, 0.1, 0, 0, 0.1][0.25, 0.25, 0.125, 0.125, 0.25]^T = 0.25$$

$$bel_p(X_2 = x_3) = [0.1, 0.8, 0.1, 0, 0][0.25, 0.25, 0.125, 0.125, 0.25]^T = 0.237$$

$$bel_p(X_2 = x_4) = [0, 0.1, 0.8, 0.1, 0][0.25, 0.25, 0.125, 0.125, 0.25]^T = 0.138$$

$$bel_p(X_2 = x_5) = [0, 0, 0.1, 0.8, 0.1][0.25, 0.25, 0.125, 0.125, 0.25]^T = 0.138$$

The complete probability distribution of prediction is

$$bel_p(X_2) = [0.237, 0.25, 0.237, 0.138, 0.138]^T$$

The correction step yields

$$bel(X_2) = \frac{[0.2, 0.2, 0.6, 0.6, 0.2]^T * [0.237, 0.25, 0.237, 0.138, 0.138]^T}{[0.2, 0.2, 0.6, 0.6, 0.2][0.237, 0.25, 0.237, 0.138, 0.138]^T}$$

$$= [0.136, 0.143, 0.407, 0.236, 0.079]^T$$

3. Similarly as in the previous two cases, the belief distribution can be obtained for time step $k = 3$:

$$bel_p(X_3) = [0.1, 0.131, 0.167, 0.363, 0.237]^T$$

$$bel(X_3) = [0.048, 0.063, 0.245, 0.528, 0.115]^T$$

4. After the third time step, the mobile system is most likely in the fourth cell, with probability 52.8%. The second most likely cell is the third cell, with probability 24.5%.

The state beliefs for all three time steps are presented graphically in Fig. 6.15. The implementation of the solution in Matlab is shown in Listing 6.9.

Listing 6.9 Implementation of the solution of Example 6.14

```
1  disp('Initial belief p(X0)')
2  bel_X0 = ones(1,5)/5
3
4  P_xxu_null = 0.8; % P(X=i|X'=j,U'=u), i=j+u
5  P_xxu_less = 0.1; % P(X=i|X'=j,U'=u), i=j+u-1
6  P_xxu_more = 0.1; % P(X=i|X'=j,U'=u), i=j+u+1
7
8  p_ZdX = [0.2 0.2 0.6 0.6 0.2]; % p(Z=dark|X)
9  p_ZbX = 1-p_ZdX;               % p(Z=bright|X)
10
11 bel_X = bel_X0;
12 for k=1:3
13     % Prediction step
14     p_xXu = [P_xxu_less 0 0 P_xxu_more P_xxu_null]; % for U=1
15     belp_X = zeros(1,5);
16     for i=1:5
17         belp_X(i) = p_xXu*bel_X.';
```

(Continued)

EXAMPLE 6.14—cont'd

```
18        p_xXu = p_xXu([end 1:end-1]);
19    end
20
21    % Correction step
22    if k==1
23        bel_X = p_ZbX.*belp_X;
24    else
25        bel_X = p_ZdX.*belp_X;
26    end
27    bel_X = bel_X/sum(bel_X);
28
29    if k==1
30        disp('Beliefs belp_X1 and bel_X1')
31        belp_X1 = belp_X
32        bel_X1 = bel_X
33    elseif k==2
34        disp('Beliefs belp_X2 and bel_X2')
35        belp_X2 = belp_X
36        bel_X2 = bel_X
37    elseif k==3
38        disp('Beliefs belp_X3 and bel_X3')
39        belp_X3 = belp_X
40        bel_X3 = bel_X
41        disp('Less likely to most likely position')
42        [m,mi] = sort(bel_X)
43    end
44 end
```

```
Initial belief p(X0)
bel_X0 =
    0.2000    0.2000    0.2000    0.2000    0.2000
Beliefs belp_X1 and bel_X1
belp_X1 =
    0.2000    0.2000    0.2000    0.2000    0.2000
bel_X1 =
    0.2500    0.2500    0.1250    0.1250    0.2500
Beliefs belp_X2 and bel_X2
belp_X2 =
    0.2375    0.2500    0.2375    0.1375    0.1375
bel_X2 =
    0.1357    0.1429    0.4071    0.2357    0.0786
Beliefs belp_X3 and bel_X3
belp_X3 =
    0.1000    0.1307    0.1686    0.3636    0.2371
bel_X3 =
    0.0484    0.0633    0.2450    0.5284    0.1149
Less likely to most likely position
m =
    0.0484    0.0633    0.1149    0.2450    0.5284
mi =
      1    2    5    3    4
```

EXAMPLE 6.14—cont'd

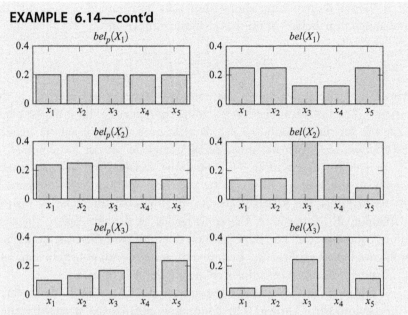

Fig. 6.15 Posterior state probability distributions in three time steps from Example 6.14.

6.5 KALMAN FILTER

Kalman filter [6] is one of the most important state estimation and prediction algorithms, which has been applied to a diverse range of applications in various engineering fields, and autonomous mobile systems are no exception. The Kalman filter is designed for state estimation of linear systems where the system signals may be corrupted by noise. The algorithm has a typical two-step structure that consists of a prediction and a correction step that are evaluated in every time step. In the prediction step, the latest system state along with state uncertainties are predicted. Once a new measurement is available, the correction step is evaluated where the stochastic measurement is joined with the predicted state estimate as a weighted average, in a way that less uncertain values are given a greater weight. The algorithm is recursive and allows an online estimation of the current system state taking into account system and measurement uncertainties.

A classical Kalman filter assumes normally distributed noises, that is, the probability distribution of noise is a Gaussian function:

$$p(x) = \frac{1}{\sqrt{2\pi\sigma^2}} e^{-\frac{1}{2}\frac{(x-\mu)^2}{\sigma^2}} \tag{6.30}$$

where μ is the mean value (mathematical expectation) and σ^2 is the variance. The Gaussian function is a unimodal (left and right from the single peak the function monotonically decreases toward zero)—more general distributions are normally multimodal (there are several local peaks). If the probability distributions of continuous variables are assumed to be unimodal, the Kalman filter can be used for optimum state estimation. In the case that the variables are not all unimodal, the state estimation is suboptimal; furthermore, the convergence of the estimate to the true value is questionable. The Bayesian filter does not have the aforementioned problems, but its applicability is limited to simple continuous problems and to discrete problems with a finite countable number of states.

In Fig. 6.16 an example of continuous probability distribution, which is not unimodal, is shown. The continuous distribution is approximated with a Gaussian function and with a histogram (domain is divided into discrete intervals). The approximation with a Gaussian function is used in the Kalman filter, and the histogram is used in the Bayesian filter.

The essence of the correction step (see Bayesian filter (6.29)) is information fusion from two independent sources, that is, sensor measurements and state predictions based on the previous state estimations. Let us use Example 6.15 again to demonstrate how two independent estimates of the same variable x can be jointed optimally if the value and variance (belief) of each source is known.

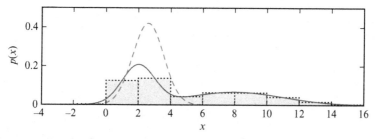

Fig. 6.16 An example of probability distribution of a continuous variable x (*solid line*), approximation with Gaussian function (*dashed line*), and approximation with a histogram (*dotted line*).

EXAMPLE 6.15

There are two independent estimates of the variable x. The value of the first estimate is x_1 and has a variance σ_1^2, and the value of the second estimate is x_2 with a variance σ_2^2. What is the optimal linear combination of these two estimates that represent the state estimate $\hat{x}$ with minimal variance?

Solution

The estimation of optimal value of the variable x is assumed to be a linear combination of two measurements:

$$\hat{x} = \omega_1 x_1 + \omega_2 x_2$$

where the parameters ω_1 and ω_2 are the unknown weights that satisfy the following condition: $\omega_1 + \omega_2 = 1$. The optimum values of the weights should minimize the variance σ^2 of the optimal estimate $\hat{x}$. Hence, the variance is as follows:

$$\sigma^2 = \mathrm{E}\{(\hat{x} - \mathrm{E}\{\hat{x}\})^2\}$$
$$= \mathrm{E}\{(\omega_1 x_1 + \omega_2 x_2 - \mathrm{E}\{\omega_1 x_1 + \omega_2 x_2\})^2\}$$
$$= \mathrm{E}\{(\omega_1 x_1 + \omega_2 x_2 - \omega_1 \mathrm{E}\{x_1\} - \omega_2 \mathrm{E}\{x_2\})^2\}$$
$$= \mathrm{E}\{(\omega_1 (x_1 - \mathrm{E}\{x_1\}) + \omega_2 (x_2 - \mathrm{E}\{x_2\}))^2\}$$
$$= \mathrm{E}\{\omega_1^2 (x_1 - \mathrm{E}\{x_1\})^2 + \omega_2^2 (x_2 - \mathrm{E}\{x_2\})^2 + 2\omega_1\omega_2(x_1 - \mathrm{E}\{x_1\})(x_2 - \mathrm{E}\{x_2\})\}$$
$$= \omega_1^2 \mathrm{E}\{(x_1 - \mathrm{E}\{x_1\})^2\} + \omega_2^2 \mathrm{E}\{(x_2 - \mathrm{E}\{x_2\})^2\} + 2\omega_1\omega_2 \mathrm{E}\{(x_1 - \mathrm{E}\{x_1\})(x_2 - \mathrm{E}\{x_2\})\}$$
$$= \omega_1^2 \sigma_1^2 + \omega_2^2 \sigma_2^2 + 2\omega_1\omega_2 \mathrm{E}\{(x_1 - \mathrm{E}\{x_1\})(x_2 - \mathrm{E}\{x_2\})\}$$

Since the variables x_1 and x_2 are independent, the differences $x_1 - \mathrm{E}\{x_1\}$ and $x_2 - \mathrm{E}\{x_2\}$ are also independent, and therefore $\mathrm{E}\{(x_1 - \mathrm{E}\{x_1\})(x_2 - \mathrm{E}\{x_2\})\} = 0$. Hence,

$$\sigma^2 = \omega_1^2 \sigma_1^2 + \omega_2^2 \sigma_2^2$$

or after introducing $\omega_2 = \omega$ and $\omega_1 = 1 - \omega$,

$$\sigma^2 = (1 - \omega)^2 \sigma_1^2 + \omega^2 \sigma_2^2$$

We are seeking the value of the weight ω that minimizes the variance that can be obtained from variance derivative:

$$\frac{\partial}{\partial \omega}\sigma^2 = -2(1 - \omega)\sigma_1^2 + 2\omega\sigma_2^2 = 0$$

which yields the solution

$$\omega = \frac{\sigma_1^2}{\sigma_1^2 + \sigma_2^2}$$

(*Continued*)

EXAMPLE 6.15—cont'd

The minimum-variance estimate is therefore

$$\hat{x} = \frac{\sigma_2^2 x_1 + \sigma_1^2 x_2}{\sigma_1^2 + \sigma_2^2} \tag{6.31}$$

and the minimum variance is

$$\sigma^2 = \frac{\sigma_1^2 \sigma_2^2}{\sigma_1^2 + \sigma_2^2} = \left(\frac{1}{\sigma_1^2} + \frac{1}{\sigma_2^2} \right)^{-1} \tag{6.32}$$

The obtained results confirm that the source with lower variance (higher belief) contributes more to the final estimate, and vice-versa.

EXAMPLE 6.16

In a particular moment in time, an initial state estimate is given $x = 2$ with variance $\sigma^2 = 4$. Then a sensor is used to measure the value of the state, which is $z = 4$ with sensor variance $\sigma_z^2 = 1$. The Gaussian probability distributions of the state and measurement are shown in Fig. 6.17.

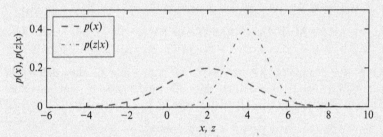

Fig. 6.17 Probability distribution of the state (*dashed line*) and the measurement (*dash-dotted line*).

What is the value of optimum state estimate that includes information from previous state estimate and current measurement? What is the probability distribution of the updated optimum state estimate?

Solution

Based on Fig. 6.17 we can foreknow that the mean value x' of the updated state will be closer to the measurement mean value, since the measurement variance (uncertainty) is lower than previous estimate variance. Using

EXAMPLE 6.16—cont'd

Eq. (6.31) the updated state estimate is obtained:

$$x' = \frac{\sigma_z^2 x + \sigma^2 z}{\sigma^2 + \sigma_z^2} = 3.6$$

Variance of the updated estimate σ'^2 is lower than both previous variances, since the integration of the previous estimate and measurement information lower the uncertainty of the updated estimate. The variance of the updated estimate is obtained from Eq. (6.32):

$$\sigma'^2 = \left(\frac{1}{\sigma^2} + \frac{1}{\sigma_z^2} \right)^{-1} = 0.8$$

and the standard deviation is

$$\sigma' = \sqrt{\sigma'^2} = 0.894$$

The updated probability distribution $p(x|z)$ of the state after the measurement-based correction is shown in Fig. 6.18.

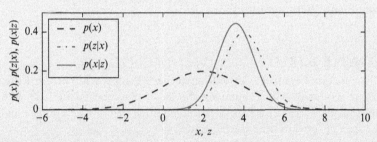

Fig. 6.18 Probability distribution of the initial state (*dashed line*), measurement (*dash-dotted line*), and updated state (*solid line*).

Let us use the findings from Example 6.15 in the derivation of the recursive state estimation algorithm. In every time step a new state measurement $z(k) = x(k) + n(k)$ is obtained by the sensor, where $n(k)$ is the measurement noise. The measurement variance $\sigma_z^2(k)$ is assumed to be known. The updated optimum state estimate is a combination of the previous estimate $\hat{x}(k)$ and the current measurement $z(k)$, as follows:

$$\hat{x}(k+1) = (1 - \omega)\,\hat{x}(k) + \omega(k)z(k) = \hat{x}(k) + \omega(z(k) - \hat{x}(k))$$

The updated state variance is

$$\sigma^2(k+1) = \frac{\sigma^2(k)\sigma_z^2(k)}{\sigma^2(k)+\sigma_z^2(k)} = (1-\omega)\sigma^2(k)$$

where

$$\omega = \frac{\sigma^2(k)}{\sigma^2(k)+\sigma_z^2(k)}$$

Therefore, given known initial state estimate x, $\hat{x}(0)$ and the corresponding variance $\sigma^2(0)$ the measurements $z(1), z(2), \ldots$ can be integrated optimally, in a way that the current state and state variance are estimated. This is the basic idea behind the correction step of the Kalman filter.

The prediction step of the Kalman filter provides the given state prediction a known input action. The initial state estimate $\hat{x}(k)$ has a probability distribution with variance $\sigma^2(k)$. In the same way, the action $u(k)$, which is responsible for transition of the state $x(k)$ to $x(k+1)$, has probability distribution (transition uncertainty) $\sigma_u^2(k)$. Using Example 6.17 let us examine the value of the state and the variance after the action is executed (after state transition).

EXAMPLE 6.17

The initial state estimate $\hat{x}(k)$ with variance $\sigma^2(k)$ is known. Then an action $u(k)$ is executed, which represents the direct transition of the state with uncertainty (variance) $\sigma_u^2(k)$. What is the value of the state estimate and the state uncertainty after the transition?

Solution

The updated state estimate after the transition is

$$\hat{x}(k+1) = \hat{x}(k) + u(k)$$

and the uncertainty of this estimation is

$$\sigma^2(k+1) = \mathrm{E}\left\{\left(\hat{x}(k+1) - \mathrm{E}\{\hat{x}(k+1)\}\right)^2\right\}$$

$$= \mathrm{E}\left\{\left(\hat{x}(k) + u(k) - \mathrm{E}\{\hat{x}(k) + u(k)\}\right)^2\right\}$$

$$= \mathrm{E}\left\{\left((\hat{x}(k) - \mathrm{E}\{\hat{x}(k)\}) + (u(k) - \mathrm{E}\{u(k)\})\right)^2\right\}$$

EXAMPLE 6.17—cont'd

$$= \mathrm{E}\big\{(\hat{x}(k) - \mathrm{E}\{\hat{x}(k)\})^2 + (u(k) - \mathrm{E}\{u(k)\})^2\big\}$$
$$+ \mathrm{E}\big\{2(\hat{x}(k) - \mathrm{E}\{\hat{x}(k)\})(u(k) - \mathrm{E}\{u(k)\})\big\}$$
$$= \sigma^2(k) + \sigma_u^2(k)$$
$$+ \mathrm{E}\big\{2\left(\hat{x}(k) - \mathrm{E}\{\hat{x}(k)\}\right)(u(k) - \mathrm{E}\{u(k)\})\big\} \qquad (6.33)$$

Since $\hat{x}$ and u are independent, it holds $\mathrm{E}\big\{2\left(\hat{x}(k) - \mathrm{E}\{\hat{x}(k)\}\right)(u(k) - \mathrm{E}\{u(k)\})\big\} = 0$, and therefore Eq. (6.33) simplifies to

$$\sigma^2(k+1) = \sigma^2(k) + \sigma_u^2(k)$$

Simplified Implementation of the Kalman Filter Algorithm

The Kalman filter algorithm for a simple case with only a single state is given in Algorithm 4, where the variables with a subscript $(\cdot)_{k|k-1}$ represent the estimated values in the prediction step, and variables with subscript $(\cdot)_{k|k}$ represent the values from the correction step. For improved readability the following notation is used: $u(k-1) = u_{k-1}$ and $z(k) = z_k$.

ALGORITHM 4   Kalman Filter for a Single State

function KALMAN_FILTER($\hat{x}_{k-1|k-1}$, u_{k-1}, z_k, $\sigma^2_{k-1|k-1}$, $\sigma_{u_{k-1}}^2$, $\sigma_{z_k}^2$)

Prediction step:
$$\hat{x}_{k|k-1} \leftarrow \hat{x}_{k-1|k-1} + u_{k-1}$$
$$\sigma^2_{k|k-1} \leftarrow \sigma^2_{k-1|k-1} + \sigma_{u_{k-1}}^2$$

Correction step:
$$\omega_k \leftarrow \frac{\sigma^2_{k|k-1}}{\sigma^2_{k|k-1} + \sigma_{z_k}^2}$$
$$\hat{x}_{k|k} \leftarrow \hat{x}_{k|k-1} + \omega_k(z_k - \hat{x}_{k|k-1})$$
$$\sigma^2_{k|k} \leftarrow (1 - \omega_k)\sigma^2_{k|k-1}$$

return $\hat{x}_{k|k}$, $\sigma^2_{k|k}$
end function

The Kalman filter has two steps (prediction and correction step) that are executed one after another in the loop. In the prediction step only the known action is used in a way that enables prediction of the state in the next time step. From the initial belief a new belief is evaluated, and the uncertainty of the new belief is higher that the initial uncertainty. In the correction step the measurement is used to improve the predicted belief in a way that the new (corrected) state estimate has lower uncertainty than the previous belief. In both steps only two inputs are required: in the prediction step the value of previous belief $\hat{x}_{k-1|k-1}$ and executed action u_{k-1} need to be known, and in the correction step the previous belief $\hat{x}_{k|k-1}$ and measurement z_k are required. The state transition variance $\sigma^2_{k-1|k-1}$, input action variance $\sigma^2_{u_{k-1}}$, and measurement variance $\sigma^2_{z_k}$ also need to be given.

EXAMPLE 6.18

There is a mobile robot that can move in only one dimension. The initial position of the robot is unknown (Fig. 6.19). Let us therefore assume that the initial position is $\hat{x}_0 = 3$ with large variance $\sigma^2_0 = 100$ (the true position $x_0 = 0$ is not known).

Then, in every time moment $k - 1 = 0, \ldots, 4$ the mobile robot is moved for $u_{0:4} = (2, 3, 2, 1, 1)$ units and the measurements of the robot positions in time moments $k = 1, \ldots, 5$ are taken, $z_{1:5} = (2, 5, 7, 8, 9)$. The movement action and measurement are disturbed with white normally distributed zero-mean noise that can be described with a constant uncertainty for movement $\sigma^2_u = 2$ and measurement uncertainty $\sigma^2_z = 4$.

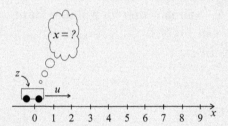

Fig. 6.19 Localization of a mobile robot in one-dimensional space with unknown initial position.

What is the estimated robot position and uncertainty of this estimate?

Solution

Let us apply Algorithm 4 to solve the given mobile robot localization problem. In the initial time step ($k = 1$) the predicted state and variance can be calculated first:

$$\hat{x}_{1|0} = \hat{x}_{0|0} + u_0 = 3 + 2 = 5$$
$$\sigma^2_{1|0} = \sigma^2_{0|0} + \sigma_u{}^2 = 100 + 2 = 102$$

EXAMPLE 6.18—cont'd

and then the correction step of the Kalman filter in the first time step $k = 1$ can be evaluated:

$$\omega_1 = \frac{\sigma_{1|0}^2}{\sigma_{1|0}^2 + \sigma_z^2} = \frac{102}{102 + 4} = 0.962$$

$$\hat{x}_{1|1} = \hat{x}_{1|0} + \omega_1(z_1 - \hat{x}_{1|0}) = 5 + 0.962(2 - 5) = 2.113$$

$$\sigma_{1|1}^2 = (1 - \omega_1)\sigma_{1|0}^2 = (1 - 0.962)102 = 3.849$$

These prediction and correction steps can be evaluated for all the other time steps. The prediction results are therefore

$$\hat{x}_{1:5|0:4} = (5.00, 5.11, 7.05, 8.02, 9.01)$$

$$\sigma_{1:5|0:4}^2 = (102, 5.85, 4.38, 4.09, 4.02)$$

and the correction results are

$$\hat{x}_{1:5|1:5} = (2.11, 5.05, 7.02, 8.01, 9.01)$$

$$\sigma_{1:5|1:5}^2 = (3.85, 2.38, 2.09, 2.02, 2.01)$$

The obtained results show that the position of the mobile robot can be determined in a few time steps with uncertainty 2, which agrees with the prediction uncertainty and measurement uncertainty $\left(\frac{1}{\sigma_{5|4}^2} + \frac{1}{\sigma_z^2}\right)^{-1} = 2.01$. The uncertainty of the predicted position estimate converges toward 4, which is in accordance with the correction uncertainty from the previous time step and measurement $\sigma_{4|4}^2 + \sigma_u^2 = 4.02$.

6.5.1 Kalman Filter in Matrix Form

Multiple input, multiple state, and multiple output systems can be represented in a matrix form for improved readability. A general linear system can be expressed in a state-space form as

$$x(k+1) = Ax(k) + Bu(k) + Fw(k)$$
$$z(k) = Cx(k) + v(k) \tag{6.34}$$

where x is the state vector, u is the input (action) vector, and z is the output (measurement); A is the state matrix, B is the input matrix, F is the input noise matrix, C is the output matrix; $w(k)$ is the process noise and v is the output (measurement) noise. In the case the noise w is added to the system input u, the following relation holds: $F = B$. The process noise $w(k)$ and measurement noise $v(k)$ are assumed to be independent (uncorrelated) white noises with zero mean value and covariance matrices $Q_k = \mathrm{E}\{w(k)w^T(k)\}$ and $R_k = \mathrm{E}\{v(k)v^T(k)\}$.

The probability distribution of the states x that are disturbed by a white Gaussian noise can be written in a matrix form as follows:

$$p(x) = \det\,(2\pi\,P)^{-\frac{1}{2}}\,e^{-\frac{1}{2}(x-\mu)^T P^{-1}(x-\mu)}$$

where P is the state-error covariance matrix.

The Kalman filter is an approach for filtering and estimation of linear systems with continuous state space, which are disturbed with normal noise. The noise distribution is represented with a Gaussian function (Gaussian noise). The input and measurement noises influence the internal system states that we would like to estimate. In the case that the model of the system is linear, the Gaussian noise propagated through the model (e.g., from inputs to the states) is also a Gaussian noise. The system must therefore be linear, since this requirement ensures Gaussian distribution of the noise on the states, an assumption used in the derivation of the Kalman filter. The Kalman filter state estimate converges to the true value only in the case of linear systems that are disturbed with Gaussian noise.

The Kalman filter for a linear system (6.34) has a prediction step:

$$\begin{aligned}\hat{x}_{k|k-1} &= A\hat{x}_{k-1|k-1} + Bu_{k-1}\\ P_{k|k-1} &= AP_{k-1|k-1}A^T + FQ_{k-1}F^T\end{aligned}$$

(6.35)

and a correction step:

$$\begin{aligned}K_k &= P_{k|k-1}C^T\left(CP_{k|k-1}C^T + R_k\right)^{-1}\\ \hat{x}_{k|k} &= \hat{x}_{k|k-1} + K_k(z_k - C\hat{x}_{k|k-1})\\ P_{k|k} &= P_{k|k-1} - K_kCP_{k|k-1}\end{aligned}$$

(6.36)

In the prediction part of the algorithm the a priori estimate $\hat{x}_{k|k-1}$ is determined, which is based on the previous estimate $\hat{x}_{k-1|k-1}$ (obtained from measurements up to the time moment $k-1$) and input $u(k-1)$. In the correction part of the Kalman filter the posteriori estimate $\hat{x}_{k|k}$ is calculated, which is based on the measurements up to time step k. The state correction is made in a way that the difference between the true and estimated measurement is calculated ($z_k - C\hat{x}_{k|k-1}$); this difference is also known as innovation or measurement residual. The state correction is calculated as a product of Kalman gain K_k and innovation. The prediction part can be evaluated in advance, in the time while we are waiting for the new measurement in the time step k. Notice the similarity of the matrix notation in Eqs. (6.35), (6.36) with the notation used in Algorithm 4.

Let us derive the equation for the calculation of the state error covariance matrix in the prediction part of the Kalman filter:

$$
\begin{aligned}
\boldsymbol{P}_{k|k-1} &= \mathrm{E}\big\{(\boldsymbol{x}_k - \hat{\boldsymbol{x}}_{k|k-1})(\boldsymbol{x}_k - \hat{\boldsymbol{x}}_{k|k-1})^T\big\} \\
&= \mathrm{cov}\big\{\boldsymbol{x}_k - \hat{\boldsymbol{x}}_{k|k-1}\big\} \\
&= \mathrm{cov}\big\{\boldsymbol{A}\boldsymbol{x}_{k-1} + \boldsymbol{B}\boldsymbol{u}_{k-1} + \boldsymbol{F}\boldsymbol{w}_{k-1} - \boldsymbol{A}\hat{\boldsymbol{x}}_{k-1|k-1} - \boldsymbol{B}\boldsymbol{u}_{k-1}\big\} \\
&= \mathrm{cov}\big\{\boldsymbol{A}\boldsymbol{x}_{k-1} + \boldsymbol{F}\boldsymbol{w}_{k-1} - \boldsymbol{A}\hat{\boldsymbol{x}}_{k-1|k-1}\big\} \\
&= \mathrm{cov}\big\{\boldsymbol{A}(\boldsymbol{x}_{k-1} - \hat{\boldsymbol{x}}_{k-1|k-1}) + \boldsymbol{F}\boldsymbol{w}_{k-1}\big\} \\
&= \mathrm{cov}\big\{\boldsymbol{A}(\boldsymbol{x}_{k-1} - \hat{\boldsymbol{x}}_{k-1|k-1})\big\} + \mathrm{cov}\big\{\boldsymbol{F}\boldsymbol{w}_{k-1}\big\} \\
&= \mathrm{E}\big\{(\boldsymbol{A}(\boldsymbol{x}_{k-1} - \hat{\boldsymbol{x}}_{k-1|k-1}))(\boldsymbol{A}(\boldsymbol{x}_{k-1} - \hat{\boldsymbol{x}}_{k-1|k-1}))^T\big\} \\
&\quad + \mathrm{E}\big\{(\boldsymbol{F}\boldsymbol{w}_{k-1})(\boldsymbol{F}\boldsymbol{w}_{k-1})^T\big\} \\
&= \mathrm{E}\big\{(\boldsymbol{A}(\boldsymbol{x}_{k-1} - \hat{\boldsymbol{x}}_{k-1|k-1}))(\boldsymbol{x}_{k-1} - \hat{\boldsymbol{x}}_{k-1|k-1})^T\boldsymbol{A}^T\big\} \\
&\quad + \mathrm{E}\big\{\boldsymbol{F}\boldsymbol{w}_{k-1}\boldsymbol{w}_{k-1}^T\boldsymbol{F}^T\big\} \\
&= \boldsymbol{A}\boldsymbol{P}_{k-1|k-1}\boldsymbol{A}^T + \boldsymbol{F}\boldsymbol{Q}_{k-1}\boldsymbol{F}^T
\end{aligned}
$$

where in line six of the derivation we took into account that the process noise $\boldsymbol{w}_k$ in time step k is independent of the state estimate error in the previous time step $(\boldsymbol{x}_{k-1} - \hat{\boldsymbol{x}}_{k-1|k-1})$.

Let us also derive the equation for computation of the state error covariance matrix in the correction part of the Kalman filter:

$$
\begin{aligned}
\boldsymbol{P}_{k|k} &= \mathrm{E}\big\{(\boldsymbol{x}_k - \hat{\boldsymbol{x}}_{k|k})(\boldsymbol{x}_k - \hat{\boldsymbol{x}}_{k|k})^T\big\} \\
&= \mathrm{cov}\big\{\boldsymbol{x}_k - \hat{\boldsymbol{x}}_{k|k}\big\} \\
&= \mathrm{cov}\big\{\boldsymbol{x}_k - \hat{\boldsymbol{x}}_{k|k-1} - \boldsymbol{K}_k(\boldsymbol{z}_k - \boldsymbol{C}\hat{\boldsymbol{x}}_{k|k-1})\big\} \\
&= \mathrm{cov}\big\{\boldsymbol{x}_k - \hat{\boldsymbol{x}}_{k|k-1} - \boldsymbol{K}_k(\boldsymbol{C}\boldsymbol{x}_k + \boldsymbol{v}_k - \boldsymbol{C}\hat{\boldsymbol{x}}_{k|k-1})\big\} \\
&= \mathrm{cov}\big\{(\boldsymbol{I} - \boldsymbol{K}_k\boldsymbol{C})(\boldsymbol{x}_k - \hat{\boldsymbol{x}}_{k|k-1}) - \boldsymbol{K}_k\boldsymbol{v}_k\big\} \\
&= \mathrm{cov}\big\{(\boldsymbol{I} - \boldsymbol{K}_k\boldsymbol{C})(\boldsymbol{x}_k - \hat{\boldsymbol{x}}_{k|k-1})\big\} + \mathrm{cov}\big\{\boldsymbol{K}_k\boldsymbol{v}_k\big\} \\
&= (\boldsymbol{I} - \boldsymbol{K}_k\boldsymbol{C})\boldsymbol{P}_{k|k-1}(\boldsymbol{I} - \boldsymbol{K}_k\boldsymbol{C})^T + \boldsymbol{K}_k\boldsymbol{R}_k\boldsymbol{K}_k^T,
\end{aligned}
$$

where in line six of the derivation we took into account that the measurement noise $\boldsymbol{v}_k$ is uncorrelated with the other terms. The obtained relation for calculation of the covariance matrix $\boldsymbol{P}_{k|k}$ is general and can be used for arbitrary gain $\boldsymbol{K}_k$. However, the expression for $\boldsymbol{P}_{k|k}$ in Eq. (6.36) is valid only for optimum gain (Kalman gain) that minimizes the mean squared correction error $\mathrm{E}\left\{\left|x_k - \hat{x}_{k|k}\right|^2\right\}$; this is equivalent to the minimization of the sum of all the diagonal elements in the correction covariance matrix $P_{k|k}$.

The general equation for $P_{k|k}$ can be extended and the terms rearranged:

$$\begin{aligned}
P_{k|k} &= (I - K_k C) P_{k|k-1} (I - K_k C)^T + K_k R_k K_k^T \\
&= P_{k|k-1} - K_k C P_{k|k-1} - P_{k|k-1} C^T K_k^T + K_k C P_{k|k-1} C^T K_k^T + K_k R_k K_k^T \\
&= P_{k|k-1} - K_k C P_{k|k-1} - P_{k|k-1} C^T K_k^T + K_k (C P_{k|k-1} C^T + R_k) K_k^T \\
&= P_{k|k-1} - K_k C P_{k|k-1} - P_{k|k-1} C^T K_k^T + K_k S_k K_k^T
\end{aligned}$$

where $S_k = C P_{k|k-1} C^T + R_k$ represents the innovation covariance matrix ($S_k = \mathrm{cov}\left\{z_k - C\hat{x}_{k|k-1}\right\}$). The sum of the diagonal terms of $P_{k|k}$ is minimal when the derivative of $P_{k|k}$ with respect to K_k is zero:

$$\frac{\partial P_{k|k}}{\partial K_k} = -2(C P_{k|k-1})^T + 2 K_k S_k = 0$$

which leads to the optimum gain in Eq. (6.36):

$$K_k = P_{k|k-1} C^T S_k^{-1} = P_{k|k-1} C^T (C P_{k|k-1} C^T + R_k)^{-1}$$

The correction covariance matrix at optimum gain can be derived if the optimum gain is postmultiplied with $S_k K_k^T$ and inserted into the equation for $P_{k|k}$:

$$\begin{aligned}
P_{k|k} &= P_{k|k-1} - K_k C P_{k|k-1} - P_{k|k-1} C^T K_k^T + K_k S_k K_k^T \\
&= P_{k|k-1} - K_k C P_{k|k-1} - P_{k|k-1} C^T K_k^T + P_{k|k-1} C^T K_k^T \\
&= P_{k|k-1} - K_k C P_{k|k-1}
\end{aligned}$$

EXAMPLE 6.19

There is a mobile robot that drives around on a plane and measures its position with a GPS sensor 10 times in a second (sampling time $T_s = 0.1\,\mathrm{s}$). The position measurement is disturbed with zero-mean Gaussian noise with variance $10\ \mathrm{m}^2$. The mobile robot drives with the speed of $1\ \mathrm{m/s}$ in the x direction and with the speed of $0\ \mathrm{m/s}$ in the y direction. The variance of Gaussian noise of the velocity is $0.1\ \mathrm{m}^2/\mathrm{s}^2$. At the beginning of observation, the mobile robot is situated in the origin $x = [0,\ 0]^T$, but our initial estimate about the mobile robot's position is $\hat{x} = [3,\ 3]^T$ with initial variance

EXAMPLE 6.19—cont'd

$$P_0 = \begin{bmatrix} 10 & 0 \\ 0 & 10 \end{bmatrix}$$

What is the time evolution of the position estimate and the variance of the estimate?

Solution

The solution of the problem can be obtained with the aid of simulation tools provided in Matlab. Let us determine the robot's movement model where the state estimate represents the robot's position in a plane $\hat{x}_{k|k-1} = [x_k, \ y_k]^T$ and input $u = [v_x, \ v_y]^T$ represents the robot's speed in x and y directions, respectively. The model for system state prediction is therefore

$$\hat{x}_{k|k-1} = \begin{bmatrix} 1 & 0 \\ 0 & 1 \end{bmatrix} \hat{x}_{k-1|k-1} + \begin{bmatrix} T_s & 0 \\ 0 & T_s \end{bmatrix} u$$

The position measurement model of a GPS sensor is

$$\hat{z}_k = \begin{bmatrix} 1 & 0 \\ 0 & 1 \end{bmatrix} \hat{x}_{k|k-1}$$

Listing 6.10 provides an implementation of the solution in Matlab. The simulation results are shown graphically in Figs. 6.20–6.22. The simulation results confirm that the algorithm provides a position estimate that is converging to the true robot position from the false initial estimate. This is expected since the mean of the measurement noise is zero. However, the estimate variance is decreasing with time and falls much below the measurement variance (the internal kinematic model is the trusted mode). The Kalman filter enables optimum fusion of data from different sources (internal kinematic model and external measurement model) if variances of these sources are known. The reader should feel free to experiment with system parameters and observe the estimate variance evolution and convergence rate.

Listing 6.10 Implementation of the solution of Example 6.19

```
1  % Linear model of the system in state-space
2  Ts = 0.1; % Sampling time
3  A = [1 0; 0 1];
4  B = [Ts 0; 0 Ts];
5  C = [1 0; 0 1];
6  F = B; % Noise is added directly to the input.
7
8  xTrue = [0; 0]; % True initial state
9  x = [3; 3]; % Initial state estimate
10 P = diag([10 10]); % Variance of the initial state estimate
11 Q = diag([1 1]/10); % Noise variance of the movement actuator
12 R = diag([10 10]); % Noise vairace of the GPS measurements
13
```

(*Continued*)

EXAMPLE 6.19—cont'd

```
14  figures_kf1_init;
15
16  % Loop
17  N = 150;
18  for k = 1:N
19      u = [1; 0]; % Movement command
20
21      % Simulation of the true mobile robot position and measurement
22      xTrue = A*xTrue + B*u + F*sqrt(Q)*randn(2, 1);
23      zTrue = C*xTrue + sqrt(R)*randn(2, 1);
24
25      % Position estimate based on known inputs and measurements
26      %%% Prediction
27      xPred = A*x + B*u;
28      PPred = A*P*A.' + F*Q*F.';
29
30      %%% Correction
31      K = PPred*C.'/(C*PPred*C.' + R);
32      x = xPred + K*(zTrue - C*xPred);
33      P = PPred - K*C*PPred;
34
35      figures_kf1_update;
36  end
```

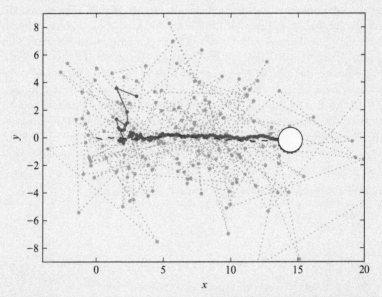

Fig. 6.20 True (*dashed line*) and estimated (*solid line*) trajectory with measurements (*dotted line*) from Example 6.19. The final position of the robot is marked with a *circle*.

EXAMPLE 6.19—cont'd

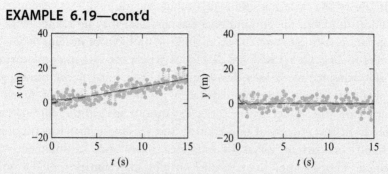

Fig. 6.21 True position (*dashed line*) and position estimate (*solid line*) of the mobile robot based on measurements (*dotted line*) from Example 6.19.

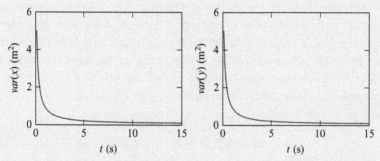

Fig. 6.22 Time evolution of mobile robot position variance from Example 6.19.

6.5.2 Extended Kalman Filter

The Kalman filter was developed for linear systems, and it is normally assumed that all disturbances and noises can be described with a zero-mean Gaussian distribution. Noise with a Gaussian distribution is invariant to linear transformations, or in other words, if Gaussian noise is transformed by a linear function, it remains Gaussian noise; it only has different noise parameters that can be calculated explicitly given the known linear function. This is the reason behind the computational efficiency of the Kalman filter. In the case the input Gaussian noise is transformed by a nonlinear function, the output noise is no longer Gaussian, but it can normally still be approximated with a Gaussian noise.

If any of the state transitions or output equations of the system are a nonlinear function, the conventional Kalman filter may no longer produce an optimum state estimate. To overcome the problem of nonlinearities, an *extended Kalman filter* (EKF) was developed where the system nonlinearities are approximated with local linear models. The local linear model can be obtained from a nonlinear system with the procedure of function linearization (first-order Taylor series expansion) around the current state estimate. In the process of linearization the sensitivity matrices (Jacobian matrices) for current values of estimated states and measurements are obtained. The obtained linear model enables approximation of the noise that is not necessarily Gaussian with a Gaussian noise distribution.

The use of noise linearization in the process of noise modeling enables a computationally efficient implementation of the EKF algorithm. Hence, the EKF algorithm is commonly used in practice. The accuracy of linear approximation depends on the noise variance (big uncertainties or noise amplitudes may lead to poor linear approximation, since the signal may be outside the linear region) and the level of nonlinearity. Due to the error that is caused by linearization the convergence of the filter can worsen or the estimation may not even converge to the true solution.

A nonlinear system may be written in a general form:

$$\begin{aligned} x_k &= f\left(x_{k-1}, u_{k-1}, w_{k-1}\right) \\ z_k &= h\left(x_k\right) + v_k \end{aligned} \tag{6.37}$$

where the noise w_k can occur at the system input or influence the states directly.

The EKF for a nonlinear system (6.37) has a prediction part:

$$\begin{aligned} \hat{x}_{k|k-1} &= f\left(\hat{x}_{k-1|k-1}, u_{k-1}\right) \\ P_{k|k-1} &= A P_{k-1|k-1} A^T + F Q_{k-1} F^T \end{aligned} \tag{6.38}$$

and a correction part:

$$\begin{aligned} K_k &= P_{k|k-1} C^T \left(C P_{k|k-1} C^T + R_k\right)^{-1} \\ \hat{x}_{k|k} &= \hat{x}_{k|k-1} + K_k(z_k - \hat{z}_k) \\ P_{k|k} &= P_{k|k-1} - K_k C P_{k|k-1} \end{aligned} \tag{6.39}$$

In the prediction part (6.38) the nonlinear state transition system is used for calculation of state prediction estimate. To calculate the noise covariance matrix, the Jacobian matrix A that describes the noise propagation from previous to current states and the Jacobian matrix F that describes the noise propagation from inputs to the states need to be determined from the state model (6.37):

$$A = \left.\frac{\partial f}{\partial \hat{x}}\right|_{(\hat{x}_{k-1|k-1}, u_{k-1})} \tag{6.40}$$

$$F = \left.\frac{\partial f}{\partial w}\right|_{(\hat{x}_{k-1|k-1}, u_{k-1})} \tag{6.41}$$

In the correction part (6.39) the measurement estimate is made based on predicted state estimate $\hat{z}_k = h\left(\hat{x}_{k|k-1}\right)$. The Jacobian matrix C that describes the noise propagation from states to outputs also needs to be determined:

$$C = \left.\frac{\partial h}{\partial x}\right|_{(\hat{x}_{k|k-1})} \tag{6.42}$$

The noise covariance matrices are $Q_k = \mathrm{E}\{w(k)w^T(k)\}$ and $R_k = \mathrm{E}\{v(k)v^T(k)\}$. Among many applications, the EKF has been successfully applied for solving the wheeled mobile robot localization problem [7, 8] and map building [9].

EXAMPLE 6.20

A wheeled mobile robot with a differential drive moves on a plane. The robot input commands are the translational velocity v_k and angular velocity ω_k that are both disturbed independently by Gaussian noise with zero mean and variances var$\{v_k\} = 0.1\,\mathrm{m}^2/\mathrm{s}^2$ and var$\{\omega_k\} = 0.1\,\mathrm{rad}^2/\mathrm{s}^2$, respectively.

The mobile robot has a sensor that enables measurement of the distance to the marker that is positioned in the origin of the global coordinate frame. The robot is also equipped with a compass that enables measurement of the mobile robot's orientation (heading). The considered setup is presented in Fig. 6.23. The distance measurement is disturbed by a Gaussian noise with zero mean and variance $0.5\,\mathrm{m}^2$, and the angle measurement is also disturbed by Gaussian noise with zero mean and variance $0.3\,\mathrm{rad}^2$.

(Continued)

EXAMPLE 6.20—cont'd

At the beginning of observation the true initial pose of the robot is $x_0 = [1, 2, \pi/6]^T$; however, the initial estimated pose is $\hat{x}_0 = [3, 0, 0]^T$ with initial variance:

$$P_0 = \begin{bmatrix} 9 & 0 & 0 \\ 0 & 9 & 0 \\ 0 & 0 & 0.6 \end{bmatrix}$$

What is the time evolution of the mobile robot pose estimate ($\hat{x}_{k|k-1} = [x_k, y_k, \varphi_k]^T$) and the variance of the estimation if the command $u = [v_k, \omega_k]^T = [0.5, 0.5]^T$ is sent to the robot in every sampling time step $T_s = 0.1$ s?

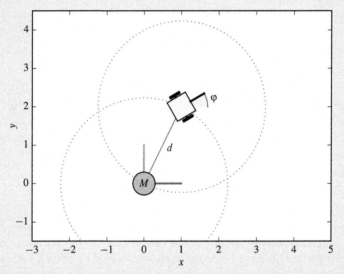

Fig. 6.23 Setup considered in Example 6.20. The mobile robot has a sensor for measuring the distance to marker M (positioned in the origin of the global frame) and a compass to determine the robot orientation in the global frame (heading).

Solution

The solution can be obtained with the aid of simulation tools provided in Matlab. Let us determine the kinematic motion model of the mobile robot. In this case the kinematic model is nonlinear:

$$\hat{x}_{k|k-1} = \hat{x}_{k-1|k-1} + \begin{bmatrix} T_s v_{k-1} \cos(\varphi_{k-1}) \\ T_s v_{k-1} \sin(\varphi_{k-1}) \\ T_s \omega_{k-1} \end{bmatrix}$$

EXAMPLE 6.20—cont'd
The distance and angle measurement model is

$$\hat{z}_k = \left[\begin{matrix} \sqrt{x_k^2 + y_k^2} \\ \varphi_k \end{matrix} \right]$$

Listing 6.11 provides an implementation of the solution in Matlab. The simulation results are shown in Figs. 6.24–6.28. Simulation results confirm that the algorithm can provide an estimate of the pose that converges to the true pose of the mobile robot although the initial estimate was biased. The innovation terms settle around zero.

Listing 6.11 Implementation of the solution of Example 6.20

```
1  Ts = 0.1; % Sampling time
2  xTrue = [1; 2; pi/6]; % True initial pose
3  x = [3; 0; 0]; % Initial pose estimate
4  P = diag([9 9 0.6]); % Initial covariance matrix of the pose estimate
5  Q = diag([0.1 0.1]); % Noise covariance matrix of movement actuator
6  R = diag([0.5 0.3]); % Noise covariance matrix of distance and
7                       % angle measurement
8  enableNoise = 1; % Enable noise: 0 or 1
9  N = 300; % Number of simulation sample times
10 figures_ekf1_init;
11
12 % Loop
13 for k = 1:N
14     u = [0.5; 0.5]; % Movement command (translational and angular velocity)
15     uNoisy = u + sqrt(Q)*randn(2, 1)*enableNoise;
16
17     % Simulation of the true mobile robot state (pose)
18     xTrue = xTrue + Ts*[uNoisy(1)*cos(xTrue(3)); ...
19                         uNoisy(1)*sin(xTrue(3)); ...
20                         uNoisy(2)];
21     xTrue(3) = wrapToPi(xTrue(3));
22
23     % Simulation of the true noisy measurements
24     zTrue = [sqrt(xTrue(1)^2 + xTrue(2)^2 ); ...
25             xTrue(3)] + sqrt(R)*randn(2, 1)*enableNoise;
26     zTrue(1) = abs(zTrue(1));
27     zTrue(2) = wrapToPi(zTrue(2));
28
29     %%% Prediction (pose and speed estimation based on known inputs)
30     xPred = x + Ts*[u(1)*cos(x(3)); ...
31                     u(1)*sin(x(3)); ...
32                     u(2)];
33     xPred(3) = wrapToPi(xPred(3));
34
35     % Jacobian matrices
36     A = [1 0 -Ts*u(1)*sin(x(3)); ...
37          0 1  Ts*u(1)*cos(x(3)); ...
38          0 0  1];
39     F = [Ts*cos(x(3)) 0; ...
40          Ts*sin(x(3)) 0; ...
```

(Continued)

EXAMPLE 6.20—cont'd

```
41        0                Ts];
42    PPred = A*P*A.' + F*Q*F.';
43
44    % Estimated measurements
45    z = [sqrt(xPred(1)^2 + xPred(2)^2); ...
46        xPred(3)];
47
48    %%% Correction
49    d = sqrt(xPred(1)^2 + xPred(2)^2);
50    C = [xPred(1)/d xPred(2)/d 0;...
51        0            0          1];
52    K = PPred*C.'/(C*PPred*C.' + R);
53    inov = zTrue - z;
54    inov(2) = wrapToPi(inov(2));
55    x = xPred + K*inov;
56    P = PPred - K*C*PPred;
57
58    figures_ekf1_update;
59 end
```

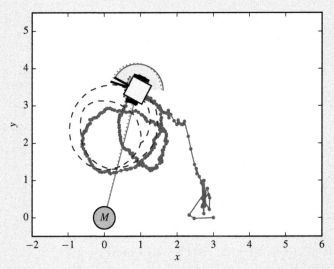

Fig. 6.24 True (*dashed line*) and estimated (*solid line*) trajectory from Example 6.20.

EXAMPLE 6.20—cont'd

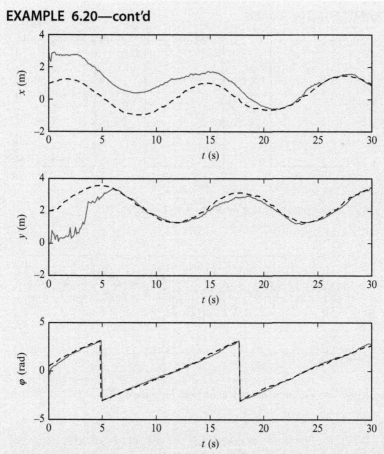

Fig. 6.25 Pose estimate (*solid line*) and true state (*dashed line*) of the mobile robot with a nonzero initial estimation error from Example 6.20.

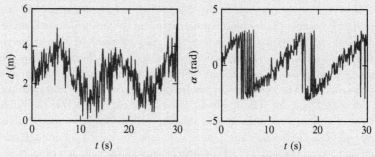

Fig. 6.26 Distance and angle measurements from Example 6.20.

(*Continued*)

EXAMPLE 6.20—cont'd

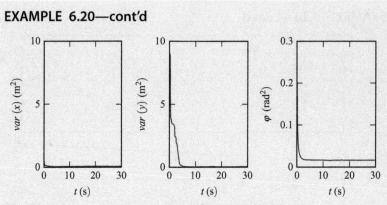

Fig. 6.27 Robot pose estimate variances from Example 6.20.

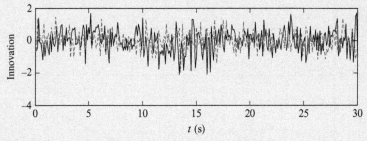

Fig. 6.28 Time evolution of innovation from Example 6.20.

When the states of the system are not measurable directly (as in this example), the question of system state observability arises. Observability analysis approaches can normally provide only sufficient conditions for system observability. However, observability analysis should be performed before designing a state estimation, since it can aid the designer choosing the appropriate set of measurement signals that make the estimation feasible. Although advanced mathematical tools could be used to check system observability, the observability analysis can also be performed with a simple graphical check that is based on the definition of indistinguishable states (see Section 6.3.3). The observability analysis of the system from this example is made in Fig. 6.29. The results in Fig. 6.29 demonstrate that the states of the system are normally distinguishable, except for some special cases (Fig. 6.29D). However, if the system states are observed long enough and mobile robot control inputs are adequately excited, the system is observable.

EXAMPLE 6.20—cont'd

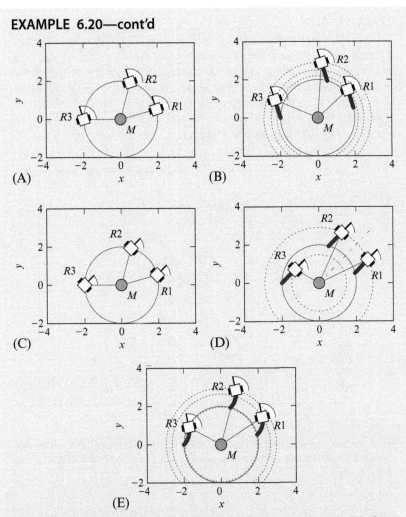

Fig. 6.29 Observability analysis of the system from Example 6.20. (A) Three particular initial robot poses that all have the same heading and distance from the marker. (B) Updated situation from Fig. (A) after the robots have traveled the same distance in a forward direction. From the measurement point of view, the robot poses are distinguishable. (C) Three particular initial robot poses that all have the same heading and distance from the marker: a special case. (D) Updated situation from Fig. (C) after the robots have traveled the same distance in a forward direction. From the measurement point of view, the robot in pose 1 is indistinguishable from the robot in pose 2, but the robot in pose 3 is distinguishable from the other two poses. (E) Updated situation from Fig. (C) after the robots have traveled the same not-straight path. From the measurement point of view, the robot poses are all distinguishable.

EXAMPLE 6.21

Let us use the same mobile robot as in Example 6.20, but let us use a slightly different sensor. Assume that the mobile robot has s sensor for measuring the distance and angle to the marker that is in the origin of the global coordinate frame (Fig. 6.30). The angle measurements are in the range $\alpha \in [-\pi, \pi]$. The distance measurement is disturbed by a Gaussian noise with zero mean and variance $0.5\,\text{m}^2$, and the angle measurement is disturbed by the Gaussian noise with zero mean and variance $0.3\,\text{rad}^2$.

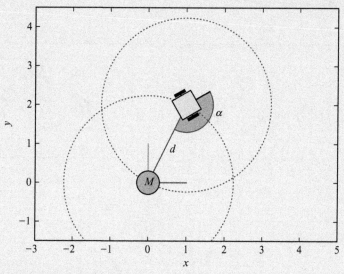

Fig. 6.30 Setup considered in Example 6.21. The mobile robot has a sensor for measuring the distance and angle to marker M (positioned in the origin of the global frame).

What is the time evolution of the robot pose estimate ($\hat{\boldsymbol{x}}_{k|k-1} = [x_k, \gamma_k, \varphi_k]^T$) and estimate variance if the input commands $\boldsymbol{u} = [v_k, \omega_k]^T = [0.5, 0.5]^T$ are sent to the robot in every sampling time step $T_s = 0.1\,\text{s}$?

Solution

The solution can be obtained with the aid of simulation in Matlab. Let us determine the mobile robot motion model. The kinematic model of the wheeled mobile robot is the same as in Example 6.20:

EXAMPLE 6.21—cont'd

$$\hat{x}_{k|k-1} = \hat{x}_{k-1|k-1} + \begin{bmatrix} T_s v_{k-1} \cos(\varphi_{k-1}) \\ T_s v_{k-1} \sin(\varphi_{k-1}) \\ T_s \omega_{k-1} \end{bmatrix}$$

In this case, the distance and angle measurement model is

$$\hat{z}_k = \begin{bmatrix} \sqrt{x_k^2 + y_k^2} \\ \text{atan2}\left(0 - y_k, 0 - x_k\right) - \varphi_k \end{bmatrix}$$

where $\text{atan2}\left(y, x\right)$ is a four-quadrant extension of $\arctan \frac{y}{x}$:

$$\text{atan2}\left(y, x\right) = \begin{cases} \arctan \frac{y}{x} & \text{if } x > 0 \\ \arctan \frac{y}{x} + \pi & \text{if } x < 0 \text{ and } y \geq 0 \\ \arctan \frac{y}{x} - \pi & \text{if } x < 0 \text{ and } y < 0 \\ \frac{\pi}{2} & \text{if } x = 0 \text{ and } y > 0 \\ -\frac{\pi}{2} & \text{if } x = 0 \text{ and } y < 0 \\ \text{undefined} & \text{if } x = 0 \text{ and } y = 0 \end{cases} \tag{6.43}$$

Listing 6.12 provides an implementation of the solution in Matlab. The simulation results in Figs. 6.31–6.35 show that the estimated states converge

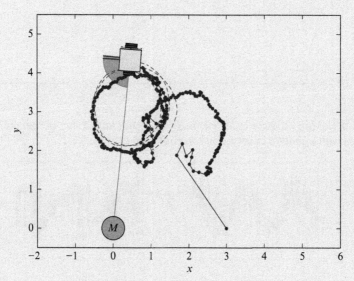

Fig. 6.31 True (*dashed line*) and estimated (*solid line*) trajectory from Example 6.21.

(*Continued*)

EXAMPLE 6.21—cont'd

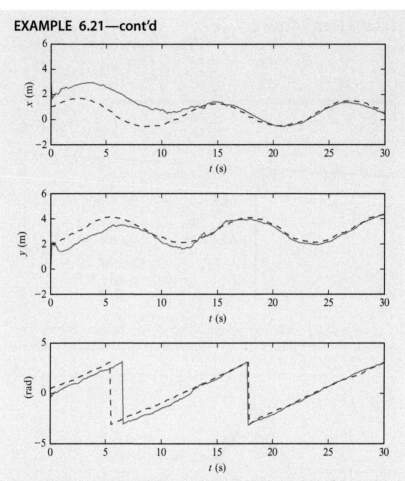

Fig. 6.32 Pose estimate (*solid line*) and true state (*dashed line*) of the mobile robot with a nonzero initial estimation error from Example 6.21.

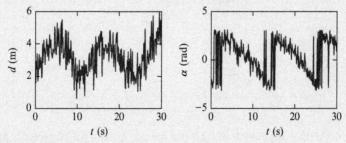

Fig. 6.33 Distance and angle measurements from Example 6.21.

EXAMPLE 6.21—cont'd

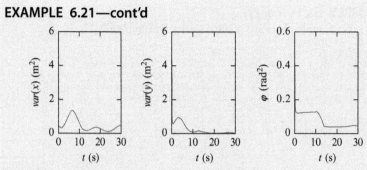

Fig. 6.34 Robot pose estimate variance from Example 6.21.

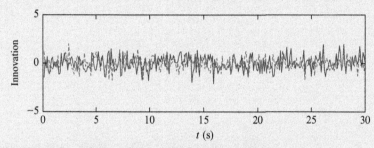

Fig. 6.35 Time evolution of innovation from Example 6.21.

to the true robot pose. Although this example is similar to Example 6.20, the estimate may not converge to the true robot pose in the presence of only slightly different environmental conditions. The case of convergence to the wrong solution is shown in Figs. 6.36–6.40. Since both sensor outputs (distance and angle) are relative measurements, the estimated states may not converge to the true solution, even though the innovation (difference between measurement and measurement prediction) is in the vicinity of a zero value (Fig. 6.40).

(*Continued*)

EXAMPLE 6.21—cont'd

Listing 6.12 Implementation of the solution of Example 6.21

```
1  Ts = 0.1; % Sampling time
2  xTrue = [1; 2; pi/6]; % True initial pose
3  x = [3; 0; 0]; % Initial pose estimate
4  P = diag([9 9 0.6]); % Initial covariance matrix of the pose estimate
5  Q = diag([0.1 0.1]); % Noise covariance matrix of movement actuator
6  R = diag([0.5 0.3]); % Noise covariance matrix of distance and
7                        % angle measurement
8  enableNoise = 1; % Enable noise: 0 or 1
9  N = 300; % Number of simulation sample times
10 figures_ekf2_init;
11
12 % Loop
13 for k = 1:N
14     u = [0.5; 0.5]; % Movement command (translational and angular velocity)
15     uNoisy = u + sqrt(Q)*randn(2, 1)*enableNoise;
16
17     % Simulation of the true mobile robot state (pose)
18     xTrue = xTrue + Ts*[uNoisy(1)*cos(xTrue(3)); ...
19                         uNoisy(1)*sin(xTrue(3)); ...
20                         uNoisy(2)];
21     xTrue(3) = wrapToPi(xTrue(3));
22
23     % Simulation of the true noisy measurements (distance and angle)
24     zTrue = [sqrt(xTrue(1)^2 + xTrue(2)^2); ...
25              atan2(0-xTrue(2), 0-xTrue(1))-xTrue(3)] + ...
26              sqrt(R)*randn(2, 1)*enableNoise;
27     zTrue(1) = abs(zTrue(1));
28     zTrue(2) = wrapToPi(zTrue(2));
29
30     %%% Prediction (pose and speed estimation based on known inputs)
31     xPred = x + Ts*[u(1)*cos(x(3)); ...
32                     u(1)*sin(x(3)); ...
33                     u(2)];
34     xPred(3) = wrapToPi(xPred(3));
35
36     % Jacobian matrices
37     A = [1 0 -Ts*u(1)*sin(x(3)); ...
38          0 1  Ts*u(1)*cos(x(3)); ...
39          0 0  1];
40     F = [Ts*cos(x(3)) 0; ...
41          Ts*sin(x(3)) 0; ...
42          0             Ts];
43     PPred = A*P*A.' + F*Q*F.';
44
45
46     % Estimated measurements
47     z= [sqrt(xPred(1)^2 + xPred(2)^2); ...
48         atan2(0-xPred(2), 0-xPred(1)) - xPred(3)];
49     z(2) = wrapToPi(z(2));
```

EXAMPLE 6.21—cont'd

```
1    %%% Correction
2    d = sqrt(xPred(1)^2 + xPred(2)^2);
3    C = [xPred(1)/d   xPred(2)/d    0; ...
4         -xPred(2)/d^2 xPred(1)/d^2 -1];
5    K = PPred*C.'/(C*PPred*C.' + R);
6    inov = zTrue - z;
7
8    % Selection of appropriate innovation due to noise and angle wrapping
9    inov(2) = wrapToPi(inov(2));
10
11   x = xPred + K*inov;
12   P = PPred - K*C*PPred;
13
14   figures_ekf2_update;
15 end
```

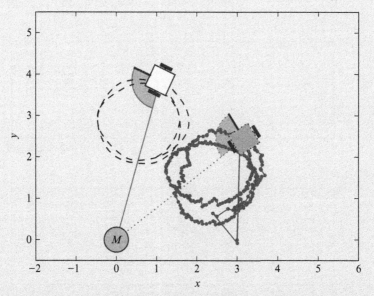

Fig. 6.36 True (*dashed line*) and estimated (*solid line*) trajectory from Example 6.21 (representative case).

A graphical check of observability made in Fig. 6.41 can provide additional insight into the reasons for estimation bias. The analyses show

(*Continued*)

EXAMPLE 6.21—cont'd

that there are states that are clearly indistinguishable from the measurement point of view, regardless of control inputs. Moreover, for every measurement there is an infinite number of states that agree with the measurement.

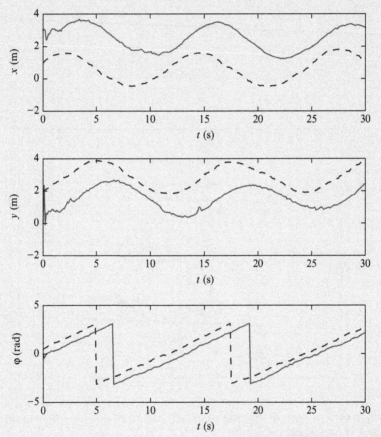

Fig. 6.37 Pose estimate (*solid line*) and true state (*dashed line*) of the mobile robot with a nonzero initial estimation error from Example 6.21 (representative case).

EXAMPLE 6.21—cont'd

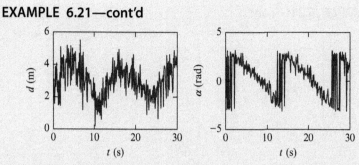

Fig. 6.38 Distance and angle measurements from Example 6.21 (representative case).

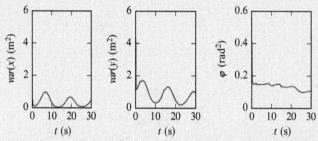

Fig. 6.39 Robot pose estimate variance from Example 6.21 (representative case).

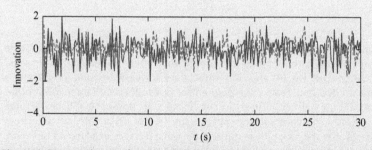

Fig. 6.40 Time evolution of innovation from Example 6.21 (representative case).

Therefore, the system is not observable. The estimation bias can be eliminated if multiple markers are observed simultaneously, since this would provide enough information to select the appropriate solution, as it is shown in Example 6.22.

(Continued)

EXAMPLE 6.21—cont'd

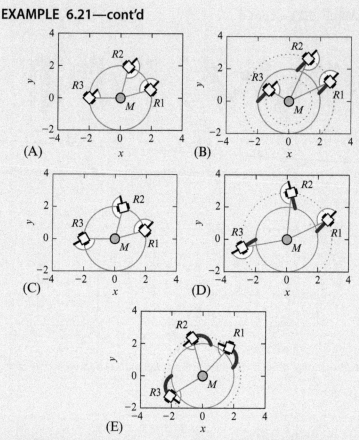

Fig. 6.41 Observability analysis of the system from Example 6.21. (A) Three particular initial robot poses with same distance from the marker. (B) Updated situation from Fig. (A) after the robots have traveled the same distance in a forward direction. From the measurement point of view, all three robot poses are distinguishable. (C) Three particular initial robot poses that all have the same angle to and distance from the marker: a special case. (D) Updated situation from Fig. (C) after the robots have traveled the same distance in a forward direction. From the measurement point of view, all three robot poses are indistinguishable. (E) Updated situation from Fig. (C) after the robots have traveled the same not-straight path. From the measurement point of view, all three robot poses are indistinguishable.

EXAMPLE 6.22

This is an upgraded Example 6.21, in a way that now two spatially separated markers, instead of only one, are observed simultaneously. In this case the sensors measure the distances and angles to two markers (with respect to the mobile robot pose), which are positioned at $x_{M1} = 0$, $y_{M1} = 0$ and at $x_{M2} = 5$, $y_{M2} = 5$ in the plane (Fig. 6.42). Otherwise, all the data are identical to Example 6.21.

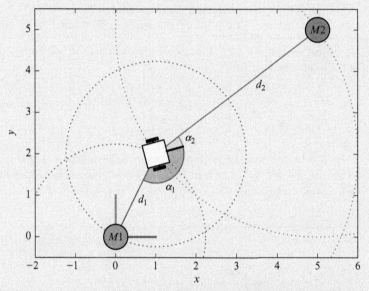

Fig. 6.42 Setup considered in Example 6.22. The mobile robot has a sensor for measuring the distance and angle to markers *M*1 and *M*2.

Solution

In this case only the correction part of the algorithm needs to be modified. The measurement vector now contains four elements, the distance and angle to the first marker, and also the distance and angle to the second marker:

$$\hat{z}_k = \begin{bmatrix} \sqrt{(x_{M1} - x_k)^2 + (y_{M1} - y_k)^2} \\ \text{atan2}\left(y_{M1} - y_k, x_{M1} - x_k\right) - \varphi_k \\ \sqrt{(x_{M2} - x_k)^2 + (y_{M2} - y_k)^2} \\ \text{atan2}\left(y_{M2} - y_k, x_{M2} - x_k\right) - \varphi_k \end{bmatrix}$$

(Continued)

EXAMPLE 6.22—cont'd

Let us determine the output matrix C (see Eq. 6.42) with linearization around the current predicted state estimate (x_k, y_k):

$$C = \begin{bmatrix} \frac{x_k}{d_1} & \frac{y_k}{d_1} & 0 \\ -\frac{y_k}{d_1^2} & \frac{x_k}{d_1^2} & -1 \\ \frac{x_k}{d_2} & \frac{y_k}{d_2} & 0 \\ -\frac{y_k}{d_2^2} & \frac{x_k}{d_2^2} & -1 \end{bmatrix}$$

where $d_1 = \sqrt{(x_{M1} - x_k)^2 + (y_{M1} - y_k)^2}$ and $d_2 = \sqrt{(x_{M2} - x_k)^2 + (y_{M2} - y_k)^2}$. Let us also determine the measurement noise covariance matrix that is

$$R = \begin{bmatrix} 0.5 & 0 & 0 & 0 \\ 0 & 0.3 & 0 & 0 \\ 0 & 0 & 0.5 & 0 \\ 0 & 0 & 0 & 0.3 \end{bmatrix}$$

Listing 6.13 provides an implementation of the solution in Matlab. The simulation results are shown in Figs. 6.43–6.47. The obtained results confirm that the estimated states converge to the true robot pose.

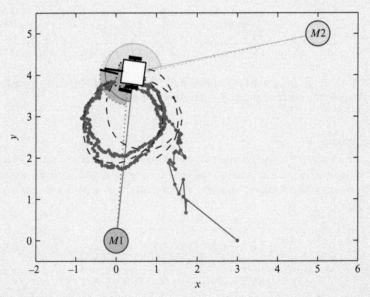

Fig. 6.43 True (*dashed line*) and estimated (*solid line*) trajectory from Example 6.22.

EXAMPLE 6.22—cont'd

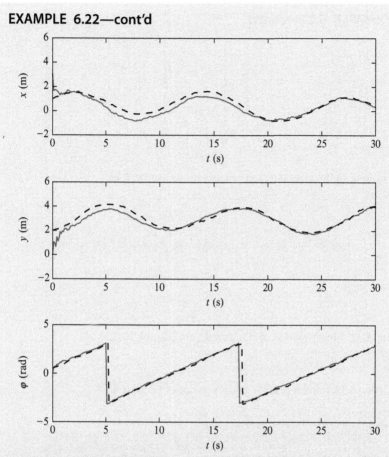

Fig. 6.44 Pose estimate (*solid line*) and true state (*dashed line*) of the mobile robot with a nonzero initial estimation error from Example 6.22.

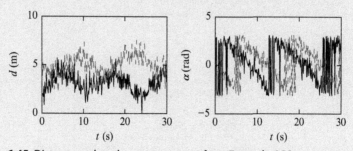

Fig. 6.45 Distance and angle measurements from Example 6.22.

(Continued)

EXAMPLE 6.22—cont'd

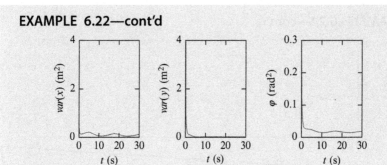

Fig. 6.46 Robot pose estimate variance from Example 6.22.

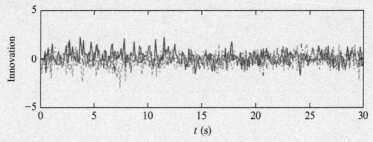

Fig. 6.47 Time evolution of innovation from Example 6.22.

Listing 6.13 Implementation of the solution of Example 6.22

```
1  Ts = 0.1; % Sampling time
2  xTrue = [1; 2; pi/6]; % True initial pose
3  x = [3; 0; 0]; % Initial pose estimate
4  P = diag([9 9 0.6]); % Initial covariance matrix of the pose estimate
5  Q = diag([0.1 0.1]); % Noise covariance matrix of movement actuator
6  R = diag([0.5 0.3]); % Noise covariance matrix of distance and
7                       %    angle measurement
8  enableNoise = 1; % Enable noise: 0 or 1
9  N = 300; % Number of simulation sample times
10 marker = [0 0; 5 5]; % Positions of markers
11
12 figures_ekf3_init;
13
14 % Loop
15 for k = 1:N
16     u = [0.5; 0.5]; % Movement command (translational and angular velocity)
17     uNoisy = u + sqrt(Q)*randn(2, 1)*enableNoise;
18
19     % Simulation of the true mobile robot state (pose)
20     xTrue = xTrue + Ts*[uNoisy(1)*cos(xTrue(3)); ...
21                         uNoisy(1)*sin(xTrue(3)); ...
22                         uNoisy(2)];
23     xTrue(3) = wrapToPi(xTrue(3));
24
25     % Simulation of the true noisy measurements (distance and angle)
```

EXAMPLE 6.22—cont'd

```
26    zTrue = [];
27    for m = 1:size(marker, 1)
28        dist = sqrt((marker(m,1)-xTrue(1))^2 + (marker(m,2)-xTrue(2))^2);
29        alpha = atan2(marker(m,2)-xTrue(2), marker(m,1)-xTrue(1))-xTrue(3));
30        zz = [dist; alpha] + sqrt(R)*randn(2, 1)*enableNoise;
31        zz(1) = abs(zz(1));
32        zz(2) = wrapToPi(zz(2));
33        zTrue = [zTrue; zz];
34    end
35
36    %%% Prediction (pose and speed estimation based on known inputs)
37    xPred = x + Ts*[u(1)*cos(x(3)); ...
38                    u(1)*sin(x(3)); ...
39                    u(2)];
40    xPred(3) = wrapToPi(xPred(3));
41
42    % Jacobian matrices
43    A = [1 0 -Ts*u(1)*sin(x(3)); ...
44         0 1  Ts*u(1)*cos(x(3)); ...
45         0 0  1];
46    F = [Ts*cos(x(3)) 0; ...
47         Ts*sin(x(3)) 0; ...
48         0            Ts];
49    PPred = A*P*A.' + F*Q*F.';
50
51    %%% Correction, measurement estimation
52    z = [];
53    C = [];
54    for m = 1:size(marker,1)
55        dist = sqrt((marker(m,1)-xPred(1))^2 + (marker(m,2)-xPred(2))^2);
56        alpha = atan2(marker(m,2)-xPred(2), marker(m,1)-xPred(1))-xPred(3));
57        zz = [dist; alpha];
58        zz(2) = wrapToPi(zz(2));
59        z = [z; zz];
60
61        % Create matrix C for correction
62        c = [xPred(1)/dist    xPred(2)/dist     0; ...
63            -xPred(2)/dist^2 xPred(1)/dist^2 -1];
64        C = [C; c];
65    end
66
67    % Measurement covariance matrix
68    RR = diag(repmat([R(1,1) R(2,2)], 1, size(marker, 1)));
69    K = PPred*C.'/(C*PPred*C.' + RR);
70
71    inov = zTrue - z;
72    % Selection of appropriate innovation due to noise and angle wrapping
73    for m = 1:size(marker, 1)
74        inov(2*m) = wrapToPi(inov(2*m));
75    end
76
77    x = xPred + K*(inov);
78    P = PPred - K*C*PPred;
79
80    figures_ekf3_update;
81 end
```

(Continued)

EXAMPLE 6.22—cont'd

Again a simple graphical observability check can be performed. In this case one particular situation that has indistinguishable states can again be found, as shown in Fig. 6.48. At first glance the considered states are indistinguishable from the measurement point of view. But since we assume that the measurement data also contains marker IDs, this particular special situation has distinguishable states, and therefore the system is observable.

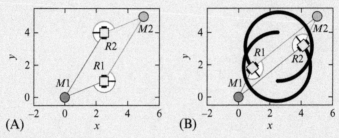

(A)

(B)

Fig. 6.48 Observability analysis of the system from Example 6.22. (A) Two particular poses that are symmetrical from the measurement point of view. (B) Updated situation from Fig. (A) after the robots have traveled the same relative path. From the measurement point of view, the both robot poses are distinguishable if measurement data contains IDs of the markers.

6.5.3 Kalman Filter Derivatives

Alongside the Kalman filter for linear systems and the EKF for nonlinear systems [10], many other extensions of the above filters have been proposed. Let us take a look at some of the most common and important approaches.

Unscented Kalman filter [11] is normally used in systems with significant nonlinearities where the EKF may not provide satisfactory results. In this case, the covariance matrices are estimated statistically based on a small subset of points that are transformed over the nonlinear function; the transformed points are used to estimate the mean value and the covariance matrix. These points are known as sigma points that are spread around the estimated value by some algorithm (normally there are $2n + 1$ points for n dimensions).

Information filter uses the information matrix and information vector instead of the covariance matrix and state estimate. The information matrix represents the inverse of the covariance matrix, and the information vector is the product of the information matrix and estimated state vector. The information filter is dual to the Kalman filter where the correction step is computationally much simpler to evaluate (no matrix summation). However, the prediction step becomes more computationally demanding.

The *Kalman-Bucy* filter is a form of the Kalman filter for continuous systems.

6.6 PARTICLE FILTER

Up until now the Bayesian filter was considered to be normally used for discrete state-space systems with a finite number of values that the state-space variables can take. If we want to apply the Bayesian filter for continuous variables, these variables can be quantified to a finite number of values. This kind of Bayesian filter implementation for continuous state variables is commonly known as the histogram filter.

For continuous state-space variables an explicit solution of the Bayesian filter (6.29) would need to be determined:

$$
\begin{aligned}
&p(\boldsymbol{x}_k|\boldsymbol{z}_{1:k}, \boldsymbol{u}_{0:k-1}) \\
&= \frac{p(\boldsymbol{z}_k|\boldsymbol{x}_k)}{p(\boldsymbol{z}_k|\boldsymbol{z}_{1:k-1}, \boldsymbol{u}_{0:k-1})} \int p(\boldsymbol{x}_k|\boldsymbol{x}_{k-1}, \boldsymbol{u}_{k-1}) p(\boldsymbol{x}_{k-1}|\boldsymbol{z}_{1:k-1}, \boldsymbol{u}_{0:k-2}) \, d\boldsymbol{x}_{k-1}
\end{aligned}
$$

$$(6.44)$$

where we took into account the complete state assumption and that the system is a Markov process (see Section 6.2). The current state probability distribution (6.44) is needed in calculation of the most probable state estimate (mathematical expectation):

$$
\mathrm{E}\{\hat{\boldsymbol{x}}_{k|k}\} = \int \boldsymbol{x}_{k|k} \cdot p(\boldsymbol{x}_k|\boldsymbol{z}_{1:k}, \boldsymbol{u}_{0:k-1}) \, d\boldsymbol{x}_{k|k}
$$

An explicit solution of Eq. (6.44) is possible only for a limited class of problems where Gaussian noise distribution and system linearity are assumed; the result is a Kalman filter. In the case of nonlinear systems, the system nonlinearity (in the motion model, actuator model, and/or sensor model) can be linearized, and this leads to the EKF.

A particle filter is a more general approach where noise distribution need not be Gaussian and the system may be nonlinear. The basic idea behind the particle filter is that the a posteriori state estimate probability distribution (6.29), after the measurement was made, is approximated with a set of N particles. Every particle in the set presents the value of the estimated state x_k^i that is randomly sampled from the probability distribution (Monte Carlo simulation approach). Every particle represents a hypothesis about the system state. The representation of the probability distribution with a set of randomly generated particles is a nonparametric description of probability distribution, which is therefore not limited to only Gaussian distributions. The description of probability distribution with particles enables modeling of nonlinear transformation of noise (actuator and/or sensor model). The particles can therefore be used to describe the noise distributions that are transformed over nonlinear functions from inputs or outputs to the system states.

Algorithm 5 presents the basic implementation of the particle filter.

ALGORITHM 5   Particle Filter Algorithm. The `Particle_filter` Function Is Called in Every Time Step With Current Values of Inputs and Measurement and the Previous State Estimate

function PARTICLE_FILTER$(\hat{x}_{k-1|k-1}, u_{k-1}, z_k)$

Initialization:

if $k > 0$ **then**

Initialize a set of N particles x_k^i based on a random sampling of the probability distribution $p(x_0)$.

end if

Prediction:

Transform every particle $\hat{x}_{k-1|k-1}^i$ based on the known movement model and known input u_{k-1}, to which add a randomly generated value based on the noise properties that are part of the movement model. The movement model is given with $p(x_k|x_{k-1}, u_{k-1})$. The obtained prediction is given as a set of particles $\hat{x}_{k|k-1}^i$.

Correction:

For every particle $\hat{x}_{k|k-1}^i$ *estimate measurement output* that would be measured if the system state would correspond to the particle state.

Based on the measurement made and comparison against estimated measurements with particles, *evaluate the particle importance*.

This is followed by *importance sampling*, a sampling approach based on the particle importance with random selection of the particles in a way that the particle selection probability is proportional to the particle importance (i.e., $p(z_k|\hat{x}^i_{k|k-1})$). More important particles should therefore be selected more times, and less important particles fewer times.

Filtered state estimate $\hat{x}_{k|k}$ can be selected as the average value of all the particles.

end function

In the initial step of the particle filter an initial population of the particles needs to be selected, $\hat{x}^i_0$ for $i \in 1, \dots, N$, which is distributed based on the belief (probability distribution) $p(x_0)$ of the system states at the beginning of observation. In the case the initial state is not known a uniform distribution should be used; therefore, the particles should be spread uniformly over the entire state-space.

In the prediction part a new state for each particle is evaluated based on the known system input. To the state estimate of every particle a randomly generated noise is added, which is expected at the system input. In this way state predictions for every particle are obtained, $\hat{x}^i_{k|k-1}$. The added noise ensures that the particles are scattered, since this enables estimation of true value in the presence of different disturbances.

In the correction part of particle filter the importance of the particles is evaluated, in a way that the difference between the obtained measurement z_k and all the estimated particle measurements $(\hat{z}^i_k)$ is calculated from estimated particle states. The difference between the measurement and estimated measurement is commonly known as innovation or measurement residual, and it can be evaluated for every particle as

$$innov^i_k = z_k - \hat{z}^i_k$$

which is low for most probable particles.

Based on measurement residual the importance of every particle, that is, the probability $p(z_k|\hat{x}^i_{k|k-1})$ can be determined, which represents the weight w^i_k for particle i. The weight can be determined with Gaussian probability distribution as

$$w^i_k = \det{(2\pi R)}^{-\frac{1}{2}} e^{-\frac{1}{2}(innov^i_k)^T R^{-1}(innov^i_k)}$$

where R is the measurement covariance matrix.

A very important step in the particle filter is importance sampling based on the weights w^i_k. The set of N particles is randomly sampled in a way that the probability of selection a particular particle from a set is proportional of the particle weight w_k^i. Therefore, the particles with larger value of the weight should be selected more times than the particles with lower value of the weight; the particles with the smallest value of the weight should not even be selected. The approach of generating a new set of particles can be achieved in various ways, one of the possible approaches is given below:

- The particle weights w_k^i are normed with the sum of all the weights $\sum_{i=1}^{N} w_k^i$; in this way the new weights $wn_k^i = \frac{w_k^i}{\sum_{i=1}^{N} w_k^i}$ are obtained.
- The cumulative sum of the normed weights is calculated in order to obtain the cumulative weights $wc_k^i = \sum_{j=1}^{i} wn_k^i$, as shown in Fig. 6.49.
- N numbers between 0 and 1 are selected randomly (from a uniform and 1 are selected randomly (from a uniform distribution) and then it is determined to which weights these random numbers correspond. The comparison between the cumulative weights wc_k^i and the randomly generated numbers is therefore made. As seen from Fig. 6.49, the

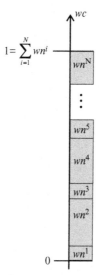

Fig. 6.49 Importance sampling in the correction part of the particle filter. The particle weights are arranged in a sequence one after another and then normed in a way that the sum of all the particles is one. More probable particles take more space on the unity interval, and vice-versa. The new population of particles is determined in a way that N numbers from the unity interval are selected randomly (with uniform distribution), and the particles that correspond to these numbers are selected.

particles with larger weights are more probable to be selected (the weights in Fig. 6.49 occupy more space).

- The selected particles are used in the correction part to determine the a posteriori state estimate $\hat{\mathbf{x}}_{k|k} = \sum_{i=1}^{N} w_k^i \hat{\mathbf{x}}_{k|k-1}^i$.

In the case the system is not moving (the current state is equal to the previous state) it is recommended that the importance sampling is not made, but the previous particles are selected as the current ones and only the weights are modified, as shown in [12].

The application of particle filter is shown in Example 6.23.

EXAMPLE 6.23

Use a particle filter for estimation of the most likely state in Example 6.22. In the implementation of particle filter use $N = 300$ particles. All the other data are the same as in Example 6.22.

Solution

Listing 6.14 provides an implementation of the solution in Matlab. The simulation results are shown in Figs. 6.50 and 6.51.

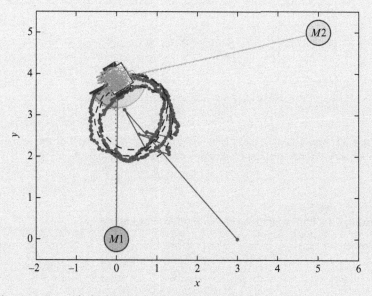

Fig. 6.50 True (*dashed line*), estimated (*solid line*) trajectory, and generated particles in the final step of simulation from Example 6.23.

(*Continued*)

EXAMPLE 6.23—cont'd

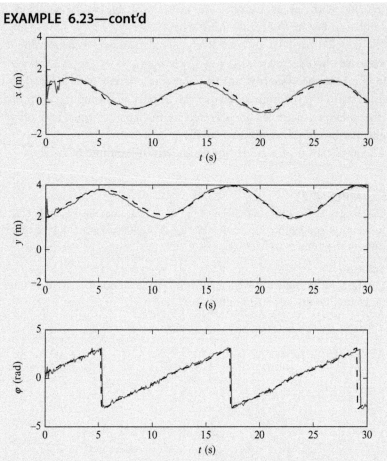

Fig. 6.51 Pose estimate (*solid line*) and true state (*dashed line*) of the mobile robot with a nonzero initial estimation error from Example 6.23.

Listing 6.14 Implementation of the solution of Example 6.23

```
1 Ts = 0.1; % Sampling time
2 xTrue = [1; 2; pi/6]; % True initial pose
3 x = [3; 0; 0]; % Initial pose estimate
4 P = diag([9 9 0.6]); % Initial covariance matrix of the pose estimate
5 Q = diag([0.1 0.1]); % Noise covariance matrix of movement actuator
6 R = diag([0.5 0.3]); % Noise covariance matrix of distance and
7                      % angle measurement
8 enableNoise = 1; % Enable noise: 0 or 1
9 N = 300; % Number of simulation sample times
10 marker = [0 0; 5 5]; % Positions of markers
11
```

EXAMPLE 6.23—cont'd

```
12  % Particle initialization
13  nParticles = 300;
14  xP = repmat(xTrue, 1, nParticles) + diag([4 4 1])*randn(3, nParticles);
15  W = ones(nParticles, 1)/nParticles; % All particles have equal probability
16
17  figures_pf1_init;
18
19  % Loop
20  for k = 1:N
21      u = [0.5; 0.5]; % Movement command (translational and angular velocity)
22      u_sum = u + sqrt(Q)*randn(2, 1)*enableNoise;
23
24      % Simulation of the true mobile robot state (pose)
25      xTrue = xTrue + Ts*[u_sum(1)*cos(xTrue(3)); ...
26                          u_sum(1)*sin(xTrue(3)); ...
27                          u_sum(2)];
28      xTrue(3) = wrapToPi(xTrue(3));
29
30      % Simulation of the true noisy measurements (distance and angle)
31      zTrue = [];
32      for m = 1:size(marker, 1)
33          dist = sqrt((marker(m,1)-xTrue(1))^2 + (marker(m,2)-xTrue(2))^2);
34          alpha = atan2(marker(m,2)-xTrue(2), marker(m,1)-xTrue(1))-xTrue(3);
35          zz = [dist; alpha] + sqrt(R)*randn(2, 1)*enableNoise;
36          zz(1) = abs(zz(1));
37          zz(2) = wrapToPi(zz(2));
38          zTrue = [zTrue; zz];
39      end
40
41      % Prediction
42      for p = 1:nParticles
43          % Particles are moved according to the noise model
44          un = u + sqrt(Q)*randn(2, 1)*1;
45          xP(:,p) = xP(:,p) + Ts*[un(1)*cos(xP(3,p)); ...
46                                  un(1)*sin(xP(3,p)); ...
47                                  un(2)];
48          xP(3,p) = wrapToPi(xP(3,p));
49      end
50
51      % Correction
52      for p = 1:nParticles
53          % Estimated measurement for every particle
54          z = [];
55          for m = 1:size(marker, 1)
56              dist = sqrt((marker(m,1)-xP(1,p))^2 + (marker(m,2)-xP(2,p))^2);
57              alpha= atan2(marker(m,2)-xP(2,p), marker(m,1)-xP(1,p))-xP(3,p);
58              zz = [dist; alpha];
59              zz(1) = abs(zz(1));
60              zz(2) = wrapToPi(zz(2));
61              z = [z; zz];
62          end
63
```

(Continued)

EXAMPLE 6.23—cont'd

```
1
2        Innov = zTrue - z; % Determine innovation
3
4        % Selection of appropriate innovation due to
5        % noise and angle wrapping
6        for m = 1:size(marker, 1)
7            iii = zTrue(2*m) - (z(2*m) + [0; 2*pi; -2*pi]);
8            [tmp, index] = min(abs(iii));
9            Innov(2*m) = iii(index);
10       end
11
12       % Determine particle weights (particle probability)
13       % Measurement covariance matrix
14       RR = diag(repmat(diag(R), size(marker, 1), 1));
15       W(p) = exp(-0.5*Innov.'*inv(RR)*Innov) + 0.0001;
16   end
17
18   iNextGeneration = obtainNextGenerationOfParticles(W, nParticles);
19   xP = xP(:,iNextGeneration);
20
21   % The new state estimate is the average of all the particles
22   x = mean(xP, 2);
23   x(3) = wrapToPi(x(3));
24   % For robot orientation use the most likely particle instead of the
25   % average angle of all the particles.
26   [gg, ggi] = max(W);
27   x(3) = xP(3,ggi);
28
29   figures_pf1_update;
30 end
```

Listing 6.15 Function used in Listings 6.14 and 6.16

```
1 function iNextGeneration = obtainNextGenerationOfParticles(W, nParticles)
2     % Selection based on the particle weights
3     CDF = cumsum(W)/sum(W);
4     iSelect  = rand(nParticles, 1); % Random numbers
5     % Indices of the new particles
6     CDFg = [0; CDF];
7     indg = [1; (1:nParticles).'];
8     iNextGeneration_float = interp1(CDFg, indg, iSelect, 'linear');
9     iNextGeneration=round(iNextGeneration_float + 0.5); % Round the indices
10 end
```

In Example 6.23 the distances and angles to the markers are measured, so the angle wrapping must be taken into account. The example of localization where only distances to the markers are measured in order to estimate the mobile robot pose is shown in Example 6.24.

EXAMPLE 6.24

Use a particle filter for estimation of the most likely state in Example 6.23, but consider only measurement of the distance to the markers. In the implementation of the particle filter use $N = 500$ particles. All the other data are the same as in Example 6.23.

Solution

Listing 6.16 provides an implementation of the solution in Matlab. The simulation results are shown in Figs. 6.52 and 6.53, where it is clear that the estimated value converges to the true value, as in Example 6.23.

Listing 6.16 Implementation of the solution of Example 6.24

```
1  Ts = 0.1; % Sampling time
2  xTrue = [1; 2; pi/6]; % True initial pose
3  x = [3; 0; 0]; % Initial pose estimate
4  P = diag([9 9 0.6]); % Initial covariance matrix of the pose estimate
5  Q = diag([0.1 0.1]); % Noise covariance matrix of movement actuator
6  R = diag([0.5 0.3]); % Noise covariance matrix of distance and
7                        % angle measurement
8  enableNoise = 1; % Enable noise: 0 or 1
9  N = 300; % Number of simulation sample times
10 marker = [0 0; 5 5]; % Positions of markers
11 R = R(1,1); % Only distance measurement
12
13 % Particle initialization
14 nParticles = 500;
15 xP = repmat(xTrue, 1, nParticles) + diag([4 4 1])*randn(3, nParticles);
16 W = ones(nParticles, 1)/nParticles; % All particles have equal probability
17
18 figures_pf2_init;
19
20 % Loop
21 for k = 1:N
22     u = [0.5; 0.5]; % Movement command (translational and angular velocity)
23     u_sum = u + sqrt(Q)*randn(2, 1)*enableNoise;
24
25     % Simulation of the true mobile robot state (pose)
26     xTrue = xTrue + Ts*[u_sum(1)*cos(xTrue(3)); ...
27                         u_sum(1)*sin(xTrue(3)); ...
28                         u_sum(2)];
29     xTrue(3) = wrapToPi(xTrue(3));
30
31     % Simulation of the true noisy measurements (distance)
32     zTrue = [];
33     for m = 1:size(marker, 1)
34         dist = sqrt((marker(m,1)-xTrue(1))^2 + (marker(m,2)-xTrue(2))^2);
35
```

(Continued)

EXAMPLE 6.24—cont'd

```
1        zz = [dist] + sqrt(R)*randn(1, 1)*enableNoise;
2        zz(1) = abs(zz(1));
3        zTrue = [zTrue; zz];
4    end
5
6    % Prediction
7    for p = 1:nParticles
8        % Particles are moved according to the noise model
9        un = u + sqrt(Q)*randn(2, 1)*1;
10       xP(:,p) = xP(:,p) + Ts*[un(1)*cos(xP(3,p)); ...
11                               un(1)*sin(xP(3,p)); ...
12                               un(2)];
13       xP(3,p) = wrapToPi(xP(3,p));
14   end
15
16   % Correction
17   for p = 1:nParticles
18       % Estimated measurement for every particle
19       z = [];
20       for m = 1:size(marker, 1)
21           dist = sqrt((marker(m,1)-xP(1,p))^2 + (marker(m,2)-xP(2,p))^2);
22           zz = [dist];
23           zz(1) = abs(zz(1));
24           z = [z; zz];
25       end
26
27       Innov = zTrue - z; % Determine innovation
28
29       % Determine particle weights (particle probability)
30       % Measurement covariance matrix
31       RR = diag(repmat(diag(R), size(marker, 1), 1));
32       W(p) = exp(-0.5*Innov.'*inv(RR)*Innov) + 0.0001;
33   end
34
35   iNextGeneration = obtainNextGenerationOfParticles(W, nParticles);
36   xP = xP(:,iNextGeneration);
37
38   % The new state estimate is the average of all the particles
39   x = mean(xP, 2);
40   x(3) = wrapToPi(x(3));
41   % For robot orientation use the most likely particle instead of the
42   % average angle of all the particles.
43   [gg, ggi] = max(W);
44   x(3) = xP(3,ggi);
45
46   figures_pf2_update;
47 end
```

EXAMPLE 6.24—cont'd

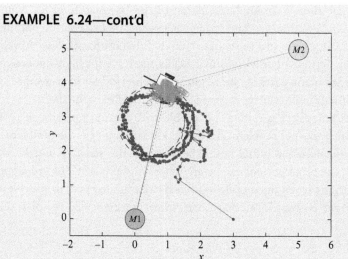

Fig. 6.52 True (*dashed line*), estimated (*solid line*) trajectory, and generated particles in the final step of simulation from Example 6.24.

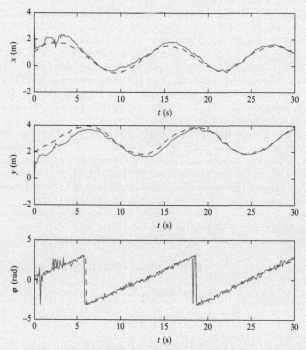

Fig. 6.53 Pose estimate (*solid line*) and true state (*dashed line*) of the mobile robot with a nonzero initial estimation error from Example 6.24.

A particle filter is an implementation of the Bayesian filter for continuous systems (continuous state space); it enables description of nonlinear systems and can take into account arbitrary distribution of noise. Applicability of the particle filter is computationally intensive for high-dimensional systems, since many particles are required to achieve filter to converge. The number of required particles increases with space dimensionality.

The particle filter is robust and is able to solve the problem of global localization and the problem of robot kidnapping. In the problem of global localization the initial pose (the value of the states) is not known. Therefore, the mobile robot can be anywhere in the space. The kidnapping problem occurs when the mobile robot is moved (kidnapped) to an arbitrary new location. Robust localization algorithms must be able to solve these problems.

REFERENCES

[1] D.C. Montgomery, G.C. Runger, Applied Statistics and Probability for Engineers, John Wiley & Sons, Englewood Cliffs, NJ, 2003.

[2] R. Hermann, A.J. Krener, Nonlinear controllability and observability, IEEE Trans. Autom. Control 22 (5) (1977) 728–740.

[3] W. Respondek, Introduction to geometric nonlinear control; linearization, observability, decoupling, in: vol. 8 of ICTP Lecture Notes Series, ICTP, 2002, pp. 173–222.

[4] A. Martinelli, Nonlinear unknown input observability: analytical expression of the observable codistribution in the case of a single unknown input, in: 2015 Proceedings of the Conference on Control and its Applications, 2015, pp. 9–15.

[5] D. Fox, J. Hightower, L. Liao, D. Schulz, G. Borriello, Bayesian filtering for location estimation, IEEE Pervasive Comput. 2 (2003) 24–33.

[6] R.E. Kalman, A new approach to linear filtering and prediction problems, Trans. ASME J. Basic Eng. 82 (Series D) (1960) 35–45.

[7] L. Teslić, I. Škrjanc, G. Klančar, Using a LRF sensor in the Kalman-filtering-based localization of a mobile robot, ISA Trans. 49 (2010) 145–153.

[8] L. Teslić, I. Škrjanc, G. Klančar, EKF-based localization of a wheeled mobile robot in structured environments, J. Intell. Robot. Syst. 62 (2010) 187–203.

[9] G. Klančar, L. Teslić, I. Škrjanc, Mobile-robot pose estimation and environment mapping using an extended Kalman filter, Int. J. Syst. Sci. (2013) 1–16.

[10] B.D.O. Anderson, J.B. Moore, Optimal Filtering, Prentice-Hall, Englewood Cliffs, NJ, 1979.

[11] S.J. Julier, Process models for the navigation of high-speed land vehicles, PhD thesis, University of Oxford, 1997.

[12] S. Thrun, W. Burgard, D. Fox, Probabilistic Robotics, vol. 4, MIT Press, Cambridge, MA, 2005.

CHAPTER 7

Autonomous Guided Vehicles

7.1 INTRODUCTION

Autonomous guided vehicles (AGVs) are mobile robots that can move autonomously on the ground within indoor or outdoor environments while performing a set of tasks. Although several locomotion principles are possible (rolling using wheels, walking, crawling, swimming, etc.). AGVs are usually wheeled mobile robots. The range of practical AGV applications as well as the number of commercially available mobile robots for various areas is increasing. Nowadays the use of mobile robots performing domestic tasks such as cleaning or lawn mowing are quite common and are about to become a must-have of every household. Quite common is also the use of mobile robots in factories, hospitals, and distribution centers for delivery service. Also very promising are future applications in agriculture, where many research groups are developing their robots that will soon be operational. Applications in public transportation are also important, such as autonomous cars. There are many other applications such as in military for unmanned reconnaissance, in space missions of planetary explorations, in disaster areas for search and rescue, and in security. Mobile robotics is an emerging field of research and development, and therefore, many new applications that are not so obvious are expected to appear in near future.

In the following a brief review of some current research activities and applications of AGVs in transportation, agriculture, industry, medical, and domestic use are presented. The purpose of this chapter is not to give a complete overview of numerous applications, but to shed some insight on how various approaches and technologies are used together in order to achieve different autonomous capabilities and novel functionalities.

7.2 AUTONOMOUS TRANSPORTATION VEHICLES

Self-driving cars are about to become common in our lives. According to Google's plans they will become available to the public by 2020 [1], and similar forecasts are also given by other autonomous car producers [2].

Wheeled Mobile Robotics
http://dx.doi.org/10.1016/B978-0-12-804204-5.00007-X

Currently there are a number of autonomous car projects going on in major companies and research institutes that develop and experiment with such prototypes, namely, Mercedes-Benz, General Motors, Continental Automotive Systems, IAV, Autoliv Inc., Bosch, Nissan, Renault, Toyota, Audi, Volvo, Tesla Motors, Peugeot, AKKA Technologies, Vislab from University of Parma, Oxford University, and Google [3]. The first really self-driving car appeared in 1980 by the Navlab group from Carnegie Mellon University's [4]. Today some countries (in the United States: Nevada, Florida, California, Michigan and in Europe: Germany, the Netherlands, and Spain) already allowed testing of autonomous vehicles in traffic and many others are about to follow.

In the following self-driving car characteristics, performance, and plans are described, which mostly are inspired from Google's car project [1].

7.2.1 About

The idea of a self-driving car is to drive people to their desired locations autonomously. During the transportation people can enjoy the ride, read magazines, or sleep because the car's intelligence is driving them safely. The car therefore needs to figure out where it is located (on which street and lane, etc.) by using sensors and maps of the environment. It needs to sense the environment to locate other objects in traffic such as other vehicles, pedestrians, cyclists, motorbikes, and the like. Then it needs to predict the intentions of all the sensed traffic participants and make appropriate decisions that result in safe, reliable, and comfortable driving. There are various scenarios that are easy to predict, but some events cannot be foreseen. Although some scenarios are highly unlikely (e.g., big crack on the road due to an earthquake, or an approaching tornado, etc.), the self-driving vehicle needs to be designed in such a way that it can take appropriate actions also in unpredicted situations. The design of autonomous vehicles also needs to be robust enough to enable safe and reliable driving in difficult weather and road conditions (e.g., robust line tracking in fog, heavy rain, or snow).

7.2.2 Setup

Self-driving cars do not need steering wheels and pedals (as for example in the latest version of the Google car shown in Fig. 7.1), hence no human interaction is possible. The on-board embedded computer systems constantly process the measured sensor data and take over the control of driving, while people are only riding. For this purpose the autonomous

Fig. 7.1 Google's self-driving car prototype unveiled in 2014. *(Photo by smoothgroover22; https://www.flickr.com/photos/smoothgroover22/15104006386, CC BY-SA 2.0.)*

vehicle needs to be equipped with an appropriate set of sensors. The sensors need to be mounted at appropriate locations on the vehicle in order to achieve maximum field of view and minimize blind spots. Automotive-grade custom embedded computer systems need to process all information gathered from the sensors in order to perceive the situation in the environment. These systems are battery powered with a backup system to enable uninterrupted operation of vital system functionalities (breaking, steering, computer). Although not necessary, autonomous vehicles can also have electric or hybrid powertrain.

7.2.3 Sensors

Fig. 7.2 shows an early self-driving car prototype, which is equipped with various sensors for perception of the environment. The vehicle is equipped with a GNSS signal receiver for long-term localization and navigation, while for short-term localization and obstacle detection several range sensors, cameras, and radars are used.

One of the main sensors that are used in autonomous vehicles is a 3D laser range finder (LIDAR), which is normally mounted on the roof of the car to achieve maximum coverage. The 3D LIDAR used in Google's car project is reported to cover the area in a radius of up to 200 m around the vehicle. The LIDAR sensor provides an ample amount of valuable data about the environment in a form of a point cloud. Advanced data processing is employed in order recognize, classify, and identify the objects from the raw point cloud data. Since all required information cannot be obtained

Fig. 7.2 Main parts of the early self-driving car prototype: (A) front side radar, (B) odometry sensor, (C) back radar, (D) GNSS receiver, (E) 3D laser range scanner (LIDAR). *(Derived from photo by Travis Wise; https://www.flickr.com/photos/photographingtravis/ 16253152591, CC BY 2.0.)*

only from the LIDAR, additional sensors are used to improve the perception of the environment. One of the important sensors is also a (normally front-facing) camera that can be used to recognize lane boundaries, other vehicles and pedestrians, traffic signs and lights, people gestures, and more. Computational capabilities of contemporary computers enable real-time implementation of advanced machine vision algorithms.

Additionally, Google's car is equipped with several radars mounted on the front and the rear bumper of the car. These sensors are used to measure the distance to the nearest car in front and to the nearest car behind the vehicle in order to maintain a safe distance and to prevent collisions with other participants on the road. Several ultrasonic sensors around the perimeter of the car are used to detect near obstacles (also on the side of the car) and therefore enable reverse driving and precise parking in tight spaces. To enable implementation of odometry the wheels of the car are equipped with rotation encoders. The autonomous vehicle is also equipped with some inertial navigation sensors, like accelerometers, gyroscopes, and an altimeter and tachymeter.

7.2.4 Localization and Mapping

The Google self-driving car prototype uses a preexisting map of the route. This map can be obtained in advance by a human driver going along all roads and collecting various sensor data on the way. The collected data needs to be processed in order to extract static background and all essential information that is required for autonomous driving from the rest of the dynamic environment. Besides the information about the physical geometry of the space a map can contain various additional information, like number and arrangement of lanes, locations of road crossings and static obstacles, placements and types of traffic signals, areas of potential hazards, and more. As an autonomous car passes along the road the newly collected data can be used to accommodate the map to some permanent or temporal (e.g., construction works) changes in the environment. An updated map can then be shared between vehicles.

On the road most of the time the vehicle can be successfully localized by GNSS at least for the purpose of navigation, but the position accuracy is not adequate for autonomous driving. Moreover, a GNSS signal may be unavailable or blocked in some areas (e.g., in tunnels and in cities between tall buildings) and also the accuracy and precision of the GNSS signal is space and time varying. To enhance precision and robustness required for autonomous driving other sensors must be used in localization, such as a laser range finder and camera. These sensors are required not only to estimate the pose of the autonomous vehicle in the environment but also all the poses of all other participants in the immediate vicinity around the vehicle (i.e., other vehicles, pedestrians, and cyclists). In Fig. 7.3 a visualization of various localization and map data is depicted.

Fig. 7.3 Localization of Google's self-driving car on the road, recognized objects in the environment, and planned path while it is approaching a railway crossing. *(Photo from video by Best Machine; https://www.youtube.com/watch?v=kaNWNtJyK4w, CC BY 3.0.)*

7.2.5 Control

A self-driving car has many control-related tasks, the main being following the path by driving between the lanes or according to the path that is planned online according to current environmental conditions. For this it uses algorithms similar to path and trajectory tracking presented in this book. It needs to observe driving objects in front and accelerate or decelerate in order to maintain the required safe distance. It needs to avoid obstacles, overtake other cars or motorists, correctly position in multilane streets, park itself, reverse drive, and much more. Although there are many scenarios that can be predicted in the car building phase there are many more that cannot be foreseen. Therefore all control actions need to be planned with safety being the main concern. To cope with unknown situations the self-driving car would also need to learn.

7.2.6 Path Planning

Long-term GNSS navigation is already present in most regular cars and in smartphones. It can therefore also be used for self-driving cars. More challenging for an autonomous car is short-term planning of its driving actions. According to sensed information, obstacles, other traffic participants, traffic lights, and other signs it needs to plan and replan its driving path, velocity profile, and the like (e.g., see Fig. 7.3).

7.2.7 Decision Making

From known (predictable) situations some decision rules and behaviors can be programmed in advance; however, traffic is highly dynamic and unpredictable. Therefore, this knowledge is not sufficient. To cope with unknown situations the self-driving car would also need to learn. The learning strategy to be successful requires a huge database of different traffic situations to which the car needs to be exposed. The Google, for example, has driven about 2 million kilometers on public roads and gathered many different traffic situations from which now the self-driving car software can learn offline and will become prepared better in future traffic scenarios [2]. However, they would still need to learn "on the fly."

Some very basic car behaviors are the following: driving on the lanes by following street markings and observing traffic, reverse driving, parking, overtaking, and driving in multilane crossroads. Any contradicting information or planned actions (e.g., correct sensor interpretations, unknown identified objects) need to be solved in a way to maximize safety.

7.3 WHEELED MOBILE ROBOTS IN AGRICULTURE

7.3.1 Introduction (About)

The use of service units (i.e., automated machinery) in agricultural or even mining applications entails several challenges from both mechatronics and control systems points of view. On one hand, it is to be noted that a service unit in agricultural applications is likely to be designed to fulfill a sole application [5–7], thus versatility is not an attribute on those kinds of machines. For example, seeding wheat does not have the same technical challenges as harvesting blueberries. Fig. 7.4 shows the general problems an agricultural process has to face when using automated machinery.

- The machinery and its capabilities are intrinsically related to the nature of the growing.
- The farmer, thus, has two options: on one hand, he/she adapts the land according to the machinery or designs a service unit for his/her own needs. The first case is a typical example of precision agriculture [8], where maneuverability issues and terrain characteristics become crucial for path planning or task planning of agricultural procedures. In the second case, a special machine is designed to face a particular problem, such as in the case of handling blueberries and harvesting grapes and olives [9].
- Nevertheless, sensors (mainly exteroceptive ones) are the ones that provide the machinery with all the necessary information to make decisions, both online or offline, according to the environment and task's needs. An example is the phenotyping that an LIDAR is able to perform in groves [10], which gives the farmer enough information for pruning, harvesting, or even disease control.

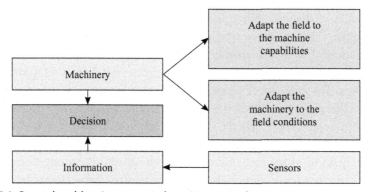

Fig. 7.4 General problem in automated precision agriculture.

(A)   (B)

Fig. 7.5 Two examples where robotized machinery still has issues to solve: how to manipulate and how to harvest from groves. (A) shows an apple grove, whereas (B) shows a vineyard. *(Left photo by Matt Jiggins, https://www.flickr.com/photos/mattjiggins/ 3942903508, CC BY 2.0; right photo in public domain, https://pixabay.com/en/grape-planting-purple-wine-fruit-1133200/, CC0.)*

An important issue to highlight, and was previously mentioned, is that service units for groves are the ones that currently face the most challenging problems. Handling the fruit, harvesting, spreading herbicide in an intelligent fashion, or even designing with the aim to compensate for the lack of human labor, are current research aims [11]. Fig. 7.5 shows two cases where two main crops (apples and grapes) are currently manually harvested; thus farmers have to face the problem of a lack of human labor.

7.3.2 Service Unit Setup

A general mechatronic architecture of a service unit design is presented in Fig. 7.6, which corresponds to the mini-loader shown in Fig. 7.7, from the *Pontificia Universidad Católica*, Chile, and partially sensorized by the *Universidad Técnica Federico Santa María*, also from Chile. Although the mini-loader is designed for mining tasks, it can be used for agricultural applications. The architecture shown in Fig. 7.6 is a general design that can be adapted to other service units. The main operating system used is ROS. The system has two computers, one for low-level processing and a one for high-level processing. At this point it is important to mention that proprioceptive sensors, actuators, and communication among the components of the vehicle should be implemented on the low-level computer, since the high-level one is left for processing sensor data and decision making. It has to be noted that such separation among computers is necessary: high-level processing usually requires larger sampling times when compared with the sampling time of an actuator. In addition, the

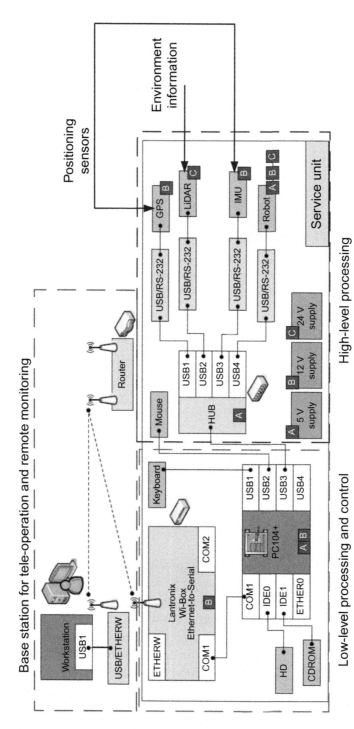

Fig. 7.6 General architecture of a service unit for agricultural (and mining) purposes. *(Scheme by Fernando Auat Cheein.)*

Fig. 7.7 Automatized mini-loader: (A) back laser range scanned, (B) side laser range scanner; although not marked in the picture, the vehicle is also equipped with GNSS and IMU systems. *(Derived from photo by Fernando Auat Cheein.)*

service unit has a tele-operation module (mandatory in some countries) for remote control of the vehicle and also remote monitoring. In the following, the main parts from Fig. 7.6 will be explained in detail.

Sensors

In the design of a service unit, the sensors, as established in [9] are closely related to the agricultural application. When phenotyping, LIDARs are the most used sensors; however, when measuring range, the Kinect sensor [10] is becoming a cheaper option. The Kinect sensor (depending on its version) delivers depth, RGB, and intensity data, and it is able to work outdoors and indoors. Due to its price, it is rapidly replacing the LIDAR in phenotyping operations.

Artificial vision systems (including monocular, binocular, and trinocular cameras) are widely used for extracting features of interest from the grove [7]. They are especially useful for predicting the crop production, although sensitivity to lighting conditions is still an unsolved issue.

In the context of positioning, GNSS systems and inertial measurements are mandatory in any service unit for two reasons: GNSS antennas offer the possibility of a (always) bounded positioning system. On the other hand, inertial units allow not only for inertial positioning, but also for monitoring vibrations of the service unit's chassis.

Some proprioceptive sensors such as odometric encoders are usually part of most service units. Dead-reckoning offers short time and yet very accurate, positioning of the vehicle.

7.3.3 Localization, Mapping, and SLAM

As stated in [9, 12, 13], localization is a crucial part of every service unit's design. A bad localization system, in automated machinery, will cause an inefficient harvesting process and loss of energy. In monitoring, phenotyping, or herbicide management tasks, errors in the position of the vehicle are propagated to all the other core tasks of the machine, as demonstrated in [9, 14]. Thus, the following scenario arises:

- When using dead-reckoning as a single localization system, the error associated with such localization of the vehicle will grow unbounded [13].
- If using GNSS as a single localization method, the GNSS signal depends on the satellites' alignment, and misalignments will occur.
- The fusion of GNSS and inertial units (through extended Kalman filter as in [14]) as well as dead-reckoning will provide the vehicle with an error bounded localization system, suitable for agricultural (and mining) operations.
- Lastly, simultaneous localization and mapping (SLAM) is a more precise technique for localizing vehicles, as shown in [15, 16], with the difference being that, due to its computational cost, it should be implemented in the high-level computer (Fig. 7.6). The main advantage of SLAM is that the vehicle only needs exteroceptive sensors, like LIDARs and a low-cost GNSS to keep the error bounded. However, the SLAM only works when the vehicle revisits previously navigated parts from the environment. Otherwise consistency and convergence of the algorithm are not guaranteed [13], and the references therein.
- When using SLAM, the algorithm itself maps the environment, since the map is part of its state vector. If other localization system is used, then the map built by the robot will be constrained to the error in such a localization system.

For a deeper comparison between different localization systems, please refer to [13, 14].

7.3.4 Control Strategies

Controlling a service unit, following the guidelines shown in [9, 17], is a threefold process. On one hand, low-level controllers should be designed and embedded on the machinery to ensure that actuators will not saturate and will response properly. On the other, a medium-level controller should be designed to achieve at least one of the three basic things a wheeled vehicle does: positioning, path following, or trajectory tracking. In many applications, as the one shown in [15] (and the references therein), a path tracking controller is implemented on the vehicle. It is to be noted that the medium-level controller is the one that actually governs the machinery motion along the agricultural (or mining) environment. There is, however, a third level, closely related to supervision. The service unit, at the end, must perform an agricultural task, and this third-level controller is the one that supervises the machinery's performance.

To design a control strategy for a service unit, the following issues are to be faced:

1. *Slippage.* A characterization of the terrain that must be considered to avoid excessive waste of energy.
2. *Maneuverability.* The dimensions of the service unit might constrain the maneuverability space [11].
3. *Energy.* The autonomy of the vehicle is strongly related to energy management. Controlling the service unit is not only related to motion, but also to an efficient use of the available resources.

7.3.5 Planning Routes and Scheduling

Path or trajectory planning is another important system part, which is executed by the high-level computer (Fig. 7.6). Planning is closely related to the agricultural task. At this stage, the path planning method incorporated on the machinery should be able to solve the agricultural problem from the robotic perspective. For example, if the aim is to monitor the olive grove shown in Fig. 7.8, then a map of the environment is required (as the geo-referenced one shown in Fig. 7.9), which will be used to plan a feasible and kinodynamically compatible path. The most common path planer is a waypoint-based system [5, 6] in which the user places some waypoints in the geo-referenced map shown in Fig. 7.9, and the service unit, using a positioning controller or a path following controller, navigates the environment reaching one waypoint at a time.

However, it is to be noted that there are several other techniques that can be used for path planning, such as A*, RRT, and RRT*, among others [8, 18–20] that only need map information. One key issue to consider is that vehicle kinematic and dynamic compatibility is not always ensured by the path planner, and thus the middle-level controller must be robust enough to overcome possible disturbances.

Fig. 7.8 Ground-based image of an olive grove. *(Photo by Fernando Auat Cheein.)*

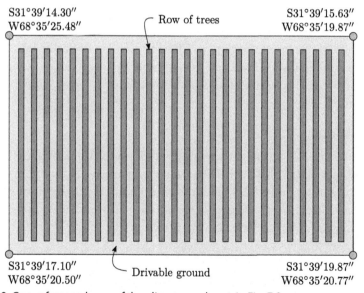

S31°39′14.30″
W68°35′25.48″

Row of trees

S31°39′15.63″
W68°35′19.87″

S31°39′17.10″
W68°35′20.50″

Drivable ground

S31°39′19.87″
W68°35′20.77″

Fig. 7.9 Geo-referenced map of the olive grove shown in Fig. 7.8.

7.4 WHEELED MOBILE ROBOTS IN INDUSTRY

Among many possible AGV applications the industry is the most common one. Wheeled mobile robots in industry are used in warehouses, factories, distribution centers, and more, mainly to transport materials. They enable a modular *fetch and carry* delivery service between material-storage facilities and assembly work-stations in a dynamically changing environment. In various applications the interaction with humans can be prevented, either with restriction of human access to the AGV's working area completely or with separate motion paths or lanes. Normally only authorized workers are allowed to be in the working area of the AGV, since they need to be aware of AGV presence (e.g., an automatic forklift can move even without the presence of a human operator).

A benefit of automated delivery is reducing production costs related to labor, space, inventory, and productivity. They improve throughput and enhance worker health and safety (heavy loads and repetitive efforts). The most important factors for the future increase of AGV's use in industry are the following: improved safety by sensing people and the environment, faster and easier reaction to changes in processes or in tasks (with no reprogramming required), enlarging knowledge of the environment by a priori given knowledge and by the use of sensors, and improvement of human-robot interaction in a more intuitive way [21].

Such mobile robot platforms are a key part of intelligent factories providing new forms of mobility as a flexible link between different working stages. They enable Industry 4.0 concepts by integrating autonomous process flow among workstations and possibilities for new human-robot interactions.

7.4.1 About

There are currently several commercially available AGVs such as KMR iiwa from KUKA [22], MiR 100 from Mobile Industrial Robots [23], Kiva from Amazon Robotics [24], TUG from Aethon [25], Lynx from Bastian Robotics [26], AGVs from Egemin Automation [27], and many others [28]. Two examples of industrial AGVs are shown in Fig. 7.10.

7.4.2 Setup

AGVs typically use wheels to move around, and some applications also use omnidirectional wheels for better maneuverability. They are designed either to pull cars or to load units (also top carrier) where the load rests on them (e.g., standard pallets) or are equipped with forks similar to those on manual fork trucks. They are equipped with sensors for safe operation and for localization. Efficiency and flexibility are increased by better knowledge of the environment (increased sensor use, SLAM), good understanding of the tasks they perform (models), ability to work in the presence of people,

(A)

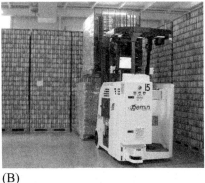

(B)

Fig. 7.10 Application of AGVs in distribution centers (A) and factory storage facilities (B). *(Left photo from video by DocumentaryVideo, https://www.youtube.com/watch? v=8of0t_tpWI0, CC BY-SA 3.0; right photo by AGVExpertJS, https://commons.wikimedia.org/ wiki/File:Hybrid_AGV,_Egemin_Automation.jpg, CC BY-SA 3.0.)*

ability to adapt to changes (in environment or in tasks), and simple use (e.g., navigation requires no special infrastructure).

7.4.3 Sensors

AGVs for transport in manufacturing apply sensors for localization purposes and to assure safe operation (Fig. 7.11). For localization they typically combine incremental encoders or inertial navigation with a laser range finder or camera. To assure safe operation they typically use a laser range finder and bumpers.

For example, Kuka's industrial mobile robot KMR iiwa is capable of picking desired boxes and delivering them to the desired location when needed. It can navigate to destinations without special markings on the floor (e.g., lines). It uses a pair of SICK laser range scanner S300 for mapping the environment and for localization purposes. The laser range scanner is therefore used for both navigation support and safety and not just for protection (e.g., detecting obstacles, preventing collisions), as usually was the case in most pioneered industrial applications.

7.4.4 Localization and Mapping

AGVs in manufacturing typically need to operate in large facilities. They can apply many features to solve localization and navigation. Quite often a robust solution is sensing induction from the electric wire in the floor or sensing magnetic tape glued to the floor. Currently the most popular solution is the usage of markers (active or passive) on known location and

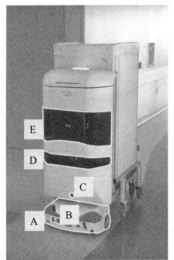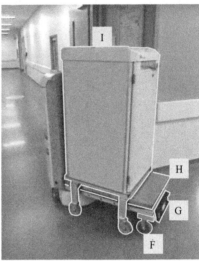

Fig. 7.11 Main parts of Aethon's TUG: (A) steerable drive unit, (B) IR range sensors, (C) camera, (D) laser range scanner, (E) front panel with ultrasonic sensors, (F) trailer caster wheel, (G) ultrasonic range sensors, (H) lift-able deck, (I) payload. *(Derived from video by Aethon, www.aethon.com, https://www.youtube.com/watch?v=REEzJfGRaZE, CC BY 3.0.)*

then AGVs localize by triangulation or trilateration. The latter is usually solved by a laser range finder and special reflecting markers. Other solutions may include wall following by range sensors or camera- or ceiling-mounted markers. All of the mentioned approaches are usefully combined with odometry. However, the recent modern solutions apply algorithms for SLAM which make them more flexible and easier to use in new and/or dynamically changing environments. They use sensors to locate usually natural features in the environment (e.g., flat surfaces, border lines, etc.). From features that were already observed and are stored in the existing map the unit can localize, while newly observed features extend the map.

Obtained maps are then used also for path planning.

7.4.5 Control

Motion control of AGVs is mostly solved by trajectory tracking, path following, and point sequence following approaches. These paths can be precomputed or better planned online using a map of the environment and path planning algorithms. In situations where magnetic tape on the floor marks desired roads of AGVs they can use simple line following algorithms with the ability to detect obstacles, stop, and move around them. Efficiency of AGVs in crowded areas greatly depends on how the obstacle avoidance problem is solved.

7.4.6 Path Planning

In environments that are mostly static the robots can operate using a priori planned routes. However, in dynamically changing environments they need to plan routes simultaneously. The most usual path planning strategy applies the combination of both mentioned possibilities where sensed markers on known locations in the environment enable accurate localization. When an unexpected obstacle is detected the AGV needs to find a way around the obstacle and then return and continue on the preplanned path.

7.4.7 Decision Making

Vehicles need to coordinate their paths to prevent intersections and possible collisions with other vehicles and humans. This coordination is usually done centrally in a well-known environment with known tasks in advance. If the latter is not the case (dynamically changing environments) then planning, coordination, and decision making are decentralized using multiagent-based methods. Correct decision making also involves the ability to recognize and predict motion of moving obstacles, other vehicles, and people.

Situations where AGVs and people are working together (in the same space) require adaptive and intelligent architectures to ensure desired safety and to have acceptable productivity.

To operate continually vehicles also need to decide when they need to recharge their batteries.

7.5 WHEELED MOBILE ROBOTS IN DOMESTIC ENVIRONMENTS

7.5.1 About

Domestic environments are normally unstructured with many dynamic obstacles. The map of the environment is normally not known and the environment is changing over time. These are some specific factors that make implementation of mobile robots into domestic environments a challenging task. Moreover, humans and other living beings cannot be excluded from their living spaces, and therefore, some level of cooperation between robots and humans is necessary to ensure safe operation. Mobile robots can already be found in millions of homes around the world, since wheeled mobile robots have proven to be useful at some domestic chores. Autonomous wheeled robotic vacuum cleaners, floor mopping, and sweeping can make our homes clean effortlessly, and autonomous wheeled robotic lawn mowers can maintain greenery around the house.

7.5.2 Setup

In order to accomplish the task of floor cleaning inside a house or lawn mowing around the house the problem of complete coverage [29] needs to be solved. Therefore the wheeled mobile robot needs to be able to autonomously reach every point in space in a limited time. There are some particular properties of indoor and outdoor environments that need to be considered or can be leveraged as an advantage in designing the autonomous mobile systems.

Indoor domestic environments are normally unstructured and cluttered. Floor-cleaning mobile robots are required to cover the entire floor in a room or multiple rooms. The floor is normally assumed to be flat, but there may be some uneven areas due to, for example, carpets. On the floor there may be some movable and immovable objects (e.g., tables, chairs, etc.). The floor is normally bounded by static objects like walls and closets, but in some cases the floor may end with a cliff (on top of stairs or on a balcony). Some floor areas may only be reached through narrow passages (e.g., doors or between dense arrangement of objects) or may not always be reachable (e.g., in the event of closed doors). The floor cleaning robot needs to be able to autonomously navigate in this kind of environment with obstacle avoidance and collision prevention. Only low-velocity collisions may be tolerated to prevent furniture damages and tip-over of objects. In order to be able to reach every part of the space (e.g., even under the sofas), the mobile cleaning robots tend to be small and compact. Cleaning robots are equipped with several brushes and a vacuum suction unit that enable collection of dirt and dust in the waste bin. Mobile cleaning robots normally come with a charging station (home base) that the robot is able to discover and also autonomously approach and initiate battery charging. Some mobile cleaning robots come with some additional hardware that can be placed in the environment to mark some forbidden zones or improve navigation between multiple rooms (e.g., infrared [IR] light barriers and lighthouses, magnetic strips).

The outdoor terrain in which autonomous robotic lawn movers operate is more challenging. The terrain is normally uneven and sometimes even inclined, and the properties of the surface are subjected to various weather conditions. The lawn is normally not separated from the rest of the terrain, with some environmental boundaries that would prevent the mobile robot from leaving the working area. For this purpose the perimeter of the working area is normally encircled with a wire in the ground that emits an electromagnetic signal, so the mobile robot can detect lawn boundaries. The wire can also be placed around the objects that the robotic lawn mower should not go to (e.g., trees, flowers, or a vegetable garden). Since the wire needs to be laid out before the first use, this requires some initial time to prepare the environment.

7.5.3 Sensors

Domestic mobile robots are normally equipped with a set of proximity sensors that enable detection of obstacles in the environment. Fig. 7.12 shows the main parts of a popular floor cleaning robot iRobot Roomba 620, which is also very developer friendly since the company released the serial communication protocol [30] that enables access to sensors and actuators. The front of the mobile robot is protected by a bumper—in normal operation the mobile robot only drives forward—that contains switches that are triggered whenever the mobile robot bumps into an obstacle. In some cases the front bumper is equipped with several IR proximity sensors that enable detection of obstacles in the immediate surrounding in front of the robot. Therefore the mobile robot can detect the presence of an obstacle before it would bump into it. Similar IR sensors are normally mounted around the perimeter of the mobile bottom,

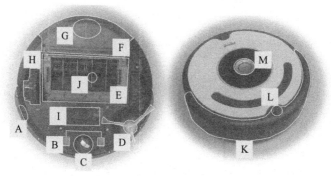

Fig. 7.12 The main parts of the iRobot Roomba 620: (A) IR cliff sensor, (B) charging contacts, (C) caster wheel, (D) side brush, (E) main brushes, (F) dust bin, (G) vacuum unit (undercover), (H) wheel, (I) battery (undercover), (J) dust sensor (inside), (K) bumper with IR proximity sensors, (L) omnidirectional IR sensor, and (M) buttons.

facing downwards, in order to detect holes in the environment and therefore prevent the robot, for example, from falling down the stairs, from the balcony, or into a hole. Robot wheels are equipped with rotation encoders to enable implementation of velocity control and wheel odometry. Moreover, the wheels are equipped with contact switches that enable detection if the mobile robot drove into a hole or if it has been picked up by the user. Some additional sensors may be used in order to verify that the mobile system is actually moving as desired (e.g., IR sensor to detect rotation of a two-colored passive caster wheel, current consumption measuring). Different models of domestic floor cleaning and lawn mowing robots use different configurations and arrangements of these sensors.

Since the wheeled robots can accomplish complete coverage of terrain with only these low-cost, robust, and simple to maintain and calibrate sensors the domestic robots can be mass produced and offered at an affordable price. Although map building and localization is not feasible with the available set of aforementioned proximity sensors, the floor cleaning task can still be accomplished, although systematic cleaning is not possible. Some indoor mobile robots are equipped with IR packet transmitters and receivers that enable communication between robots or some other external devices. Floor cleaning mobile robots are normally also equipped with an acoustic and/or optical sensor for detection of dust and dirt. Some robots are also equipped with magnetic field detectors that enable detection of magnetic strips that are laid down on the floor. Lawn mowing robots (e.g., Husqvarna Automower Fig. 7.13C) have a sensor that enables detection of signals emitted by the wire in the ground that marks the boundary of the working area. Some lawn mowers, which have been designed for large

(A) (B) (C)

Fig. 7.13 (A) Neato Robotics Neato XV, (B) Miele Scout RX1, and (C) Husqvarna Automower. *(Left photo by Autopilot, https://commons.wikimedia.org/wiki/File:Neato_XV-11.jpg, CC BY-SA 3.0; middle photo by Karlis Dambrans, https://www.flickr.com/photos/janitors/20591412974, CC BY 2.0; right photo by Tibor Antalóczy, https://commons.wikimedia.org/wiki/File:Husqvarna_Automower_305.jpg, CC BY-SA 3.0.)*

mowing areas, are also equipped with a GNSS sensor to enable mapping of perimeter wire and already covered areas.

Although laser range scanners are considered relatively expensive for domestic applications, developers of the Neato robot implemented a low-cost 360-degree laser scanner into their vacuum cleaning mobile robot (Fig. 7.13A). A laser-range scanner can provide necessary information about the environment that enables implementation of localization, map building, and systematic complete coverage of the space. The latest models of the indoor mobile robots use one or more cameras for environment perception (e.g., Miele Scout RX1 in Fig. 7.13B, iRobot Roomba 980, or LG Hom-Bot). The latest models also come equipped with a WiFi module that enables parameterization and remote control even from a smartphone.

7.5.4 Localization and Mapping

Since some domestic wheeled mobile robots are equipped only with proximity sensors these system are not capable of global localization and environment mapping. These mobile robots can only rely on the odometry obtained from wheel encoders. Some systems include external devices that are placed into the environment and operate as IR lighthouses (each device emits an IR signal with a unique ID). The lighthouses can assist the system when moving between different areas (e.g., rooms). Odometry drift over small distances in indoor environments can be relatively low, and therefore, this information can be useful for short-term navigation. However, in an outdoor application of a robotic lawn mower the odometry is less useful due to uneven terrain and weather dependent wheel slippage. Some robotic lawn mowers are equipped with GNSS sensors that enable mobile robot localization, but with limited precision. Therefore, localization might not be precise enough for mapping and implementation of navigation and systematic lawn mowing.

Mobile robots that are equipped with a laser range scanner can use the approach of SLAM in order to determine the pose of the robot in space. The latest vacuum cleaning robots use the camera in order to solve the problem of robot localization. The camera is mounted in a way that it is observing the ceiling of the scene in front of the robot. Machine vision algorithms are used in order to detect natural visual features in images. The approach of visual simultaneous localization and mapping (VSLAM) is used for fusion of the detected features into a map of the environment. Some models of robotic vacuum cleaners are also equipped with a down-facing camera. This camera is used to detect optical flow and implement visual odometry.

7.5.5 Path Planning

Path planning can only be applied when a map of the environment is known. Only the robots that are capable of SLAM can therefore use optimum coverage path planning approaches [29, 31, 32] in order to achieve systematic covering of the entire free space. The *complete coverage path* problem differs from the problem of optimum path planning. If in optimum path planning the goal is to find the optimum path between the initial and goal point, the goal of complete coverage is to find the optimum path so the robot covers the entire space. If the space is divided into a grid of cells (cell size depends on the robot dimensions), the goal of optimum coverage is to visit every cell at least once, and in an optimal case only once. This problem is also known as the *traveling salesman problem*. Once the optimum path is found the robot can systematically traverse the space and therefore be more time and energy efficient.

However in floor cleaning tasks it might be desired that the robot covers some parts of the space more than once or in a specific way (e.g., more dust can be expected near the edges and in the corners) to achieve better cleaning results. These additional requirements could also be considered in the path planning phase. However, the planned path could also be accommodated online, if dynamic obstacles are encountered or dirt is detected.

7.5.6 Control

If the path is available and the robot is capable of localization (map–based navigation), the robot can travel along the path with an appropriate path following algorithm. The control algorithms must be capable to adapt to the changes in the environment and unforeseen obstacles. Therefore the control algorithms at a low level must be designed in a way that safe and collision-free control is achieved. The control algorithms are normally designed in a way that the mobile robot slows down in the vicinity of the obstacles, and if the obstacles is encountered the robots stops and an alternative control action is found. If the robot is raised or drives into a hole the task is terminated and human intervention is required. Whenever the desired path is obstructed the mobile robot should try to find an alternative way or try to go around the obstacle.

If the mobile robot is not able to localize in the environment and build a map of the environment, some basic behaviors can be implemented that can also achieve complete coverage of the space. One of the possible control approaches is to drive straight forward until an obstacle is encountered, then rotate in place for a random angle or until no obstacles are detected. In a bounded space the complete coverage of the space is eventually achieved if only the robot drives around the environment long enough. This simple control approach could be extended with some additional rules to achieve better performance (e.g., make only left turns, but after a particular number of obstacle is encountered start making only right turns, and vice versa). Another behavior could be wall following. Once the obstacle is encountered try to follow the obstacle envelope; for example, the goal of the control is to ensure that only a single-side proximity sensor is active, since in this case the wall is on the side of the robot and the robot should drive forward. If no obstacles are detected, the robot should turn in one direction, but if multiple sensors become active the robot should turn to the other direction. Another behavior would be to follow the obstacle for some short distance and then try to go toward the free space in an orthogonal direction with respect to the obstacle. For large open spaces without obstacles the mobile robot can move in a spiral based on odometry data (problem of drift).

Some mobile robots are also able to detect IR signals from external devices. These external devices can make invisible barriers that the system should not cross. These devices emit different IR light beams in different directions, to enable determination where the mobile robot is with respect to the device. In this way the mobile robot can also return to base station for charging. The home base emits an IR beam with one ID to the left of the base station and a beam with a different ID to right of the base station. These two beams can be used to enable the mobile robot to reach the base station from the forward direction. The proximity sensors cab be used in order to detect distance from the base station.

Lawn mowing robots have implemented control algorithms for following the wire. This enables them to follow the perimeter of the working area. The wire can be placed into the ground also to guide the robot to the base station or through narrow passages.

7.5.7 Decision Making

Mobile robots that are unable to perform map building can ensure complete coverage with a set of limited behaviors (e.g., edge following, random walk, moving in a pattern, etc.). Different behaviors are more or less appropriate for a particular arrangement of the environment. With a random walk approach the majority of the space may be quickly but sparsely visited.

Fig. 7.14 An example of floor coverage with iRobot Roomba. *(Photo by Andreas Dantz, https://www.flickr.com/photos/szene/8649326807, CC BY 2.0.)*

With this approach finding the way through narrow passages might be difficult. When the robot is following the obstacle edges it might find a way between different areas that are connected by narrow passages, but this kind of behavior cannot cover free space. For dense coverage of free space, moving in a pattern (e.g., spiral motion) might be the best option.

To achieve optimum results in solving the problem of optimum coverage, domestic mobile robots normally switch between different behaviors. The switching can be done in an indeterministic way, that is, random switching between different behavior. The transition to a new behavior can be made based on time or traveled path. A more intelligent way of switching between behaviors can be made on analysis of signals from sensors. For example, if the robot tries to follow the wall, but it keeps losing the signal of the wall, then this might not be the best behavior in a given situation, so a different behavior should be used. If the robot detects that some part of space needs more attention (e.g., dirty floor in cleaning applications), it should switch to the behavior that can cover the particular part of space more thoroughly (e.g., motion in a spiral). An example of space covering with a floor cleaning robot is shown in Fig. 7.14.

7.6 ASSISTIVE MOBILE ROBOTS IN WALKING REHABILITATION THERAPY

Assistive robotics has huge potential in daily life assistance and its application growth is closely following industrial robotics. Robot assistants share a common working environment with humans, therefore harmless human-machine interactions need to be ensured. These solutions represent the new field of technology, tasked to improve the human-machine interaction in medical and domestic environments with the inclusion of technology that is able not only to detect, but also to perceive the environment [33].

The main topics of research and applications include assistance in daily life, remote presence (e.g., telepresence in remote diagnosis, social interactions with elderly people and those with physical impairments), rehabilitation robotics (e.g., physical therapy after injuries or neural impairments), and wearable robots and exoskeletons for elderly and physically challenged people.

Wheeled mobile robots are used in walking rehabilitation and as a walking aid [34–36]. Such systems are aimed to improve and restore the freedom of motion for people with reduced motor abilities by taking over or assisting in motion tasks [37]. They typically appear in the form of a smart wheelchairs and hoists where humans are directly interacting with the machine.

7.6.1 About

Several walking support and rehabilitation systems are commercially available and many more are in a research phase. Their purpose is to improve the rehabilitation process by enabling repetition of the exercise many times, possibilities to measure and evaluate progress, and to ease physical strain of the therapist. Common technological solutions include treadmill-based devices, exoskeletons, and movable support platforms. The latter solutions use wheeled mobile robots and an example of it is illustrated in the sequel.

Two examples of commercially available systems in rehabilitation of walking are e-go from THERA-Trainer [38] and Andago from Hocoma [39], which are shown in Fig. 7.15.

In the following some more insight will be illustrated on the THERA-Trainer e-go and on its extended functionality version (Hoist project) that was researched by Bošnak and Škrjanc [37].

7.6.2 Setup

The moving platform THERA-Trainer e-go is meant for rehabilitation of walking in clinical use. The platform consists of two motor-driven wheels using differential drive kinematics and four caster wheels for stability. The patient's waist is fastened to the vertical frame and the latter is at the bottom attached to the wheeled platform using elastic joints with limited range

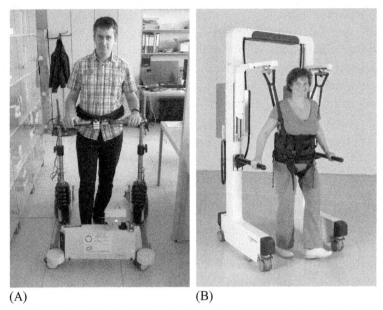

(A)   (B)

Fig. 7.15 Examples of wheeled robots dedicated to rehabilitation for walking: (A) modified e-go from THERA-Trainer (Hoist project) and (B) Andago from Hocoma. *(Right photo by Hocoma, Switzerland.)*

of angular motion to prevent the patient from falling. The therapist can manually control the platform motion and the patient should follow. The main goals of the therapy are [38] improved walking quality and increased walking speed, increased walking distance, and improved gait safety.

The main idea behind the Hoist project ([37], Fig. 7.16) is to equip the platform with additional sensors and algorithms to obtain autonomous (platform moves autonomously and the patient follows) or semiautonomous motion (platform moves considering to patient's intentions) of the platform and adapt rehabilitation strategies accordingly. The patient intentions are predicted by measuring the angle of the vertical frame according to the base platform. For additional safety and for localization and mapping (SLAM) a laser range scanner is also included.

7.6.3 Sensors

In the Hoist project DC motors that drive the wheels are equipped with incremental encoders in order to control the wheels' speed and enable implementation of odometry.

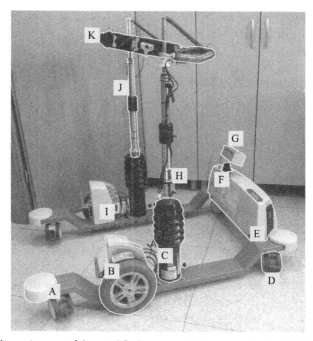

Fig. 7.16 The main parts of the modified THERA-Trainer e-go (Hoist project): (A) bumper, (B) motor drive with a wheel, (C) elastic joint, (D) caster wheel, (E) control unit and power supply (covered), (F) laser range scanner, (G) projector (part of future development), (H) tilt sensor on the strait, (I) tilt sensor on the base, (J) strait, and (K) bar with patient harness.

It turns out that patient intentions can be estimated from the tilt of the two rods that offer support to the patient. In the particular case, the tilt of each of the two straits is measured with a three-axis accelerometer and a three-axis gyroscope. A Kalman filter is used in estimation of the tilt (i.e., estimation of the direction of the gravity vector with respect to the local coordinate frame of the sensor) from the data provided by the accelerometer and gyroscope. Based on the measurement of the tilt of the two rods the displacement of the patient from the mobile robot center and also the orientation of the patient's pelvis (around the vertical axis) with respect to the robot's base can be determined. The mobile platform is equipped with an additional tilt sensor on the mobile robot base to enable estimation of the stait tilt with respect to the mobile robot base, which enables the use of the platform even on the inclined surface (Fig. 7.16).

To provide safe motion without collisions a laser range scanner is used and additional four bumpers to detect collisions (Fig. 7.16). A laser range scanner is also used for localization and mapping.

7.6.4 Localization and Mapping

Basic operation of the Hoist does not require localization and mapping. However, if the platform's pose and environment map are known the patient's walking routes can be preplanned and the therapists can monitor patient progress. According to the therapy goals the therapists can prescribe different walking routes (in the map of the environment) that are the most suitable to a particular patient. Based on the recorded performance data of the patient the therapist can gradually increase or decrease the level of difficulty to achieve optimal therapy results.

Capability of localization and mapping is required to enable advanced control capabilities that can enhance rehabilitation therapy. The mobile platform can autonomously guide the patient between arbitrary goals in the environment. It can notify the patient when he/she leaves the desired walking path prescribed by the therapist, or even help the patient to stay on the desired path whenever the patient needs some support.

7.6.5 Control

There are multiple levels of control that enable different modes of operation. At a low level, a control of the speed of each wheel based on the readings from the incremental encoder on the motor shaft is implemented. Since this mobile platform has differential drive, the reference tangential and angular speed of the mobile platform (or equivalently reference speeds of both wheels) can therefore be used as control actions in higher level control algorithms (cascade control).

One of the primary enhancements of the conventional robotic rehabilitation trainer is the capability of semiautonomous following the patient during the rehabilitation therapy. In this mode the patient controls the speed and turning of the platform. Based on the tilt measurements of both rods the control actions (reference wheel speeds) are calculated in a way that the mobile robot follows the patient. Determination of control signals is not straight-forward, since the control algorithm needs to be robust enough to filter out oscillations in tilt measurements that occur due to walking dynamics. In this control mode a negligible strain from the patient is required to move the platform, the patient determines the level of difficulty (speed) to his/her own capabilities, and the platform follows the patient seemingly. The therapist therefore no longer needs to drive the patient in the platform manually, but can only observe the patient and provide the patient with instructions.

The mobile platform in the Hoist project has been equipped with a laser range scanner that enables detection of obstacles in the range of 270 degrees up to a few meters in front of the mobile platform. The measured distances to the nearest obstacles in different directions are used implicitly in the control of the mobile platform in a way that collisions with obstacles are prevented. The control algorithm allows the patient to drive the mobile platform only in the directions that are free of obstacles. To enable soft touching (i.e., with zero velocity in the direction perpendicular to the obstacle) and also navigation through narrow passages (e.g., doors) the control actions are modified according to the estimated time to collide (based on predicted motion). In this way the patient can carry out the rehabilitation therapy safely alone without constant monitoring from the therapist.

Since a laser range scanner enables implementation of SLAM, this enables the design of some more advanced control strategies that can further enhance the rehabilitation therapy. Therefore a path following and trajectory tracking control algorithms can be implemented. In a particular case different control modes have been considered. The mobile platform is capable of autonomously driving the patient along the desired path. In this mode the patient has no control over the mobile platform and needs to follow it. Although if the mobile platform detects the patient is not capable of following (based on tilt measurements), the speed is reduced or the task is terminated. In another control mode, the patient should walk along the path designed by the therapist and the mobile platform is only following the patient and prevents too large deviations from the desired path. In yet another control mode, the mobile platform can also provide the patient some assistance in following the path whenever it detects (from the measurement of the pose error and straits tilt) help is required. Therefore, new rehabilitation therapy approaches can be achieved that were previously not possible.

7.6.6 Path Planning

In order to enable autonomous navigation of the mobile platform between arbitrary goals in the environment the path planning problem needed to be addressed. For this purpose the A* algorithm is used where the footprint of the mobile platform shape is taken into account. Due to the particular shape of the mobile platform, the assumption of a round robot for the purpose of path planning simplification is not preferred. Since the smallest circle around the axis of rotation that encircles the whole mobile platform is much larger than the actual outer convex envelope of the robot, the simplified path planning algorithm would fail to find the path between narrow (and also between relatively wide) passages. To overcome this problem the boundary of the mobile platform is approximated with a rectangle that is used during the path planning.

The result of A^* path planning is a sequence of points (states) that represent the feasible path that connect the start pose with the goal pose. To enable smooth trajectory tracking the path needs to be smoothed. For this purpose a smooth spline of parametric curves is fitted to the obtained sequence of path points. In path planning not only is it important to design the shape of the path, but also to design an appropriate path velocity and acceleration profile. The goal of path planning is therefore to also take into account some velocity and acceleration constraints that are imposed by the particular patient's capabilities. Path planning algorithms can therefore also enable the therapist to customize the rehabilitation therapy in order to achieve optimum results.

REFERENCES

[1] Google self-driving car project, 2016, http://www.google.com/selfdrivingcar/ (accessed 25.05.16).
[2] Driverless car market watch, 2016, http://www.driverless-future.com/ (accessed 25.05.16).
[3] Autonomous car, 2016, https://en.wikipedia.org/wiki/Autonomous_car (accessed 25.05.16).
[4] Navlab: The Carnegie Mellon University Navigation Laboratory, 2016, http://www.cs.cmu.edu/afs/cs/project/alv/www/index.html (accessed 25.05.16).
[5] D. Bochtis, H.W. Griepentrog, S. Vougioukas, P. Busato, R. Berruto, K. Zhou, Route planning for orchard operations, Comput. Electron. Agric. 113 (2015) 51–60.
[6] K. Zhou, A.L. Jensen, C.G. Sørensen, P. Busato, D.D. Bothtis, Agricultural operations planning in fields with multiple obstacle areas, Comput. Electron. Agric. 109 (2014) 12–22.
[7] S. Nuske, K. Wilshusen, S. Achar, L. Yoder, S. Narasimhan, S. Singh, Automated visual yield estimation in vineyards, J. Field Rob. 31 (5) (2014) 837–860.
[8] B. Zion, M. Mann, D. Levin, A. Shilo, D. Rubinstein, I. Shmulevich, Harvest-order planning for a multiarm robotic harvester, Comput. Electron. Agric. 103 (2014) 75–81.
[9] F.A.A. Cheein, R. Carelli, Agricultural robotics: unmanned robotic service units in agricultural tasks, IEEE Ind. Electron. Mag. 7 (2013) 48–58.
[10] J.R. Rosell-Polo, F.A. Cheein, E. Gregorio, D. Andújar, L. Puigdomènech, J. Masip, A. Escolà, Advances in structured light sensors applications in precision agriculture and livestock farming, in: D.L. Sparks (Ed.), Advances in Agronomy, vol. 133, Academic Press, 2015, pp. 71–112, Chapter 3.

[11] F.A.A. Cheein, SLAM-based maneuverability strategy for unmanned car-like vehicles, Robotica 31 (2013) 905–921.

[12] F.A.A. Cheein, J. Guivant, SLAM-based incremental convex hull processing approach for treetop volume estimation, Comput. Electron. Agric. 102 (2014) 19–30.

[13] F.A.A. Cheein, J. Guivant, R. Sanz, A. Escolà, F. Yandún, M. Torres-Torriti, J.R. Rosell-Polo, Real-time approaches for characterization of fully and partially scanned canopies in groves, Comput. Electron. Agric. 118 (2015) 361–371.

[14] F.A. Cheein, G. Steiner, G.P. Paina, R. Carelli, Optimized EIF-SLAM algorithm for precision agriculture mapping based on stems detection, Comput. Electron. Agric. 78 (2) (2011) 195–207.

[15] F.A. Cheein, G. Scaglia, Trajectory tracking controller design for unmanned vehicles: a new methodology, J. Field Rob. 31 (6) (2014) 861–887.

[16] L. de Paula Veronese, F.A. Cheein, T. Bastos-Filho, A.F. De Souza, E. de Aguiar, A computational geometry approach for localization and tracking in GPS-denied environments, J. Field Rob. 33 (7) (2015) 946–966.

[17] F.A. Cheein, D. Herrera, J. Gimenez, R. Carelli, M. Torres-Torriti, J.R. Rosell-Polo, A. Escolà, J. Arnó, Human-robot interaction in precision agriculture: sharing the workspace with service units, in: 2015 IEEE International Conference on Industrial Technology (ICIT), 2015, pp. 289–295.

[18] D.D. Bochtis, C.G. Sørensen, S.G. Vougioukas, Path planning for in-field navigation-aiding of service units, Comput. Electron. Agric. 74 (1) (2010) 80–90.

[19] N. Noguchi, H. Terao, Path planning of an agricultural mobile robot by neural network and genetic algorithm, Comput. Electron. Agric. 18 (2–3) (1997) 187–204.

[20] M. Spekken, S. de Bruin, J.P. Molin, G. Sparovek, Planning machine paths and row crop patterns on steep surfaces to minimize soil erosion, Comput. Electron. Agric. 124 (2016) 194–210.

[21] M. Shneier, R. Bostelman, Literature review of mobile robots for manufacturing, Technical Report, National Institute of Standards and Technology, U.S. Department of Commerce, 2015.

[22] Discover new horizons for flexible automation solutions—KMR iiwa, 2016, http://www.kuka-robotics.com/en/products/mobility/KMR_iiwa/start.htm (accessed 07.06.16).

[23] Easily & flexibly automate your internal transport, 2016, http://mobile-industrial-robots.com/en/ (accessed 07.06.16).

[24] Amazon robotics, 2016, https://www.amazonrobotics.com/ (accessed 07.06.16).

[25] TUG for industrial use, 2016, http://www.aethon.com/industrialtug/ (accessed 07.06.16).

[26] Lynx mobile robot, 2016, https://www.bastiansolutions.com/solutions/techno logy/industrial-robotics/industrial-robotic-solutions/mobile-robotics/lynx (accessed 07.06.16).

[27] Automated guided vehicles and AGV systems, 2016, http://www.egemin-auto mation.com/en/automation/material-handling-automation_ha-solutions/ agv-systems (accessed 07.06.16).

[28] Automated guided vehicle, 2016, https://en.wikipedia.org/wiki/Automated_ guided_vehicle (accessed 07.06.16).

[29] E. Galceran, M. Carreras, A survey on coverage path planning for robotics, Robot. Auton. Syst. 61 (12) (2013) 1258–1276.

[30] iRobot, iRobot(R) Create(R) 2 Open Interface (OI) Specification based on the iRobot(R) Roomba(R) 600, Technical Report, iRobot, 2015.

[31] E.U. Acar, H. Choset, Y. Zhang, M. Schervish, Path planning for robotic demining: robust sensor-based coverage of unstructured environments and probabilistic methods, Int. J. Robot. Res. 22 (7–8) (2003) 441–466.

[32] M. Dakulović, S. Horvatić, I. Petrović, Complete coverage D* algorithm for path planning of a floor-cleaning mobile robot, in: {IFAC} Proceedings Volumes, 18th {IFAC} World Congress, vol. 44, 2011, pp. 5950–5955.

[33] R.A. Cooper, H. Ohnabe, D.A. Hobson (Eds.), An introduction to rehabilitation engineering, in: Medical Physics and Biomedical Engineering, vol. 306, Taylor & Francis, 2006.

[34] P. Wenger, C. Chevallereau, D. Pisla, H. Bleuler, A. Rodić (Eds.), New trends in medical and service robots, human centered analysis, control and design, in: Mechanisms and Machine Science, vol. 39, Springer International Publishing, 2016.

[35] Z.Z. Bien, D. Stefanov (Eds.), Advances in rehabilitation robotics: human-friendly technologies on movement assistance and restoration for people with disabilities, in: Lecture Notes in Control and Information Sciences, vol. 306, Springer Science & Business, 2004.

[36] C.-Y. Lee, J.-J. Lee, Walking-support robot system for walking rehabilitation: design and control assistance device, Artif. Life Robot. 4 (4) (2000) 206–211.

[37] M. Bošnak, I. Škrjanc, Embedded control system for smart walking assistance device, IEEE Trans. Neural Syst. Rehabil. Eng. 2016, http://dx.doi.org/10.1109/TNSRE.2016.2553369.

[38] THERA-trainer e-go, 2016, https://www.thera-trainer.de/en/thera-trainer-products/gait/thera-trainer-e-go/ (accessed 15.07.16).

[39] Andago, 2016, https://www.hocoma.com/world/en/media-center/media-images/andago/ (accessed 15.07.16).

CHAPTER 8

Project Examples for Laboratory Practice

8.1 INTRODUCTION

This chapter contains several examples that demonstrate the applications of the methods presented in the book for solving different problems in autonomous mobile robotics. This chapter contains several practical examples that demonstrate somewhat more involved applications of the methods presented in the book on wheeled mobile robots. Each example is about a specific problem and/or particular problem solving methodology. Every example is structured as follows: at the beginning is a description of the problem, then one of the possible solutions is presented in more detail, and at the end some experimental results are shown. The examples are given in a way that the interested reader should be able to implement the presented algorithms in a simulation environment and also on a real wheeled mobile system. Therefore, the experiments are designed in a way that the proposed methods can be implemented on some common educational and research wheeled mobile robots. Although every example contains the recommended robotic platform for algorithm implementation, other robotic platforms could also be considered. Wheeled mobile robots used in the examples to obtain the real-world experimental results were the following: vacuum cleaning mobile robot iRobot Roomba, research mobile robot Pioneer 3-AT with SICK laser range-finder, Lego Mindstorms EV3 set, and soccer robots that are normally used in FIRA Micro Robot World Cup Soccer Tournaments.

The examples are intended to guide the reader through the process of solving different problems that occur in implementation of autonomous mobile robots. The examples should motivate the reader to implement the proposed solutions in a simulation environment and/or real mobile robots. Since most of the examples can be applied to various affordable mobile robots, the examples should give the reader enough knowledge to repeat the experiments on the same robotic platform or modify the example for a

Wheeled Mobile Robotics
http://dx.doi.org/10.1016/B978-0-12-804204-5.00008-1
419

particular custom mobile robot. The examples could be used as a basis for preparation of lab practice in various autonomous mobile robotic courses.

8.2 LOCALIZATION BASED ON BAYESIAN FILTER WITHIN A NETWORK OF PATHS AND CROSSROADS

8.2.1 Introduction

This demonstration exercise shows how the Bayesian filter can be applied for localization of a wheeled mobile system within a known map of the environment. The exercise also covers some basic principles of line-following, map building, and navigation within the map. Consider that the map of the environment is represented with lines that are drawn on the ground and represent the paths that the mobile system can drive on. The intersection of lines represent crossroads where the mobile system can select different paths. For the sake of simplicity, only two lines can intersect in a single crossroad and there should be some minimum distance between every pair of crossroads. The goal of the exercise is to develop a mobile system that is able to localize within the network of interconnected lines (map) and navigate between arbitrary locations in the map. The algorithms should be implemented on a two-wheeled differentially driven mobile platform that is equipped with an infrared light reflection sensor, which is used for detection of a line that is drawn on the ground. Through the rest of this exercise we will consider that a Roomba mobile system is used for implementation. However, any other mobile system that is capable of line following and crossroad detection could also be used. Although this is only an exercise, an appropriate solution to the problem has some practical applications in real-life environments, such as automated warehouses and distribution centers, where wheeled mobile platforms are used to transfer objects (parcels) between different stations.

The exercise is composed of several sections, where in every section a small subsystem is developed that is required for implementation of the complete localization solution. The first section is devoted to implementation of line following and crossroad selection control algorithm. The second section is about implementation of the odometry algorithm. In the third section an approach for map building is presented, and the final section presents the method for localization within the map.

8.2.2 Line Following

The first task consists of implementation of a control algorithm for line following and crossroad detection. In our case, we consider using an infrared light sensor for line detection. Consider which other sensors do you think could also be used for line detection, and how.

Exercise. First, write a simple program that enables reading the data from the light sensors. Then, read the sensor values at different colors of the ground and at different distances from the ground. Based on the sensor readings try to answer the following questions. Under which circumstances can the light sensors be used for detection of a black line on a white background? Could the sensor be used for detection of holes in the ground and detection of black and white ground at the same time?

Exercise. Write a program for line following based on light sensors. How many sensors are required for implementation of the line following control algorithm? Which arrangements of one, two, and more light sensors can be used for implementation of line following control algorithm? Which setup do you find to be the most robust? Based on the arrangement of the light sensors on a mobile system, select the most appropriate control algorithm for line following and implement it. Try to implement a simple on-off control algorithm and some version of a PID controller. Explain the advantages and disadvantages of the implemented control algorithm with respect to some other possible solutions.

Exercise. Extend the line following algorithm with crossroad detection and path selection algorithm. Once the mobile system reaches the crossroad, it should stop. Based on the command (continue forward, left, right, backward) the mobile system should take the appropriate path. Implement an algorithm for selection of four different paths in a crossroad.

Exercise. Based on the available sensors, implement some safety mechanisms that enable safe operation of the mobile system in the environment.

The bottom of the iRobot Roomba (series 500) mobile system is equipped with four light sensors that are normally used for detection of holes (cliffs) in the ground. The front two light sensors can be used for line detection and the side sensors for crossroad detection. Fig. 8.1 shows how the IR sensors can be used for line tracking (front sensors) and for crossroad detection (side sensors).

8.2.3 Odometry

Observing the rotation of the wheels, the position of the mobile platform in the environment can be tracked using the approach that is known as odometry. Assume that the rotation of a wheel is measured with an encoder with counter γ. Let us assume that the value of the encoder counter is increasing when the wheel rotates in a forward direction and the value of

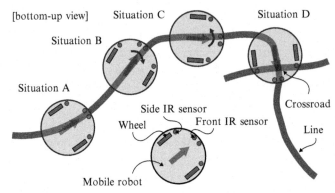

Fig. 8.1 Line-following and crossroad-detecting mobile system based on IR sensors.

the counter is decreasing when the wheel rotates in the opposite direction. The temporal difference of counter values $\Delta\gamma(k) = \gamma(k) - \gamma(k-1)$ is proportional to the relative rotation of the wheel. Based on the temporal difference of counter values of the left $\Delta\gamma_L(k)$ and right $\Delta\gamma_R(k)$ encoder, the relative rotations of the left $\Delta\varphi_L(k)$ and right $\Delta\varphi_R(k)$ wheel can be calculated, respectively:

$$\Delta\varphi_L(k) \propto \Delta\gamma_L(k)$$
$$\Delta\varphi_R(k) \propto \Delta\gamma_R(k)$$

(8.1)

The pose $(x(k), y(k), \varphi(k))$ of the mobile system can be approximated with a discrete solution of the direct kinematic model of the wheeled mobile system, using the Euler integration method of first differences:

$$x(k+1) = x(k) + \Delta d(k) \cos\varphi(k)$$
$$y(k+1) = y(k) + \Delta d(k) \sin\varphi(k)$$
$$\varphi(k+1) = \varphi(k) + \Delta\varphi(k)$$

(8.2)

The relative distance $\Delta d(k)$ and relative rotation $\Delta\varphi(k)$ of the mobile system can be determined from relative rotations of the wheels:

$$\Delta d(k) = \frac{r}{2}(\Delta\varphi_R(k) + \Delta\varphi_L(k))$$
$$\Delta\varphi(k) = \frac{r}{L}(\Delta\varphi_R(k) - \Delta\varphi_L(k))$$

(8.3)

where L is the distance between the wheels and r is the radius of the wheels.

Exercise. Implement the algorithm for odometry. Read the encoder values of the left and right wheel, implement the odometry algorithm, and calibrate the odometry model parameters. Propose a calibration procedure, that is based on measurement of the mobile system pose at two moments in time with an external measurement device. If required, take into account the encoder counter overflow. Determine the maximum allowable tangential and angular velocities of a mobile system that do not cause wheel slippage.

Exercise. Evaluate the performance of an implemented odometry algorithm in terms of accuracy. The odometry model (8.2) was obtained using the discrete Euler integration of first differences. Try to use one of the more accurate discrete odometry models, reimplement the odometry algorithm, and compare the results.

Exercise. The odometry enables estimation of the mobile system pose; therefore, it also enables estimation of the rotation angle when the mobile system is spinning about the vertical axis of rotation. Because the four paths in crossroads are assumed to be intersecting at 90-degree angles, the odometry can be used in the path selection algorithm. If required, improve the path selection algorithm implemented in the previous section with the rotation measurement provided by odometry.

In Fig. 8.2 an example of a path obtained by odometry is shown.

8.2.4 Map Building

In this section a method for a semiautomated supervised method of map building is presented. The lines drawn on the ground intersect at several crossroads, for example, in Fig. 8.3. For the sake of simplicity, we assume that all the crossroads connect four paths and that all paths are connected to crossroads at both ends. The shapes of the paths between the crossroads are assumed to be arbitrary. We label all the K crossroads with a set $\mathcal{K} = \{K_i\}_{i=1,2,...,K}$ and all the P direct paths between the crossroads with a set $\mathcal{P} = \{p_i\}_{i=1,2,...,P}$. Every path is given a dominant direction in a way that if direction of path p_i is from crossroad K_j to K_l, the path $\bar{p}_i$ is in the opposite direction, that is from crossroad K_l to K_j. The nondominant paths are gathered in a set $\bar{\mathcal{P}} = \{\bar{p}_i\}_{i=1,2,...,P}$. In the case in Fig. 8.3 there are $K = 3$ crossroads, which are labeled with a set $\mathcal{K} = \{A, B, C\}$, and $P = 6$ direct paths, which are labeled with a set $\mathcal{P} = \{a, b, c, d, e, f\}$. The spatial representation of the crossroads and paths can therefore be represented as a graph, like shown on the right-hand side of Fig. 8.3. When redrawing spatial representation of the map into a graph form, the connections (paths) into nodes (crossroads) should be drawn in the same order as in the real map.

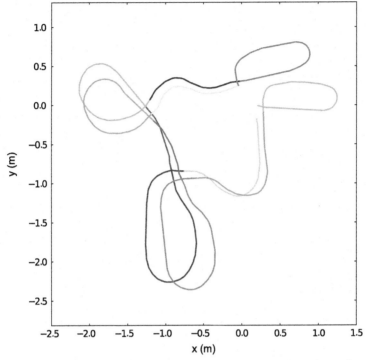

Fig. 8.2 An example of path obtained by odometry when traveling along all possible direct paths in a map with three crossroads. The accumulation of odometry error is clearly visible.

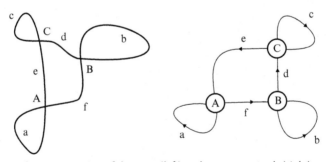

Fig. 8.3 Spatial representation of the map (*left*) and map as a graph (*right*).

This is required to be able to determine which path in a crossroad is on the left, right, and straight ahead if the mobile system enters the crossroad from an arbitrary path. An alternative is to save the order of paths in a crossroad in a table, like it is shown in Table 8.1, where the locations of the paths in

Table 8.1 Connections (paths) in the nodes (crossroads)

Node	A	B	C
Connections	$a, f, \bar{e}, \bar{a}$	$b, d, \bar{f}, \bar{b}$	$c, e, \bar{d}, \bar{c}$

Table 8.2 Ending nodes (crossroads) of the connections (paths)

Connection	a	b	c	d	e	f
Nodes	A, A	B, B	C, C	B, C	C, A	A, B

a crossroad can be determined from the ordered list of paths. The ending nodes of all the direct connections can also be represented in the form of a table (Table 8.2), where the order of nodes is important to distinguish between dominant and nondominant path direction. Therefore the entire map can be represented in a form of a graph or an equivalent table; these data structures are simple to implement in a computer's memory.

With the current representation of the map in a form of a graph (or table) the spatial relations between the crossroads (nodes) are lost as are lost the shapes of the paths (connections). Since we assume all the crossroads are of the same shape (four paths perpendicular to adjacent paths), the entire map can be built when the shapes of all the direct paths between crossroads are known. The shapes of all the direct paths can be recorded with the process of odometry (Fig. 8.4). Although the process of odometry has some obvious disadvantages, it should be sufficient to determine the

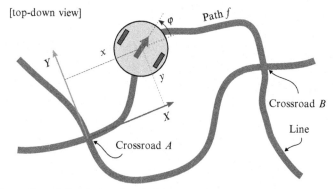

Fig. 8.4 Recording of the shape of path *f* during the line following from crossroad *A* to crossroad *B*.

approximate shape of the path between two nodes. Therefore, if the shapes of at least some paths are different enough, it should be possible to determine the location of the mobile system within a map, based on the stored map data (graph representation of the map and shapes of all the direct paths). A map of the environment is represented with a set of crossroads $\mathcal{K}$, a set of paths $\mathcal{P}$, and recorded shapes of the paths with the process of odometry. In the process of map building all this information needs to be written into appropriate data structures and saved in a nonvolatile memory area.

Exercise. Assign node labels to all the crossroads and mark all the paths with connection labels. Draw a graphical representation of the map and present the graph in the form of a table. Choose appropriate data structures to represent the graphical representation of the map and insert the map into the code. The data structures should be selected in a way that enables storing a map with different numbers of crossroads and paths.

Exercise. Record the shapes of all the direct paths with the odometry (Fig. 8.5) as follows: put the mobile system at the beginning of path in the starting crossroad, let the mobile system follow the line until the next (ending) crossroad, save the recorded shape of the path, and repeat for all the other paths. The map of the environment should be saved into a nonvolatile memory area (e.g., as a text file on a hard disk).

Exercise. Write a program that automates the process of map building to an appropriate level.

8.2.5 Localization

In this section an algorithm for localization within a map built in the previous section is presented. The goal of localization is to determine the current path followed by the mobile system, the direction of driving (from the starting node to the ending node), and the distance from the starting or ending node.

Assume the mobile system is traveling along the paths by line following and randomly choosing the next path at the crossroads. Comparing the current odometry data with the odometry data stored in a map, the mobile system should be able to localize within the map. The basic concept of localization is as follows. The current odometry data from paths between crossroads is compared to all the odometry data stored in a map. Based on predefined goodness of fit, the probabilities of all the paths are calculated.

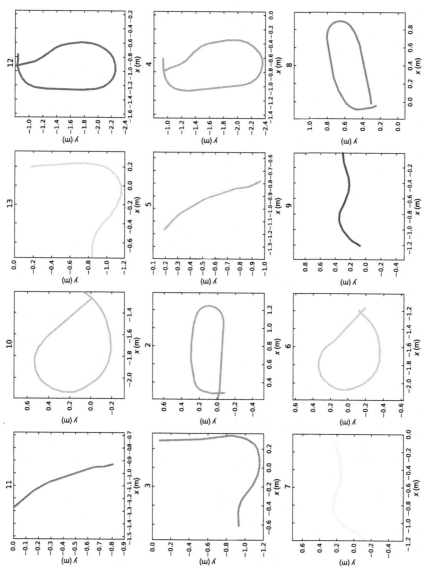

Fig. 8.5 An example of all recorded paths obtained by odometry when traveling along all possible direct paths in a map with three crossroads (see Fig. 8.2). The accumulation of odometry error is clearly visible.

Once the probability of a specific path is significantly larger than the probabilities of all the other paths, the location of the mobile system is considered to be known (with some level of probability).

The localization problem can be solved in a framework of Bayesian filter. For the sake of implementation simplicity, assume that the localization algorithm is executed only once the mobile system reaches a new crossroads, just before the next path is selected. The location of the mobile system is therefore known, if it is known from which path the mobile system reached the crossroad. Since every path has two directions, we need to consider the probabilities of all the direct paths in both directions. For every path $x_k \in \mathcal{P} \bigcup \bar{\mathcal{P}}$ the following belief functions need to be calculated:

$$bel_p(x_k) = \sum_{x_{k-1} \in \mathcal{P} \bigcup \bar{\mathcal{P}}} p(x_k|x_{k-1}, u_k)bel(x_{k-1}) \qquad (8.4)$$

$$bel(x_k) = \frac{1}{\eta_k}p(z_k|x_k)bel_p(x_k) \qquad (8.5)$$

where $\eta_k = \sum_{x_k \in \mathcal{P} \bigcup \bar{\mathcal{P}}} p(z_k|x_k)bel_p(x_k)$.

Assume that the localization algorithm is first executed when we arrive at the first crossroad. In the initial step the location of the mobile system is unknown; therefore, all the paths have the same level of probability. The mobile system therefore leaves the crossroad at an arbitrary direction based on the given control action (go forward, left, right, or back). The path selection control algorithm is assumed to select the desired path with some level of probability. All the exiting paths in a crossroad can be marked with labels s_1, s_2, s_3, and s_4 as shown in Fig. 8.6. The probabilities of selecting the path s_1 if the current path is one of the paths $\bar{s}_1$, $\bar{s}_2$, $\bar{s}_3$, and $\bar{s}_4$ need to be determined for all the possible control actions. For example, for control action *forward* the probabilities could be as follows:

$$p(X_k = s_1|X_{k-1} = \bar{s}_1, U_k = forward) = 0.05$$
$$p(X_k = s_1|X_{k-1} = \bar{s}_2, U_k = forward) = 0.1$$
$$p(X_k = s_1|X_{k-1} = \bar{s}_3, U_k = forward) = 0.75 \qquad (8.6)$$
$$p(X_k = s_1|X_{k-1} = \bar{s}_4, U_k = forward) = 0.1$$

Exercise. Based on the robustness of the path selection algorithm determine the probabilities $p(x_k|x_{k-1}, u_k)$ of selecting path $X_k = s_1$ if the current path is one of the paths $X_{k-1} \in \{\bar{s}_1, \bar{s}_2, \bar{s}_3, \bar{s}_4\}$ for all the possible control actions $U_k \in \{forward, left, right, back\}$.

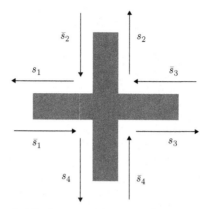

Fig. 8.6 The directions of paths in a general crossroad.

Since the crossroads are symmetric, the following relations are assumed to hold:

$$
\begin{aligned}
p(X_k = s_1 | X_{k-1} = \bar{s}_3, u_k) &= p(X_k = s_2 | X_{k-1} = \bar{s}_4, u_k)\\
&= p(X_k = s_3 | X_{k-1} = \bar{s}_1, u_k)\\
&= p(X_k = s_4 | X_{k-1} = \bar{s}_2, u_k)\\
p(X_k = s_1 | X_{k-1} = \bar{s}_2, u_k) &= p(X_k = s_2 | X_{k-1} = \bar{s}_3, u_k)\\
&= p(X_k = s_3 | X_{k-1} = \bar{s}_4, u_k)\\
&= p(X_k = s_4 | X_{k-1} = \bar{s}_1, u_k)\\
p(X_k = s_1 | X_{k-1} = \bar{s}_4, u_k) &= p(X_k = s_2 | X_{k-1} = \bar{s}_1, u_k)\\
&= p(X_k = s_3 | X_{k-1} = \bar{s}_2, u_k)\\
&= p(X_k = s_4 | X_{k-1} = \bar{s}_3, u_k)\\
p(X_k = s_1 | X_{k-1} = \bar{s}_1, u_k) &= p(X_k = s_2 | X_{k-1} = \bar{s}_2, u_k)\\
&= p(X_k = s_3 | X_{k-1} = \bar{s}_3, u_k)\\
&= p(X_k = s_4 | X_{k-1} = \bar{s}_4, u_k)
\end{aligned}
\tag{8.7}
$$

For all the exiting paths a table of corresponding entering paths can be written in a table form that proves to be convenient when implementing the localization algorithm (see Table 8.3).

The mobile system then leaves the crossroad, follows the line and records the odometry data until the next crossroad is detected. Once the mobile system arrives in the next crossroad the recorded odometry between the previous and current crossroad is compared to all the odometry data stored in a map. To determine the probability $p(z_k | x_k)$ some distance measure for

Table 8.3 Entering paths for all the exit paths from example in Fig. 8.3

s_1	a	b	c	d	e	f	$\bar{a}$	$\bar{b}$	$\bar{c}$	$\bar{d}$	$\bar{e}$	$\bar{f}$
$\bar{s}_1$	$\bar{a}$	$\bar{b}$	$\bar{c}$	$\bar{d}$	$\bar{e}$	$\bar{f}$	a	b	c	d	e	f
$\bar{s}_2$	a	b	c	$\bar{b}$	$\bar{c}$	$\bar{a}$	e	f	d	$\bar{e}$	$\bar{f}$	d
$\bar{s}_3$	e	f	d	b	c	a	$\bar{f}$	$\bar{d}$	$\bar{e}$	$\bar{c}$	$\bar{a}$	b
$\bar{s}_4$	$\bar{f}$	$\bar{d}$	$\bar{e}$	f	d	e	$\bar{a}$	$\bar{b}$	$\bar{c}$	c	a	b

comparing the shapes of odometry data needs to be defined. The most probable path is therefore the path with the smallest distance value.

Exercise. Define a distance measure for comparing the paths obtained by odometry. Take into account that the mobile system may not be traveling along the path at constant velocity. Experimentally evaluate the performance of the selected distance measure.

In Fig. 8.7 an idea for definition of path comparison measure is shown. Different distance measures can be defined, for example: $SASE = \frac{1}{m}\sum_{i=1}^{m}\sqrt{(x_i - x_{o,i})^2 + (y_i - y_{o,i})^2}$ or $SSSE = \frac{1}{m}\sqrt{\sum_{i=1}^{m}(x_i - x_{o,i})^2 + (y_i - y_{o,i})^2}$, where m is the number of comparison points (the measures are also shown in Fig. 8.8).

The belief function in Eq. (8.5) can now be updated. The evolution of belief distribution is shown in Fig. 8.9. In the initial step the probabilities of all the paths are equal (Fig. 8.9A). When the mobile system travels the first path between two crossroads, the probability of paths based on measured data is calculated (Fig. 8.9B) and the belief function is updated (Fig. 8.9C). After several steps, when the mobile system visits several crossroads the most probable path according to the belief function should be the real path.

Exercise. Implement the localization algorithm based on Bayesian filter.

Exercise. Extend the localization in a way that enables reporting of mobile system position on the path as a relative distance from the starting crossroad.

Exercise. Try to improve the localization algorithm in a way that the path belief distributions are updated not only in crossroads but also during the path tracking.

Exercise. Implement an algorithm for path finding from an arbitrary location within a map to the desired end location. Implement the control algorithm for driving the mobile system across the map to the desired

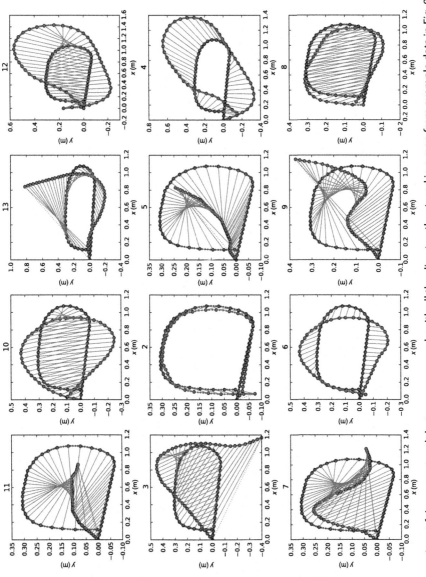

Fig. 8.7 Comparison of the current path between crossroads with all the direct paths stored in a map from example data in Fig. 8.5.

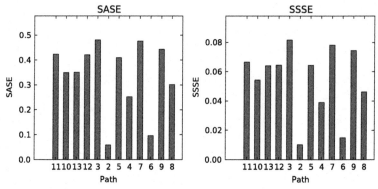

Fig. 8.8 Two distance measures calculated from path comparison in Fig. 8.7. The *SASE* (*SSSE*) measure is a sum off all the absolute (*squared*) lengths of the *lines* between the shapes of two paths in Fig. 8.7.

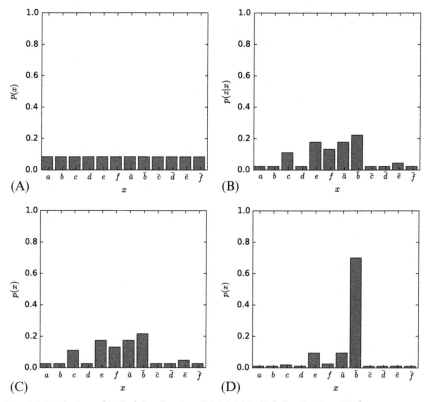

Fig. 8.9 Evolution of belief distribution. (A) Initial belief distribution, (B) first measurement probability distribution, and (C) belief distribution after first update and (D) after several updates.

location. Try to implement the localization algorithm in a way that enables finding an alternative route in case of path obstruction.

8.3 LOCALIZATION BASED ON EXTENDED KALMAN FILTER USING INDOOR GLOBAL POSITIONING SYSTEM AND ODOMETRY

8.3.1 Introduction

Mobile robots are normally required to be able to localize in the environment in order to perform some desired tasks autonomously. In order to give a mobile system the localization capability the mobile system must be equipped with an appropriate sensor that can determine the position of the mobile system with respect to the external (global) coordinate frame. This can be achieved in various ways. In this exercise a localization approach that acquires the position from an external measurement system is considered. In an outdoor environment the global position can be determined with a global navigation satellite system (GNSS) that is nowadays available all over the world (the accuracy may vary). Since the GNSS docs not work in an indoor environment, some special localization systems have been developed for indoor localization. Some are based on measuring wireless network signal strength or time of flight between several static transmission points, while others use image-based object tracking technology. Global positioning systems normally measure only the position but not also the orientation of the tracked object. However, for implementation of autonomous navigation, the wheeled mobile system must be aware of its position and orientation.

8.3.2 Experimental Setup

This exercise covers how the pose of the wheeled mobile robot in space can be estimated based on odometry data and measurement of the position of the mobile system within the environment. The considered setup (Fig. 8.10) consists of a wheeled mobile system equipped with wheel encoders and a measurement unit that is able to determine the position of the tag (mounted on top of a mobile system) with respect to some anchor nodes with fixed and known positions in the environment. A mobile robot with differential drive (Section 2.2.1) will be used; however, the presented approach can be used on any other wheeled mobile drive. The experimental setup consists of a wheeled mobile system that has rotation encoders on each wheel.

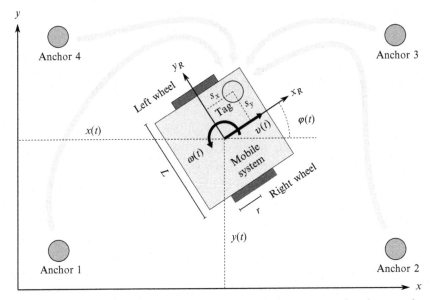

Fig. 8.10 Wheeled mobile system with a differential drive equipped with a tag that measures its position with respect to the external stationary anchors.

Based on the encoder readings, an odometry approach can be used for estimation of mobile system pose. Since the pose obtained by odometry is susceptible to drift, the pose estimate must be corrected by external position measurements. The position of the mobile system is obtained using an indoor positioning system, but this measurement system has limited precision. This is a sensor fusion problem, that for the given setup can be solved in the framework of an extended Kalman filter (EKF). Once the localization algorithm is implemented and tuned, the mobile system can be extended with the capability of autonomous navigation between desired locations within the environment.

8.3.3 Extended Kalman Filter

Here an extended Kalman filter (EKF) is considered for solving the mobile system localization problem. The EKF (Section 6.5.2) is an extension of a conventional Kalman filter to nonlinear systems with linear approximation of covariance matrices. The algorithm has a two-step structure, prediction of the states based on the current action (Eq. 6.38), and correction of the predicted states when a new system output measurement becomes available (Eq. 6.39). To implement the EKF, a model of the system and model of the noise must be given. The nonlinear model of the system (6.37) has to be linearized around the current estimated state in order to obtain the matrices (6.40)–(6.42). The most challenging part in implementation of the EKF is an appropriate modeling of system kinematics and noise, which come as a result from various processes. Noise is normally time or state dependent, and is hard to measure and estimate.

Prediction Step

Let us develop a kinematic odometry model that takes into account system uncertainties. The robot wheels are equipped with encoders γ_L and γ_R that track the orientation of left and right wheel, respectively. The relative encoder readings of the left and right wheel, $\Delta\gamma_{\{L,R\}}(k) = \gamma_{\{L,R\}}(k) - \gamma_{\{L,R\}}(k-1)$, are assumed to be proportional to the angular velocities of the wheels, $\omega_{\{L,R\}}(k) \propto \Delta\gamma_{\{L,R\}}(k)$. Let us assume that the readings of rotation encoders of each wheel, γ_L and γ_R, are disturbed by additive white noise $w^T(k) = [w_{\Delta\gamma_L}(k) \; w_{\Delta\gamma_L}(k)]$. Distinguish between angular velocities, annotated with Greek letter ω, and noise, annotated with letter w. This noise is assumed to encapsulate not only the measurement noise, but also the effects of all the noises that are due to different sources, including wheel slippage. The encoder readings $\Delta\hat{\gamma}_{\{L,R\}}(k)$ are therefore assumed to be related to the true encoder values $\Delta\gamma_{\{L,R\}}(k)$ as follows:

$$\begin{bmatrix} \Delta\hat{\gamma}_L(k) \\ \Delta\hat{\gamma}_R(k) \end{bmatrix} = \begin{bmatrix} \Delta\gamma_L(k) \\ \Delta\gamma_R(k) \end{bmatrix} + \begin{bmatrix} w_{\Delta\gamma_L}(k) \\ w_{\Delta\gamma_R}(k) \end{bmatrix} \tag{8.8}$$

Assume that the distribution of noise is normal (Gaussian) with zero mean, $w(k) \sim \mathcal{N}(0, Q)$. For the application of EKF the distribution of the noise need not be normally distributed, but the noise should be unbiased. The noise covariance matrix Q is assumed to be diagonal, since no cross-correlation between the wheel noise is assumed:

$$Q = \begin{bmatrix} \sigma^2_{\Delta\gamma_L} & 0 \\ 0 & \sigma^2_{\Delta\gamma_R} \end{bmatrix} \tag{8.9}$$

Based on the relative encoder readings, $\Delta\hat{\gamma}_{\{L,R\}}(k)$, the relative longitudinal traveled distance $\Delta\hat{d}(k)$ and relative rotation $\Delta\hat{\varphi}(k)$ of the wheeled mobile system can be determined:

$$\Delta\hat{d}(k) = r\alpha \frac{\Delta\hat{\gamma}_R(k) + \Delta\hat{\gamma}_L(k)}{2}$$

$$\Delta\hat{\varphi}(k) = r\alpha \frac{\Delta\hat{\gamma}_R(k) - \Delta\hat{\gamma}_L(k)}{L} \tag{8.10}$$

The discrete kinematic model of a differential drive (Eq. 2.3) can be written in the following form:

$$\hat{x}(k + 1) = \hat{x}(k) + \Delta\hat{d}(k) \cos\hat{\varphi}(k)$$
$$\hat{y}(k + 1) = \hat{y}(k) + \Delta\hat{d}(k) \sin\hat{\varphi}(k) \qquad (8.11)$$
$$\hat{\varphi}(k + 1) = \hat{\varphi}(k) + \Delta\hat{\varphi}(k)$$

Since the wheels are disturbed by noise, the estimated pose of the wheeled mobile system using the recursive estimation approach (Eq. 8.11) is susceptible to drift.

Exercise. Combine Eqs. (8.8), (8.10) into Eq. (8.11). Substitute the terms $\Delta\gamma_{\{L,R\}}(k)$ in Eq. (8.10) for Eq. (8.8), and the terms $\Delta d(k)$ and $\Delta\varphi(k)$ in Eq. (8.11) for Eq. (8.10).

Exercise. Based on Eqs. (8.10), (8.11) implement the odometry algorithm. For some practical implementation advices and notes, follow the description and go through the exercises given in Section 8.2.3. Design and describe a procedure for calibration of odometry algorithm parameters.

The parameters in the odometry algorithm (T, r, L, and α) could all be measured. However, to implement the odometry algorithm only the parameters $Tr\alpha$ and L need to be known. These two parameters can be obtained with an appropriate calibration procedure. For example, the mobile system can be drawn along a circular arc, and the two parameters $Tr\alpha$ and L can be obtained from two equations (8.10) if the length and relative rotation of the circular arc are known.

The kinematic model of wheel odometry (Eq. 8.11) without noise can be used to make predictions about the mobile system pose and can therefore be used in the prediction step of the EKF. In the prediction step of the EKF, the new state of the system (mobile system pose) is predicted based on the state estimate from the previous time step (using the nonlinear system model, Eq. 8.11 without noise in our case), and also the new covariance matrices are calculated (from linearization of the system around the previous estimate).

Exercise. Using model (8.10), derive the Jacobian matrices (6.40), (6.41), and calculate the matrix $F(k)QF^T(k)$ that is required in calculation of the prediction step (6.38) in the EKF algorithm.

The Jacobian matrices required in the prediction step of the EKF are determined from Eqs. (6.40), (6.41), as follows:

$$A(k) = \begin{bmatrix} 1 & 0 & -T r\alpha \frac{\Delta\gamma_D(k)+\Delta\gamma_L(k)}{2} \sin\varphi(k) \\ 0 & 1 & T r\alpha \frac{\Delta\gamma_D(k)+\Delta\gamma_L(k)}{2} \cos\varphi(k) \\ 0 & 0 & 1 \end{bmatrix}$$

$$F(k) = \frac{T r\alpha}{2L} \begin{bmatrix} L\cos\varphi(k) & L\cos\varphi(k) \\ L\sin\varphi(k) & L\sin\varphi(k) \\ -2 & 2 \end{bmatrix}$$

$$(8.12)$$

Taking into account the input noise covariance matrix (8.9), the matrix FQF^T that occurs in Eq. (6.38) is then

$$F(k)QF^T(k) = \frac{T^2 r^2\alpha^2 \left(\sigma_{\Delta\gamma_R}^2 + \sigma_{\Delta\gamma_L}^2\right)}{4} \begin{bmatrix} \cos^2\varphi(k) & \cos\varphi(k)\sin\varphi(k) & 0 \\ \cos\varphi(k)\sin\varphi(k) & \sin^2\varphi(k) & 0 \\ 0 & 0 & \frac{4}{L^2} \end{bmatrix}$$

$$+ \frac{T^2 r^2\alpha^2 \left(\sigma_{\Delta\gamma_R}^2 - \sigma_{\Delta\gamma_L}^2\right)}{2L} \begin{bmatrix} 0 & 0 & \cos\varphi(k) \\ 0 & 0 & \sin\varphi(k) \\ \cos\varphi(k) & \sin\varphi(k) & 0 \end{bmatrix}$$

$$(8.13)$$

Considering that the variance of all the diagonal elements in Eq. (8.9) are equal to each other, the last term in Eq. (8.13) is zero.

Fig. 8.11 shows simulation results, where constant input velocities were applied to the robot. The bottom row in Fig. 8.11 shows time evolution of the state predictions with state error covariance boundaries of one

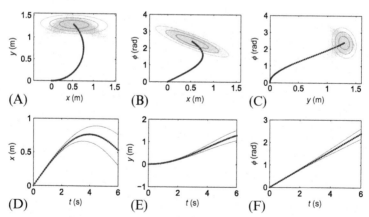

Fig. 8.11 Model prediction simulation. *Top row*: 2D state projections with final state error covariance matrix ellipses and simulated cloud of final states: (A) phase plot $y(x)$, (B) phase plot $\varphi(x)$ and (C) phase plot $\varphi(y)$. *Bottom row*: time evolution of the states with error variance margins of one standard deviation: (D) signal $x(t)$, (E) signal $y(t)$ and (F) signal $\varphi(t)$.

standard deviation. The top row in Fig. 8.11 shows the predicted path (2D projections of the generalized robot coordinates) and contours of the projected state error covariance matrix in the final state. The plots in the top row also show the predicted clouds of final state after many repetitions of the model simulation with added simulated noise. Notice the dissimilar shape between the final state clouds and the estimated error covariance matrices (ellipses).

Exercise. Implement the prediction step of the EKF in a simulation environment and on a real mobile system. In a simulation environment evaluate the performance of state predictions, plot system pose, and plot the covariance elements as a function of time, as shown in Fig. 8.11.

Exercise. Assume that a white noise is independently added to the signals Δd and $\Delta \varphi$ instead of the signals $\Delta \gamma_L$ and $\Delta \gamma_R$, that is, $\Delta \hat{d} = \Delta d + w_{\Delta d}$ and $\Delta \hat{\varphi} = \Delta \varphi + w_{\Delta \varphi}$. Calculate the new state error covariance equations and modify the state prediction algorithm. Notice that, in this case, the values in covariance matrix $\mathbf{Q}$ have different meanings.

Correction Step

Once a new measurement of the system output state is available, a current state estimate can be updated in the correction step (6.39) of the EKF. To be able to update the system states a model of the system output needs to be determined.

$$\boldsymbol{h}(k) = \begin{bmatrix} \cos \varphi(k) & -\sin \varphi(k) \\ \sin \varphi(k) & \cos \varphi(k) \end{bmatrix} \begin{bmatrix} s_x \\ s_y \end{bmatrix} + \begin{bmatrix} x \\ y \end{bmatrix} + \begin{bmatrix} v_x \\ v_y \end{bmatrix} \qquad (8.14)$$

The measurement noise $\boldsymbol{v} = [v_x \; v_y]$ can be modeled as white noise with zero mean and Gaussian distribution, $\boldsymbol{v} \in \mathcal{N}(\boldsymbol{0}, \boldsymbol{R})$.

Exercise. Based on the measurement model (8.14), calculate the Jacobian matrix $\boldsymbol{C}(k)$ as defined in Eq. (6.42). Consider three forms of the covariance matrix $\boldsymbol{R}$: positive and symmetric matrix, diagonal matrix with positive elements, and diagonal matrix with all the elements positive and equal to each other. Which form of the covariance matrix is the most appropriate for your problem? How would the model change, if the noise covariance matrix would be given in the mobile robot local coordinate frame instead of the global coordinate frame?

The Jacobian matrix $\boldsymbol{C}(k)$ is determined from the nonlinear measurement model (8.14), as follows:

$$\boldsymbol{C}(k) = \begin{bmatrix} 1 & 0 & -s_x \sin \varphi(k) - s_y \cos \varphi(k) \\ 0 & 1 & s_x \cos \varphi(k) - s_y \sin \varphi(k) \end{bmatrix} \qquad (8.15)$$

Normally we can assume a diagonal measurement noise covariance matrix:

$$\mathbf{R}(k) = \begin{bmatrix} \sigma_{v_x}^2 & 0 \\ 0 & \sigma_{v_y}^2 \end{bmatrix} \qquad (8.16)$$

where $\sigma_{v_x}^2$ and $\sigma_{v_y}^2$ are the measurement noise variances along the x and y directions, respectively, and are normally equal to each other. If there is significant correlation between noise signals $v_x(k)$ and $v_y(k)$ a general form of the noise covariance matrix $\mathbf{R}$ should be used. In a special case, when the sensor is mounted in the rotation center of the mobile system ($s_x = 0$ and $s_y = 0$) the model (8.14) simplifies, the last column in Jacobian matrix $\mathbf{C}(k)$ becomes zero, and therefore all the measurement Jacobian matrices become constant unity matrices.

Exercise. Implement the correction step of the EKF as described by Eqs. (6.39) based on evaluation of position sensor characteristics determine the values of the measurement noise covariance matrix.

Accurate noise modeling is one of the most crucial parts in implementation of EKF. Assume that the variance of the noise on each wheel is dependent on the speed of the wheel, so that the following relation holds $\sigma_{\omega\{L,R\}}^2(k) = \delta \omega_{\{L,R\}}^2(k)$ that yields [1]:

$$\mathbf{Q}(k) = \begin{bmatrix} \delta \omega_L^2(k) & 0 \\ 0 & \delta \omega_R^2(k) \end{bmatrix} \qquad (8.17)$$

In this noise model, the parameter δ defines the proportion of the noise with respect to the wheel velocity.

Exercise. Modify the EKF algorithm to use the noise model (8.17). Tune the model parameters according to the system noise properties. Evaluate the performance of the modified EKF algorithm.

An example of the estimated path is shown in Fig. 8.12. Although the initial pose estimate is wrong, the localization algorithm converges to true mobile robot pose. The filter also estimates mobile robot heading, which cannot be obtained from the global positioning system directly. The heading could be determined from differentiation of position measurements, but in the presence of noise this is extremely unreliable without appropriate signal prefiltering, which introduces lag. A Kalman filtering approach as shown in this example is therefore preferred.

Exercise. As an additional exercise, implement a simple control algorithm that enables autonomous navigation between arbitrary goals in the environment. Use additional proximity sensors to detect obstacles and try to implement an obstacle avoidance algorithm.

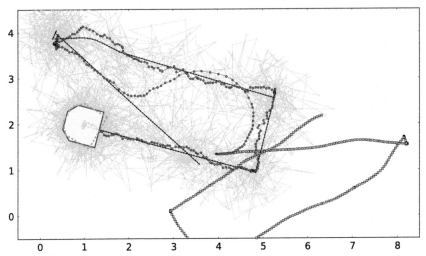

Fig. 8.12 Odometry, measured positions, and estimated path.

8.4 PARTICLE-FILTER-BASED LOCALIZATION IN A PATTERN OF COLORED TILES

8.4.1 Introduction

Localization is one of the most fundamental problems in wheeled autonomous robotics. It is normally solved in the framework of Bayesian filter. Among various filtering approaches (Kalman filter, EKF, UKF, etc.) particle filter [2] is gaining more and more attention. In this exercise it will be shown how the particle filter can be used to solve the localization of a wheeled mobile robot on a plane made of colored tiles.

8.4.2 Experimental Setup

The wheeled mobile robot is equipped with wheel encoders and a color sensor that can determine the color of the surface below the robot (Fig. 8.13). The mobile robot is assumed to have differential drive with axle length L and wheel radius r. The mounting of the color sensor on the robot is displaced from the robot's center of rotation for s_x and s_y in robot longitudinal and lateral direction, respectively. The ground plane consists of square, colored tiles of equal sizes that cover a rectangular area of $M \in \mathbb{N}$ times $N \in \mathbb{N}$ tiles. The mobile robot is only allowed to move inside the area of tiles that is enclosed with a wall boundary. There are $C \in \mathbb{N}$ distinct colors of the tiles. The arrangement of the colored tiles on the ground (i.e., the map) is assumed to be known. Fig. 8.14 shows a map of the

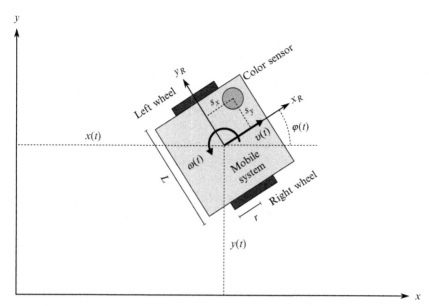

Fig. 8.13 Wheeled mobile robot with differential drive equipped with a color sensor.

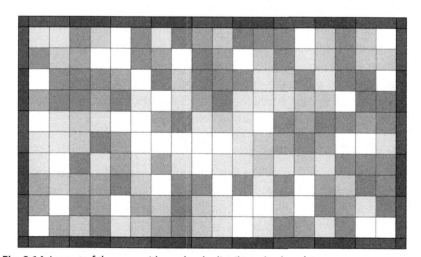

Fig. 8.14 Layout of the map with randomly distributed colored squares.

environment with dimensions of 180×100 cm^2, where $M = 10$, $N = 18$, and $L = 6$. The considered setup can be built with the elements provided in the Lego Mindstorms EV3 kit, which is also used in this example to obtain some real-world data.

Exercise. Prepare the flat surface with colored tiles on which the mobile robot can move. Make a rectangular grid in which square tiles can be placed. Enclose the rectangular area with a wall boundary in order to prevent the mobile robot from leaving the working area. Choose a set of several distinguishable colors for colored tiles and distribute colors randomly across the grid on a flat surface (Fig. 8.14). The colored tiles can be printed or painted on the ground, or square color stickers can be used.

Exercise. Assemble a mechanically working mobile robot that will be able to move around the flat environment; use of differential drive is recommended. Mount a color sensor in an appropriate location and distance from the ground to enable reliable color detection of the tiles on the floor. An example of mobile robot design is shown in Fig. 8.13.

8.4.3 Manual Control

This exercise is about implementation of the localization algorithm. Therefore, it suffices to drive the mobile robot only manually, but once the localization algorithm is implemented and evaluated also some automatic driving control algorithm can be developed.

Exercise. In order to drive the mobile robot around the surface write a simple program for remote control of the mobile robot using a keyboard or some other operator-input device. Implement a control of wheel velocity based on the feedback provided by wheel encoders. Design the control algorithm in a way that the desired tangential and angular reference velocities of the robot can be set instead of only the desired reference velocities of the wheels. In the case of EV3 motors, motor speed control is already implemented and only needs to be enabled.

8.4.4 Wheel Odometry

Wheel odometry is one of the basic localization approaches that can be used to estimate mobile robot pose, given a known initial pose. It is known that the approach is susceptible to drift, due to wheel slippage, but the error over short distances is normally relatively small. The drift can be eliminated if additional information about position is available; in this case a color sensor is considered. Estimated motion based on the internal kinematic model of the wheeled mobile robot is normally used in the prediction step of every localization algorithm that is based on Bayesian filtering, including the particle filter. Wheel odometry is normally implemented based on incremental encoder measurements (motors in the Lego Mindstorms EV3 kit are all equipped with encoders). In order to fine tune the parameters (wheel radius and axle length) of the odometry algorithm to a particular wheeled mobile system design, an appropriate calibration is normally employed.

Exercise. Implement odometry based on wheel encoder readings. Be sure to properly calibrate and evaluate the precision of odometry. Follow the instructions and tips for implementation of odometry given in Section 8.2.3.

8.4.5 Color Sensor Calibration

The odometry drifting can be eliminated with the readings of the tile color below the wheeled mobile robot, since the arrangement of the colored tiles (map of the environment) is assumed to be known. For detection of the surface color a color sensor needs to be used (Mindstorms EV3 kit contains a color sensor). Suppose that a single color sensor measurement is encoded as RGB triple, that is, value of red, green, and blue components in the RGB color space. The color sensor is required to distinguish between C different colors of the tiles that make up the ground surface. For this purpose the raw sensor readings (RGB values) need to be classified into one of the C color classes (an additional class can be defined for any other color. The color sensor needs to be calibrated for the particular mounting of the sensor on the mobile robot and the ground surface. Robust color classification is normally simpler to achieve in HSV (hue–saturation–value) that in the RGB color space. Therefore RGB to HSV color conversion (Eq. 5.87) needs to be used. Fig. 8.15 shows that color readings of six different colored tiles are clearly separable in the HSV color space.

Exercise. Write a program for reading color sensor data. Convert the obtained raw RGB values to HSV color space using Eq. (5.87). Make several measurements of the same color and represent the readings of all colors in a graph like shown in Fig. 8.15. Evaluate susceptibility of sensor readings to illumination changes and small pose perturbations with respect to the ground surface. Based on the sensor readings reposition the sensor if necessary. Also a cover that protects the sensor from exterior changes in illumination may be added. Based on the collected color measurement data implement a color clustering algorithm. A set of rules can be determined that enable clustering of colors, that is, rules should define cluster boundaries. Ensure that the rules are given in a way that cluster boundaries are wide enough to enable detection of color in different illumination conditions, but also narrow enough to prevent misclassification during color transitions.

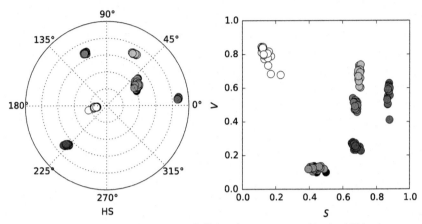

Fig. 8.15 Readings of the color sensor of all the tiles represented in the HSV color space.

8.4.6 Particle Filter

Color sensor measurements and odometry data provide partial information about the global pose of the wheeled mobile robot. A particle filter can be used in order to join information from different sources to obtain a better estimate about the robot pose. A color sensor alone cannot provide information about the mobile robot heading directly, but in the framework of a particle filter it can be used to remove the odometry drift and enable global localization, that is, the initial pose of the mobile robot need not be known in advance. The basic idea behind the particle filter is that the probability distribution of the state estimate is approximated with a set of particles. In this case particles represent hypotheses about the pose of the wheeled mobile robot. The algorithms consists of several steps: particle initialization, prediction, importance resampling, and evaluation of the most plausible wheeled mobile robot pose based on particle distribution. The steps of the algorithm are depicted in Fig. 8.16 (working area is smaller than in Fig. 8.14 to improve visibility).

Exercise. Although not necessary, write a program for recording of all sensor data (odometry and color measurements) and control actions into a file. This can simplify the development of the particle filter, since the algorithm can be developed offline. The same data can be used multiple times during algorithm testing, debugging, and tuning. A simple comma-separated file format is sufficient for this purpose. Then drive the robot

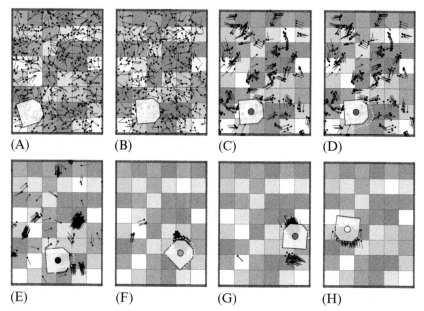

Fig. 8.16 Evolution of the particle filter: (A) particle initialization, (B) prediction step, (C) importance resampling, (D) evaluation of most plausible robot pose, (E) second iteration, (F) third iterations, (G) fourth iteration, and (H) after a few more iterations. A *solid polygon* represents true robot pose (*colored circle* is current sensor reading), and a *dashed polygon* is the estimated robot pose and *black* vectors are particles.

around the playground to collect sensor data. Write a program that enables loading of the stored data.

In the initial step of particle filter initial population of $N \in \mathbb{N}$, the particles need to be determined. In the particular case each particle represents a hypothesis about the wheeled robot pose, that is, the ith particle represents the possible robot position and heading $q_i^T = [x_i, \ y_i, \ \varphi_i]$. The particles should be distributed in accordance with the initial probability distribution of the wheeled robot pose. In a case that the initial robot pose is not known, a uniform distribution can be used, that is, the particles should be spread uniformly over the entire state space. Fig. 8.16A shows the initial uniform distribution of particles.

Exercise. Implement the initialization step of the particle filter.

The wheeled mobile robot can then make a move. The first step of the algorithm is prediction of the particles. Each particle is updated based on the assumed change in mobile robot pose. This is normally made according to the estimated pose change based on odometry data $\Delta d(k)$ and $\Delta \varphi(k)$, that is, for the ith particle the update equations are (a simple Euler integration of the differential drive kinematic model is used) the following:

$$x_i(k+1) = x_i(k) + \Delta d(k) \cos(\varphi_i(k))$$
$$y_i(k+1) = y_i(k) + \Delta d(k) \sin(\varphi_i(k)) \qquad (8.18)$$
$$\varphi_i(k+1) = \varphi_i(k) + \Delta\varphi(k)$$

To the predicted pose estimate of each particle a randomly generated noise with appropriate distribution can be added to model disturbances, for example, wheel slippage. The predicted distribution of the particles after prediction of the particles is shown in Fig. 8.16B.

Exercise. Implement the prediction step of the particle filter. Observe evaluation of the particles as the mobile robot drives around. Evaluate the influence of different noise levels of the random noise that is added to the predicted particle states.

The color of the tile below the mobile robot is then measured. For the sake of simplicity assume that the color sensor is positioned in the rotation center of the mobile robot ($s_x = 0$ and $s_y = 0$). Sensor measurement is used to determine the importance of each particle based on comparison of the measured color with the tile color below the particular particle. The particles that agree with the measurement are given a high importance score and the others are given a small score. The score of a particular particle can be determined according to the distance to the nearest tile that is of the same color as measured by the sensor.

Exercise. Use the measured sensor data in order to determine importance of each particle.

Assigned importance to each particle is used in the particle resampling step. Particles are randomly sampled in a way that the probability of selecting a particular particle from a set is proportional of the particle importance score. Therefore, the particles with larger importance score are expected to be selected more times. The updated distribution of the particles after the resampling step is shown in Fig. 8.16C.

Exercise. Implement particle resampling based on the determined importance scores.

The final step consists of particle evaluation in order to estimate the wheeled mobile robot pose from the distribution of the particles. This can be made in various ways, as a (weighted) average of all the particles or based on the probabilities of the particles. The estimated pose is shown in Fig. 8.16D.

Exercise. Evaluate distribution of the particles in order to determine the current estimate of the wheeled mobile robot.

The mobile robot then makes another move, and in the next time-step the algorithm proceeds to the next prediction step. Fig. 8.16E–G shows the estimated pose after three more particle filter iterations, and Fig. 8.16H shows the estimated pose after several more iterations. In the demonstrated case the particle filter was updated only when the value of the color measurement changed. More details about the particle filter algorithm can be found in Section 6.6.

Exercise. Evaluate the performance of the particle filter and experiment with algorithm parameters. Evaluate the performance of the algorithm in the case of *robot kidnapping* (repositioning the robot within the field while the estimation algorithm is paused) and in the presence of large wheel slippage.

Exercise. As an additional task, once the localization is working as expected, implement a simple control algorithm that enables autonomous driving of the mobile robot among arbitrary goals in the environment. Add additional sensors to enable detection of obstacles (field boundaries).

8.5 CONTROL OF VEHICLES IN A LINEAR FORMATION

8.5.1 Introduction

Future intelligent transportation systems are hard to imagine without autonomous wheeled vehicles. One important application is an automated platoon of vehicles on highways that can drive autonomously in so-called virtual train formation. Vehicles in this platoon should precisely and safely follow their leader vehicles with minimal safety distance required. This would increase vehicle density on highways, avoid traffic jams, and increase safety.

In this exercise a control algorithm of mobile robots driving in linear formation will be designed. To obtain automated driving formation precise sensor systems are required, measuring global information such as GPS or relative information such as distance and bearing or both. Here we will implement a decentralized control strategy relying only on a relative sensor—laser range finder (LRF). Suppose that each vehicle has LRF to measure the distance and the azimuth of its leading vehicle. This information is then used to record and track the path of the leader vehicle. The leading vehicle path is recorded in local coordinates of the follower using odometry and LRF measurement (distance D and azimuth α) as

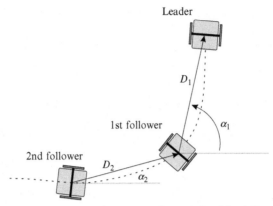

Fig. 8.17 Robots driving in a linear formation where each of the followers estimate the path of its leader and drive that path using trajectory tracking control.

shown in Fig. 8.17. The follower vehicles than follow the estimated paths of their leader vehicles using trajectory tracking control approaches as presented in Section 3.3. It is well known that robot localization using odometry is pruned to accumulation of different errors (e.g., wheel sleep, sensor and actuator noise, and the like). Odometry is therefore usable only for short-term localization. As illustrated a latter absolute pose error made using odometry is not significant in linear formation control because only the relative position (D and α) among the vehicles is important. The latter is measured precisely from LRF and only short-term odometry localization is used to estimate the part of the leader path that the follower needs to drive in the near future. Some more information can be found in [3].

The exercise contains three sections, where subsystems are developed which are later integrated in the implementation of vehicle formation control. The section deals with localization using odometry, and follows the second section describes estimation of the leading vehicle path. In the third section trajectory tracking control is adopted for the formation control task.

8.5.2 Localization Using Odometry

Odometry is the simplest localization approach where the pose estimate is obtained by integration of the robot kinematic model at known velocities of the robot. If differential robot velocities change at discrete time instants

$t = kT_S$, $k = 0, 1, 2, \ldots$ where T_s is the sampling interval then the next robot pose (at $(k+1)$) is obtained from the current pose (k) and the current velocities (see Section 2.2.1).

$$x(k + 1) = x(k) + v(k) T_s \cos(\varphi(k))$$
$$y(k + 1) = y(k) + v(k) T_s \sin(\varphi(k)) \tag{8.19}$$
$$\varphi(k + 1) = \varphi(k) + \omega(k) T_s$$

Exercise. Implement localization of a robot with differential drive using odometry, where robot initial pose is not important. Therefore, the initial pose can be in the origin. Validate localization by comparing the estimated path of the robot with the known true path.

8.5.3 Estimating Reference Trajectory

Knowing the current robot pose in global coordinates obtained by odometry, the robot can also define the location of the leader robot from a known relative position among them. This relative position is obtained from LRF measurement containing distance $D(k)$ and azimuth to the leader $\alpha(k)$. The main idea is to record positions of the leader, estimate the leader trajectory, and then the follower robot can use this trajectory as the reference and follows it.

The leader position $(x_L(k), y_L(k), \varphi_L(k))$ is therefore estimated as follows:

$$x_L(k) = x(k) + D(k) \cos(\varphi(k) + \alpha(k))$$
$$y_L(k) = y(k) + D(k) \sin(\varphi(k) + \alpha(k)) \tag{8.20}$$

If the position of the leader robot at some time $t \neq kT_s$ is required then interpolation is used:

$$x_L(t) = x_L(kT_s) + \frac{t - kT_s}{Ts} (x_L (k + 1) T_s - x_L (kT_s))$$
$$y_L(t) = y_L(kT_s) + \frac{t - kT_s}{Ts} (y_L (k + 1) T_s - y_L (kT_s)) \tag{8.21}$$

Exercise. Estimate the path of the leading vehicle by the known position of the follower vehicle and measured distance and azimuth to the leader. To validate the code you can drive the leading vehicle on some path (manually or automatically by some controller) and then compare the true and estimated robot paths' shape. Note that origins of the estimated leader coordinates may not be the same as the true leader pose due to the unknown

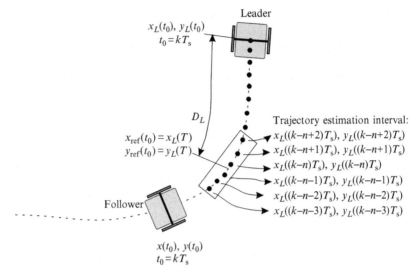

Fig. 8.18 The follower needs to follow the leader at distance D_L measured on the trajectory. Reference trajectory is estimated using six points around the current follower reference $x_{ref}(t_0)$, $y_{ref}(t_0)$.

(e.g., set to zero) initial pose of the follower in localization with odometry. However, the shape of the estimated leader trajectory must be the same as the true one.

Suppose each follower must track its leader at distance D_L measured on the trajectory of the leading robot. So at current time t_0 the position of the leading vehicle at time $t = T$ ($x_L(T)$, $y_L(T)$) needs to be found such that the distance between the current leader position at t_0 and previous leader position at $t = T$ is D_L. This position presents the reference for the follower (see Fig. 8.18).

Exercise. Estimate the reference position for the follower robot so that the follower would be distance D_L (measured on the path) behind the leader.

To follow the leader's path the follower needs to know its reference trajectory not only the current reference position. The estimated trajectory of the leader $\hat{x}_L(t)$, $\hat{y}_L(t)$ is therefore estimated in the parametric polynomial form:

$$\hat{x}_L(t) = a_2^x t^2 + a_1^x t + a_0^x$$
$$\hat{y}_L(t) = a_2^y t^2 + a_1^y t + a_0^y$$

(8.22)

taking six leader positions, three before the reference position and three after as shown in Fig. 8.18. The coefficients of the polynomials a_i^x and a_i^y ($i = 0, 1, 2$) are calculated using the least-squares method.

The reference pose of the following vehicle at current time t_0 is determined using

$$
\begin{bmatrix} x_{\text{ref}}(t_0) \\ y_{\text{ref}}(t_0) \\ \varphi_{\text{ref}}(t_0) \end{bmatrix} = \begin{bmatrix} \hat{x}_L(T) \\ \hat{y}_L(T) \\ \hat{\varphi}_L(T) \end{bmatrix} = \begin{bmatrix} a_2^x T^2 + a_1^x T + a_0^x \\ a_2^y T^2 + a_1^y T + a_0^y \\ \text{atan2}\left(2a_2^y T + a_1^y, 2a_2^x T + a_1^x\right) \end{bmatrix} \quad (8.23)
$$

where atan2 $(\cdot, \cdot)$ is a four-quadrant version (see Eq. 6.43) of the inverse tangent (in Matlab use atan2 function).

Exercise. Estimate referenced trajectory as a second-order polynomial function of time. Use six samples around the estimated reference position to identify parameters of polynomials (Eq. 8.23) using the least squares method.

8.5.4 Linear Formation Control

Each follower vehicle needs to follow the estimated reference trajectory (8.23) using trajectory tracking control. For a trajectory controller use the nonlinear controller defined in Section 3.3.6 as follows:

$$
v_{\text{fb}} = v_{\text{ref}} \cos e_\varphi + k_x e_x
$$
$$
\omega_{\text{fb}} = \omega_{\text{ref}} + k_y v_{\text{ref}} \frac{\sin e_\varphi}{e_\varphi} e_y + k_\varphi e_\varphi \quad (8.24)
$$

where the feedforward signals (see Section 3.3.2) are obtained from the estimated reference trajectory defined in Eq. (8.23) by

$$
v_{\text{ref}}(t_0) = \sqrt{\dot{x}_{\text{ref}}^2 + \dot{y}_{\text{ref}}^2} = \sqrt{(2a_2^x T + a_1^x)^2 + (2a_2^y T + a_1^y)^2} \quad (8.25)
$$

and

$$
\omega_{\text{ref}}(t_0) = \frac{\dot{x}_{\text{ref}}(t)\ddot{y}_{\text{ref}}(t) - \dot{y}_{\text{ref}}(t)\ddot{x}_{\text{ref}}(t)}{\dot{x}_{\text{ref}}^2(t) + \dot{y}_{\text{ref}}^2(t)} = \frac{(2a_2^x T + a_1^x)2a_2^y - (2a_2^y T + a_1^y)2a_2^x}{(2a_2^x T + a_1^x)^2 + (2a_2^y T + a_1^y)^2} \quad (8.26)
$$

A control tracking error is computed considering the follower's actual posture $(x(t_0),\ y(t_0),\ \varphi(t_0))$ and its reference posture $(x_{\text{ref}}(t_0),\ y_{\text{ref}}(t_0),\ \varphi_{\text{ref}}(t_0))$ using Eq. (3.35).

Exercise. Implement a linear formation off three robots. The first robot can be driven manually or it automatically follows some trajectory.

While each follower observes its leader (e.g., robot 2 observes robot 1 and robot 3 observes robot 2), estimates its reference trajectory, and computes control actions according to control law (8.24). Test the platoon performance on simulation and/or on real robots.

8.6 MOTION AND STRATEGY CONTROL OF MULTIAGENT SOCCER ROBOTS

8.6.1 Introduction

This exercise illustrates how wheeled mobile robots can be programmed to play a soccer game autonomously. Each robot is presented as an agent that can sense the environment and has some knowledge about the environment, its own goals, and ability to act according to sensed information using reactive behaviors. When more agents are applied to reach a common goal (e.g., play a robot soccer game) they need to cooperate and coordinate their actions. And each agent has knowledge how to solve basic tasks such as motion, ball kicking, defending goals, and the like.

The exercise starts with implementation of simple motion control algorithms on a two-wheeled differentially driven mobile platform such as trajectory tracking, reaching of reference pose, and obstacle avoidance. These control algorithms are then used to program different behaviors of agents (e.g., goalie, attacker, defense player). Each behavior needs to consider current sensed information and map this information to actions. For example, the goalie needs to position itself in a goal line to block the ball shots, and the attacker should include skills on how to approach the ball and kick it to the goal and how to avoid obstacles (playground boundaries and other players). Behaviors can be reactive or they can also include some cognitive capabilities such as prediction of ball motion, and actions should be planned accordingly to reach the ball more optimally and in a shorter time.

The exercise will be illustrated on a robot soccer test bed shown in Fig. 8.19, which consists of small two-wheeled robots (10 robots generating 2 teams) of size 7.5 cm cubed, an orange golf ball, rectangular playground of size 2.2×1.8 m, digital color camera, and personal computer. The camera is mounted above the playground and is used to track objects (ball and players) from their color information [4]. Basic operation of the agent program is shown in Fig. 8.19. A personal computer first assigns behavior to each agent. Behavior contains an algorithm for defining the direction and velocity of an agent. The direction controller calculates the desired translational and angular velocity of the robot and sends it as a

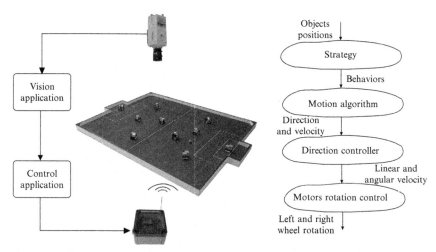

Fig. 8.19 Robot soccer setup (*left*) and principle of soccer agent control (*right*).

reference command to the robot via a radio connection. An onboard PID controller achieves desired robot wheel velocities. Note that any other kind of wheeled mobile robot that is capable of sensing its position and position of the ball can be used in this exercise.

The exercise comprises three sections, where in each section a subsystem is developed that is required for implementation of the complete multiagent system. In the first section the desired agents' simple motion capabilities are programmed to retrieve the ball and kick it in the opponent's goal, to defend the goal, and to avoid collisions. The second section introduces a behavior-based operation of a single agent. The last section introduces a multiagent system where more agents are playing cooperatively in a team considering a simple game strategy.

8.6.2 Motion Control

The first task requires coding a motion control algorithm in order to approach the ball from the right direction and push it toward the goal. On the other side the goalkeeper should control its motion in front of the goal line to block the goal shuts.

Attacker Motion Control

The basic operation of an attacking robot is to move to the ball and push it toward the goal. Therefore its reference pose contains the reference position, which is the position of the ball (x_{ref}, y_{ref}), while the reference

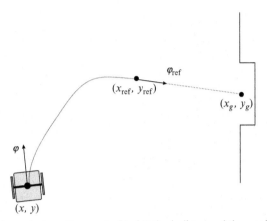

Fig. 8.20 Attacking robot motion control to hit the ball toward the goal.

orientation (φ_{ref}) is the desired direction of the ball motion after the kick (see Fig. 8.20). To hit the goal location (x_g, y_g) the reference orientation is

$$\varphi_{ref} = \text{atan2}\left(y_g - y_{ref}, x_g - x_{ref}\right)$$

where atan2 ($\cdot, \cdot$) is the four-quadrant version (see Eq. 6.43) of the inverse tangent (in Matlab use `atan2` function).

Exercise. Write a control algorithm that would drive the robot in the reference pose. For controller implementation you can use one of the basic approaches described in Section 3.2.3. Test the algorithm for various robot and ball positions.

Is the performance of the control algorithm sufficient in cases when the ball (x_b, y_b) moves? For moving the ball a robot should predict the ball's future position and intercept it there. A general predicted ball position solution for nonstraight robot motion is difficult to compute analytically. Therefore, suppose the robot can move on a straight line as shown in Fig. 8.21. From the dot product,

$$\frac{[v_b \cos \varphi_b , v_b \sin \varphi_b]^T \cdot [x - x_b , y - y_b]^T}{v_b \sqrt{(x - x_b)^2 + (y - y_b)^2}} = \cos \alpha$$

angle α is computed. Then angle β that defines the required relative robot driving direction is expressed using law of sines $\frac{v_b}{\sin \beta} = \frac{v}{\sin \alpha}$ and angle $\gamma = \pi - \alpha - \beta$. Finally from $\frac{d}{\sin \gamma} = \frac{v_b t}{\sin \beta}$ the time needed for the robot to reach the ball reads

$$t = \frac{d \sin \beta}{v_b \sin \gamma} \tag{8.27}$$

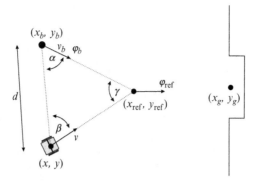

Fig. 8.21 Predicting future ball motion and estimating the point where the robot with traveling velocity v can intercept the ball.

And the predicted ball or reference position is

$$x_{\text{ref}} = x_b + v_b t \cos \varphi_b$$
$$y_{\text{ref}} = y_b + v_b t \sin \varphi_b$$

Note, however, that this calculated predicted reference position can only be successfully used for controlling robot motion if the robot initially is facing the reference position and the desired kicking direction is close to φ. This, however, is rarely true because the true robot path is usually longer than the direct line supposed in the calculation. The robot initially is not facing the predicted reference position and the final kick direction needs to be toward the goal. If one can approximately estimate the true path length l to reach point x_{ref}, y_{ref} with orientation φ_{ref} then the robot speed can be adjusted to reach the predicted ball in time t, as follows $v_n = \frac{l}{t}$.

Exercise. Extend the algorithm so that the robot will not move toward the current ball position but it will predict the future ball position where it can reach the ball. Test the control algorithm and try to adjust parameters so that robot will hit the moving ball.

Straight Accelerated Kick Motion Control

As already mentioned the moving ball can be kicked quite accurately in the goal if the robot's initial position faces toward the goal x_g, y_g with accelerated motion. The situation is illustrated in Fig. 8.22. An unknown reference point can be calculated from current ball position and its known motion:

$$x_{\text{ref}} = x_b + v_b \cos \varphi_b t$$
$$y_{\text{ref}} = y_b + v_b \sin \varphi_b t \tag{8.28}$$

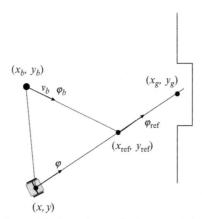

Fig. 8.22 Straight accelerated kick to the goal. The required robot acceleration and predicted ball position need to be calculated.

or from the following robot position,

$$x_{\text{ref}} = x + (x_g - x)p$$
$$y_{\text{ref}} = y + (y_g - y)p \tag{8.29}$$

where t is time when the robot can reach the ball with the requested straight motion and p scalar parameter. Combining Eqs. (8.28), (8.29) results in the following matrix equation:

$$\begin{bmatrix} x_g - x & -v_b \cos \varphi_b \\ y_g - y & -v_b \sin \varphi_b \end{bmatrix} \begin{bmatrix} p \\ t \end{bmatrix} = \begin{bmatrix} x_b - x \\ y_b - y \end{bmatrix} \tag{8.30}$$

which is in form $\boldsymbol{Au} = \boldsymbol{b}$ and its solution is $\boldsymbol{u} = \boldsymbol{A}^{-1}\boldsymbol{b}$. A feasible solution requires $0 < p < 1$, meaning that the interception ball point is between the robot and the goal. The reference point then reads

$$x_{\text{ref}} = x_b + v_b t \cos \varphi_b$$
$$y_{\text{ref}} = y_b + v_b t \sin \varphi_b$$

and the required robot acceleration a considering its current velocity v reads

$$a = \frac{2\left(\sqrt{(x_{\text{ref}} - x)^2 + (y_{\text{ref}} - y)^2} - vt\right)}{t^2} \tag{8.31}$$

Exercise. Write a motion control algorithm that will kick a moving ball in the goal using straight accelerated motion. If a solution is possible compute the required acceleration using Eq. (8.31). This is computed in each control loop iteration and a new robot translational speed is set to $v(k) = v(k-1) + a(k-1)T_s$, where $v(k)$ is the current velocity and $v(k-1)$ is the previous velocity. Angular velocity needs to be controlled so that the robot will drive in the desired direction $\varphi_{\text{ref}} = \arctan\frac{y_g - y}{x_g - x}$. Test the algorithm's performance.

Goalie Motion Control

Goal keeping robot motion control could be as follows. Suppose the robot drives on a goal line between points $\overline{T}$ and $\underline{T}$, as shown in Fig. 8.23. Its current reference position x_{ref}, y_{ref} on the dotted line needs to be calculated according to the ball position x_b, y_b and its velocity vector v_b in direction φ_b to predictively move to the location where the ball will pass the dotted line. From known data compute the time in which the ball will cross the goal line and the reference position

$$t = \frac{x_g - x_b}{v_b \cos \varphi_b}$$

$$x_{\text{ref}} = x_g$$

$$y_{\text{ref}} = y_b + v_b t \sin \varphi_b$$

$$\varphi_{\text{ref}} = \frac{\pi}{2}$$

where x_g is the coordinate of the goal line.

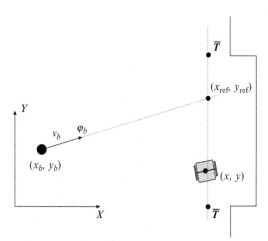

Fig. 8.23 Motion control of goal-defending robot.

To drive the robot along the goal line a linear trajectory tracking controller described in Section 3.3.5 could be used, where reference velocities (v_{ref}, ω_{ref}) in the reference position are zero. The control law therefore simplifies to

$$v = k_x e_x$$
$$\omega = k_\varphi e_\varphi + \text{sign}(v) k_y e_y$$

where gains k_x, k_y, and k_φ can be constants determined by experimenting. Local errors e_x, e_y, and e_φ are calculated using Eq. (3.35) and e_φ must be in the range $-\pi < e_\varphi \leq \pi$.

Exercise. Write a control algorithm that would position the goal keeping robot on the goal line (dotted line in Fig. 8.23) to defend the goal. Test the robot's performance by pushing the ball from different locations toward the goal. Observe the controller performance in the case of large tracking errors $|e_\varphi| > \frac{\pi}{2}$; can this be improved?

Exercise. Modify the goalkeeper motion control algorithm so that it will never rotate for more than $\frac{\pi}{2}$. This can be obtained by driving the robot in the opposite direction when $|e_\varphi| > \frac{\pi}{2}$, and modify the orientation error as follows:

$$e_\varphi = \begin{cases} e_\varphi - \pi & e_\varphi > \frac{\pi}{2} \\ e_\varphi + \pi & e_\varphi < -\frac{\pi}{2} \end{cases}$$

Obstacle Avoidance

If a playground is surrounded by a fence or there are more players on the playground, then the robot's motion control algorithm should also consider obstacle avoidance. One way to achieve such behavior is the use of the potential field method, similarly as introduced in Section 4.2.4.

Exercise. Extend one of the attacker control algorithms also with obstacle avoidance capabilities.

8.6.3 Behavior-Based Agent Operation

Suppose you have three robots playing in the same team. The first robot is the goalkeeper, the second is the attacker, and the third is the midfield player. For each of the players or agents there are different behaviors possible inside their role (e.g., goalkeeper) according to the current situation in the game (their position, position of the ball, or other players). In the current moment each agent needs to decide which behavior it will perform.

This behavior selection can be realized by reactive behavior-based architecture, which means that from the currently sensed environment information and internal agent knowledge the most appropriate behavior is selected autonomously. If there are more behaviors possible, then the one with the higher priority is chosen.

For instance the goalkeeper needs to be positioned in the goal area; if this is not the case then it needs to drive in the goal. This goal-arriving behavior can be set as the highest priority. If it is in the goal then it needs to defend the goal. It needs to calculate where the ball would pass the goal line and position itself to this location before the ball arrives. If the ball is not rolling toward the goal it can position itself in the center of the goal line.

Exercise. Program the goalkeeper agent with at least two different behaviors as suggested. Verify its performance by moving the ball manually.

The attacker agent can use basic motion control without ball prediction if its velocity is too small to reach the moving ball; otherwise, it can use predictive motion. If the conditions for accelerated straight kick are valid, then it should perform accelerated straight-kick behavior. Additionally it should detect possibility of collision and try to avoid collision. However, in some situations (e.g., when it possessed the ball) it should not try to avoid other players or opponents.

Exercise. Program the attacker agent with more different behaviors as suggested. Verify its performance by simulating the goalkeeper manually and passing the ball away from the goal.

The third player is a midfield player, which means that it should try to place itself at some (predefined) locations on the playground and orient itself toward the opponent goal. It therefore needs to arrive to the defined location, and if it is at the location then it needs to rotate to face toward the goal. This agent currently seems useless if agents' roles are static, but if they can change dynamically (agent can decide during the game to be attacker, goalie or midplayer) it becomes important, as seen in the next section.

8.6.4 Multiagent Game Strategy

Running all previously defined agents together should result in an autonomous soccer-like game. Here, make the agents choose autonomously which role (goalie, attacker, or midplayer) they will play. For each role define some numerical criteria such as the following examples: distance to the ball and distance to the goal. Each agent needs to evaluate those criteria and choose the role that they can currently perform with the highest success of all other agents. First you can evaluate agents for the goalkeeper role (e.g., which agent is closest to the goal) and the most successful agent becomes the goalkeeper. The goalkeeper is the role with the highest priority because it is assigned first.

The remaining two agents should verify which one can be the best attacker (e.g., which is closest to the ball). And the last agent is the midplayer. The agent that is currently the midplayer can take the role of attacker if the condition for accelerated straight kick becomes valid and the current attacker is not successful (e.g., not close to the ball). To negotiate which agent occupies a certain role the agents need to communicate their numerical criteria to others and then decide. Role assignment can also be done globally.

Exercise. Program game strategy for a multiagent system playing robot soccer. The agents can choose their role (e.g., attacker, goalkeeper, midplayer) dynamically during the game. After each agent knows its role it chooses the behavior inside the role that suits the current situation the most.

An example of dynamic role assignment during a robot soccer game in [5] is shown in Figs. 8.24 and 8.25. Here, five robots are playing against five opponent robots and the game strategy consists of 11 behaviors, where some of them are very similar to those presented in this exercise.

In Fig. 8.24, a long breakthrough of agent 1 is shown. It starts approaching the ball by playing attacker behavior (behavior index 0) and then continuing with approaching to the future interception point of the ball (behavior index 8) in a predictive manner. Finally, agent 1 switches to guider behavior (index 4) and scores a goal. It should be mentioned that opponent players' positions drawn in Fig. 8.24 correspond to a screen shot time at 12.08 s. The opponent player that seems to block agent 1 and the ball direction has therefore moved away, leaving agent 1 a free pass. From the behavior index plot for the first agent, a sudden jump of behavior index from 4 (ball guider) to 3 (predictive attacker) and then back again can be observed. This happens because the ball moved some distance away from agent 1 when pushing it.

A nice assist of agent 3, which rebounds the ball from the boundary by playing behavior 0, can be seen in Fig. 8.25. After the rebound, the straight ball motion is intercepted by agent 5 playing behavior 6, and the action is finished by behavior 4.

Screen shot time = 12.08 s

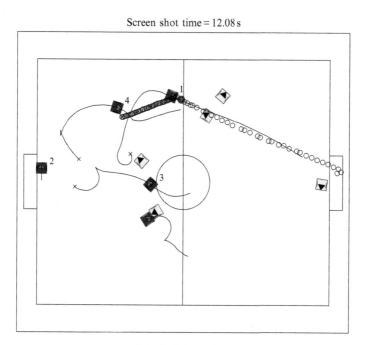

Behavior indexes of agents

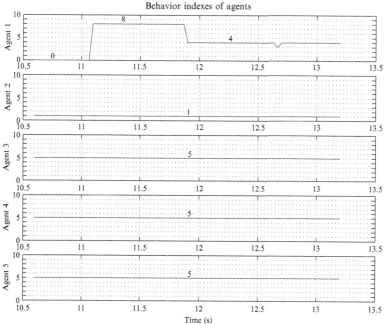

Fig. 8.24 Example of strategy operation in a robot soccer game. Robot motion and dynamic role assignment graph.

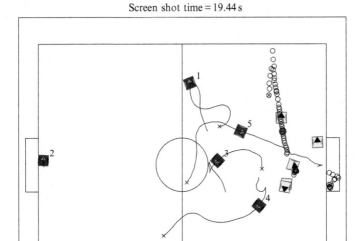

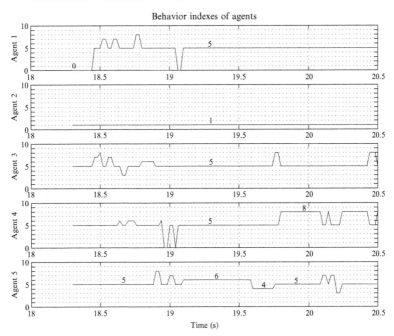

Fig. 8.25 Example of strategy operation in a robot soccer game. Robot motion and dynamic role assignment graph.

8.7 IMAGE-BASED CONTROL OF A MOBILE ROBOT

8.7.1 Introduction

Usage of machine vision in control of mobile robots is an emerging field of research and development. A camera is an attractive sensor for use in mobile systems due to its nice properties: noninvasive sensor, a lot of data, relatively small, and a lightweight and affordable sensor. Contemporary capabilities of computer processors enable real-time implementation of computationally intensive image processing algorithms, even on battery-powered embedded systems. However, image processing is considered a difficult problem, due to various reasons: the image is formed as a projection of 3D space, various image distortions and abbreviations, illumination changes, occlusions, and other issues. The required information for control is normally not easy to extract from the huge amount of data the image provides, especially if real-time performance is required.

In this laboratory practice an approach of visual control of a wheeled mobile robot equipped with a camera to a desired location is presented.

8.7.2 Experimental Setup

Usage of a two-wheeled mobile robot with differential drive is assumed. On-board of the mobile robot a front-facing camera is mounted (Fig. 8.26); this is known as eye-in-hand configuration. The considered control algorithms will require in-image tracking of at least four coplanar points. To somewhat simplify image processing some color markers or special patterns can be introduced into the environment; later some more advanced machine vision approaches can be used in order to track some features already present in the environment. The considered configuration is depicted in Fig. 8.26. The pose of the camera C on the mobile robot R is given with rotation matrix $\boldsymbol{R}_C^R$ and translation vector $\boldsymbol{t}_C^R$. The desired mobile robot pose is given with respect to the pose of the world frame W, that is, with rotation matrix $\boldsymbol{R}_D^W$ and translation vector $\boldsymbol{t}_D^W$.

8.7.3 Position-Based Visual Servoing

To implement visual control some image processing is required in order to determine control error. There are two basic approaches: the control error can be defined in the 3D frame—this is known as position–based visual servoing (PBVS)—or the control error is determined directly as a 2D error in the image; this is known as image-based visual servoing (IBVS). In this

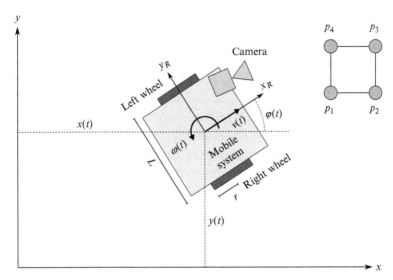

Fig. 8.26 Wheeled mobile robot with differential drive equipped with a camera.

case a PBVS control scheme will be considered. A pinhole camera model (5.48) will be used. Visual servoing normally requires camera calibration, where radial and tangential lens distortion need to be eliminated.

Exercise. Mount the camera on-board of the mobile system and implement an image grabbing program that enables real-time access to the acquired image data. Evaluate camera lens distortion and select an appropriate method for camera calibration. Determine the values of the camera intrinsic parameter matrix S, given in Eq. (5.49).

Once the camera is calibrated an appropriate image feature tracking algorithm needs to be implemented. Without loss of generality assume that the camera is observing four corners of a square, which is placed on the ground in the world frame, where the corners of the square are color coded in colors that are unique in the observed environment. Under controlled illumination conditions color patches can be obtained with simple image processing (see Section 5.3.4). It is assumed that the tracked features are always in front of the camera image plane and always visible, that is, in the camera field of view.

Exercise. Implement a color-based, or some other feature, tracking algorithm that enables image tracking of at least four image features. The algorithm needs to be able to robustly detect features and also identify features across multiple camera views.

Assume that the positions of the four corners of the square, which are placed in the ground plane, in the world frame W are known: $\boldsymbol{p}_{W,i}^T = [x_{W,i}, \; y_{W,i}]$, $i = 1, 2, 3, 4$. The control goal is to drive the wheeled mobile robot to the origin of the world frame, that is, align the robot frame R with the world frame W. Since the points $\boldsymbol{p}_{W,i}$, $i = 1, 2, 3, 4$, in the world frame are constrained to a plane and images of these points are also on a plane, the transformation between the world and image (picture) point can be described by homography (5.56), denoted as $\boldsymbol{H}_P^W$. A point in the image can be transformed to the plane with the origin in the robot frame R. This transformation can be described by homography $\boldsymbol{H}_P^R$.

Exercise. Determine the homography between the robot plane R and the image plane P based on the known transformation between the coordinate frames and camera intrinsic parameters.

Once the positions of the tracked points in the image frame P are transformed to the robot frame R, the control error between the robot frame R and the world frame W can be determined. Using an appropriate control algorithm that is defined directly in the robot frame the wheeled mobile robot can be driven to the goal pose. During execution of the visual control algorithm some image features may leave the camera field of view, and the visual control would fail. Therefore, the feature square should be positioned in an appropriate position in the world plane that prevents this situation. A more appropriate solution would be to design the control algorithm in a way that prevents this situation.

Exercise. Transform the tracked image features to the robot frame R, determine the control error, and implement an appropriate pose control algorithm.

Notice that the tracked features normally cannot be positioned near the origin of the world frame (reference pose), since the camera normally cannot observe the features that are below the mobile robot.

8.7.4 Image-Based Visual Servoing

In this section the same control problem as in previous section will be solved in the IBVS control scheme. Let us initially assume that the camera is mounted on the wheeled mobile robot in a way that the origin of the robot frame R is in front of the image plane, although it may not be visible in the image due to limited field of view.

In this case also the transformation between the image plane P and robot plane R again needs to be determined. This could be made in the same way as already suggested in the previous section. However, here an alternative approach will be considered. The wheeled mobile robot can be positioned into the origin of the world frame W in a way that the robot frame R is aligned with the world frame W. Given known positions of the feature points in the world frame $\{p_{W,i}\}_{i=1,2,3,4}$ and measured positions of their projections in the image frame $\{p_{P,i}\}_{i=1,2,3,4}$, the homography matrix H_P^R can be estimated using the least-squares method based on point correspondences.

Exercise. Estimate homography H_P^R using the least-squares method based on point correspondences between world and image points.

Once the homography H_P^R is known, the origin of the world frame can be moved around to define the desired reference pose. Square pattern feature points can then be transformed into the image plane to obtain the desired location of the feature points in the image. Comparing current locations of the tracked points with the desired points in the image can be used in order to define control error and design a control algorithm.

Exercise. Based on the known homography calculate the desired image positions of the features and use these features to calculate control error and implement IBVS control algorithm.

Notice that the desired positions of the image features can also be determined without the knowledge of homography H_P^R. This can be achieved if the wheeled mobile robot is positioned to the desired pose, and image locations of the tracked features are captured. This approach is sometimes referred to as *teach-by-showing*.

8.7.5 Natural Image Features

The control algorithms designed in the previous sections required that a special marker was introduced into the environment. However, some features that are inherently present in the environment can also be used in order to implement visual control. This enables more useful application, since no special preparation of the environment is required. If the ground is flat and contains enough visual cues that can be robustly detected with some machine vision algorithm, the similar control approaches as designed in the previous sections can be used, except that a different feature tracking algorithm is used. Implementations of various feature tracking algorithms can be found in the open source computer vision library OpenCV [6].

Exercise. Implement some machine vision algorithm that enables tracking of features that are inherently present in the environment. Modify the visual control algorithms to use the selected feature tracking algorithm.

8.8 PARTICLE-FILTER-BASED LOCALIZATION USING AN ARRAY OF ULTRASONIC DISTANCE SENSORS

8.8.1 Introduction

This exercise deals with localization of differentially driven wheeled robots equipped with three ultrasonic distance sensors. An environment where the robot moves is known and described by the sequence of straight line segments. Localization is performed using a particle filter that fuses odometry information and measured sensor information to obtain robot location in the map.

An example of such a robot build from the Lego Mindstorm EV3 kit is shown in Fig. 8.27. This figure also contains a view of the robot environment. Any other robot with distance sensors can also be used in this exercise or if preferred the whole exercise can be done using simulation. Read Section 6.5.3 to get better insight to the localization using a particle filter. This exercise is an upgrade of Example 6.24, which can be used and modified for the purpose of this exercise.

In the following, basic steps that will guide you to the final solution are outlined: prediction of particles' location using odometry, measurement model for estimation of simulated particle measurement, and correction of particle filter by particle restamping based on their success (importance).

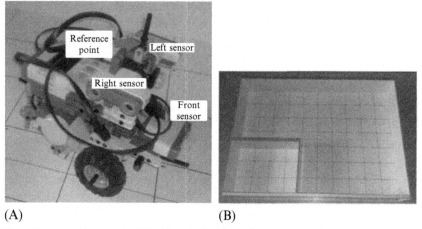

(A)   (B)

Fig. 8.27 Lego Mindstorm EV3 robot with three ultrasound distance sensors and an example of an *L-shaped* environment.

8.8.2 Predicting Particles' Locations From Known Robot Motion

A particle filter consists of a prediction and correction part. First we will implement a basic framework of the particle filter and implement prediction of the particles' locations.

Exercise. Write the framework of the particle filter. If you are doing simulation in Matlab then your starting point can be the source code of Example 6.24.

Robot pose estimation using odometry needs to be implemented to predict each particle's future location based on known motion commands sent to the true robot. Suppose motion commands sent to the robot in current time instant k are v_k and ω_k. With this command all particles are moved from their current pose $\hat{x}_k^i = [\hat{x}_k^i, \hat{y}_k^i, \hat{\varphi}_k^i]^T$ to the new predicted pose $\hat{x}_{k+1|k}^i = [\hat{x}_{k+1}^i, \hat{y}_{k+1}^i, \hat{\varphi}_{k+1}^i]^T$, where i is the index of ith particle. The predicted particle pose considering sifting of the time instant $k \to (k-1)$ and sampling time T_s reads

$$\hat{x}_{k|k-1}^i = \hat{x}_{k-1|k-1}^i + \begin{bmatrix} T_s v_{k-1} \cos(\varphi_{k-1}^i) \\ T_s v_{k-1} \sin(\varphi_{k-1}^i) \\ T_s \omega_{k-1} \end{bmatrix}$$

Exercise. Implement motion of the particles considering robot commands (v_k and ω_k). Initial poses of particles can be set randomly inside the selected environment while future particle poses are calculated using Eq. (8.8.2).

When simulating particle motion also add expected (estimated) system noise (normal distribution) to known commands v_k and ω_k. If this noise is unknown just select some small standard variance similarly as in Example 6.24.

8.8.3 Sensor Model for Particle Measurement Prediction

The robot in Fig. 8.27 has three sensors that measure distances to obstacles in three defined directions according to the local robot frame. Each particle presents a hypothesis of the robot position and therefore needs to calculate its measurement. Those measurements are needed to evaluate which particle is the most important (its pose is most likely to be the robot pose). This evaluation is based on comparison of robot and particle measurements.

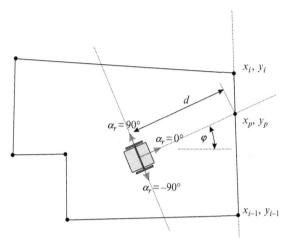

Fig. 8.28 Simulated distance sensor measurement. The robot has three distance sensors measuring the distance to the obstacle in directions 0 degree, 90 degree, or −90 degree. Sensor measurement is obtained by calculating the intersection of two straight lines: one belonging to the sensor and the other to the environment.

To calculate simulated particle measurements the environment is first described by straight line segments where each jth segment can be defined by two neighboring points x_{j-1}, y_{j-1} and x_j, y_j (see Fig. 8.28) as follows:

$$Ax + By + C = 0 \qquad (8.32)$$

where $A = y_j - y_{j-1}$, $B = -x_j + x_{j-1}$ and $C = -x_{j-1}A - y_{j-1}B$. The distance sensor measurement direction is described by the equation

$$A_s x + B_s y + C_s = 0 \qquad (8.33)$$

where $A_s = \sin \alpha$, $B_s = -\cos \alpha$, $C_s = -xA_s - yB_s$ where x, y are robot center coordinates and α is a sensor measurement angle obtained by summing the robot (or particle) orientation and relative sensor angle (α_r which in Fig. 8.28 can be 0 degree, 90 degree, or −90 degree) according to the robot ($\alpha = \varphi + \alpha_r$). If the straight lines (8.32), (8.33) are not parallel ($det = A_s A - AB_s \neq 0$) then they intersect in

$$x_p = \frac{B_s C - BC_s}{det}$$

$$y_p = \frac{C_s A - CA_s}{det}$$

and measured distance of the simulated sensor (between the robot position and the intersection point) is

$$d = \sqrt{(x - x_p)^2 + (y - y_p)^2}$$

This computed distance of course makes sense only if the object is in front of the distance sensor and not behind and if the intersection point is in between the straight line segment, which can be checked by the condition

$$\left[\left(x_{j-1} < x_p < x_j\right) \text{ or } \left(x_j < x_p < x_{j-1}\right)\right] \text{ and } \left[\left(y_{j-1} < y_p < y_j\right) \text{ or } \left(y_j < y_p < y_{j-1}\right)\right]$$

To obtain the final simulated sensor distance all line segments of the environment need to be checked and minimal distance to one of the lines needs to be found.

Exercise. Implement the distance sensor model and estimate robot measurements (the robot has three sensors as in Fig. 8.28). If the robot measurement is simulated also some sensor noise can be added.

8.8.4 Correction Step of Particle Filter and Robot Pose Estimate

The first step in the correction part of the particle filter is estimation of the particles' simulated measurements. For each particle position and orientation in the simulated environment its measurements are calculated using a derived sensor model. Those measurements are used to evaluate particles by comparing them to the true robot measurement.

Exercise. A program simulated sensor is used to estimate particles' measurements. Evaluate the particles' importance by comparing true robot measurements with particle measurements. Estimate particle weights. Use particle weights to generate a new generation of particles for the next filter iteration as shown in Section 6.5.3. The best estimate of robot pose is the average value of all particles' poses.

Exercise. Test the performance of the particle filter by driving the robot in the environment and observing its estimated pose in comparison to the true pose.

First suppose that the robot's initial pose is known; therefore, set initial particle poses around the known robot pose with small uncertainty.

Then compare filter performance in case the robot's initial pose is unknown, meaning that initial particle set is uniformly distributed in the environment.

Is the robot pose estimation problem observable, and does it always converge to the true robot pose? Can you identify situations when the robot pose is not observable (see environment example in Fig. 8.27)? How can the observability and convergence can be improved (by changing the environment or changing the measuring system)?

8.9 PATH PLANNING OF A MOBILE ROBOT IN A KNOWN ENVIRONMENT

In order to find the optimum path between the obstacles from the current mobile robot pose to the desired goal pose a path planning problem needs to be solved. This is an optimization problem that requires a known map of the environment and definition of optimality criterion. We can look for the shortest path, the quickest path, or the path that satisfies some other optimality criterion. This exercise covers the A* algorithm that is one of the commonly used approaches of solving the path planning problem.

8.9.1 Experimental Setup

This exercise does not require any special hardware. The task is to develop an algorithm that can find an optimum path between an arbitrary start and goal location given a map of the environment. The map can be represented in various ways, but in this experimental setup we will assume that the map is available as an occupancy grid. An example of an occupancy grid is represented as a grayscale image in Fig. 8.29A, which consists of 170×100 square pixels (width of each pixel corresponds to 10 cm in the real world). The particular occupancy map has the following color coding: white pixels (value 255)—free space, light-gray pixels (value 223)—unknown space, dark-gray pixels (value 159)—border. The map shown Fig. 8.29A has been acquired automatically with an SLAM approach. However, the map can also be drawn manually in any image manipulation software or even hard-coded into a file.

For the purpose of path planning we will assume that the robot can only move inside a free space. In this case it is therefore adequate to represent an occupancy grid with only two color levels, as a binary image where white pixels correspond to free space and black pixels correspond to a nontraversable space. Thresholding the occupancy map in Fig. 8.29A with the threshold level 160 results in a binary image (map) shown in Fig. 8.29B.

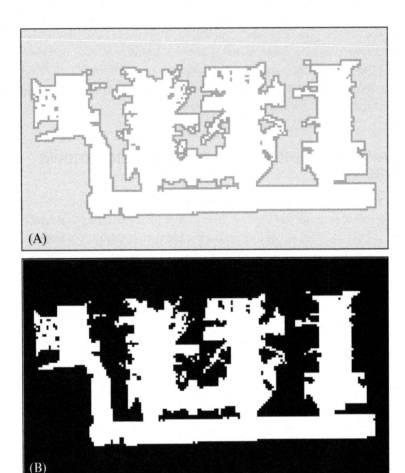

Fig. 8.29 (A) Occupancy grid of an indoor space and (B) corresponding binary map that represents only free and occupied space.

8.9.2 A* Path Searching Algorithm

For the purpose of algorithm development we will use a simple and small map of the environment, with and without obstacles. Examples of small maps that are appropriate for algorithm development are shown in Fig. 8.30; the dimensions of the map grids are 6×4 cells. The map in Fig. 8.30A is without obstacles and the map in Fig. 8.30B contains a single L-shaped obstacle. The start and end locations are marked with letters S and G, respectively.

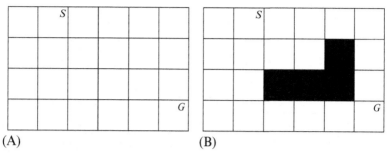

Fig. 8.30 (A) A simple map without obstacles and (B) a map with an L-shaped obstacle. Start and goal locations are marked with S and G, respectively.

Exercise. Choose an appropriate data structure for storing the map data. Map data is a 2D matrix of elements that can only take two values: for example, values 1 and 0 represent free and occupied space, respectively. Define variables for storing the start and goal position, in the form of map coordinates or a map cell index. Hard code the maps, initial, and goal positions shown in Fig. 8.30 into the selected data structures. Write a test function that checks that the start and goal cell are inside the free space.

In the A* algorithm a direct distance from any cell in the map to the goal needs to be determined (neglecting any obstacles on the way). For this purpose different distance measures can be used, for example, Manhattan or Euclidean distance.

Exercise. Write a function that takes as an input the arbitrary cell position or index in the map and goal location and returns a Euclidean distance of the input cell to the goal.

The A* search algorithm begins exploring the map in the start cell (node) and expands the search looking at the connected neighbors. In the A* algorithm immediate neighbors of the currently active cell need to be determined. In a regular grid map a four-cell or eight-cell neighborhood around the cell is normally used (Fig. 8.31). For each cell in the neighborhood a transition cost Δg from the center cell needs to be defined. In Fig. 8.31 Euclidean distance is used, although some other measure could also be used (e.g., cost d for any vertical and horizontal move and cost 3 for any diagonal move).

Exercise. Write a function that returns a list of all the neighbors around the current cell (use an eight-cell neighborhood). Pay special attention to boundary and corner cells that do not have neighbors in all directions.

In an A* search algorithm an open and closed list of cells are used. The open list contains all the cells that need to be searched, and the closed list

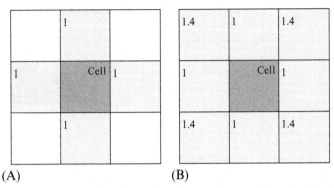

Fig. 8.31 (A) Four-cell and (B) eight-cell neighborhood around a cell; all cells have a marked transition cost from the center cell.

contains the cells that have already been processed. All the cells in the open list are ordered according to cost-of-the-whole-path, which is a sum of the cost-to-here and cost-to-goal (also known as heuristic). Initially only the start cell is in the open list. Until there are cells in the open list and the goal cell is not reached, the cell with the lowest cost-of-the-whole-path is set as the active cell. The active cell is removed from the open list and added to the closed list. The cost-to-here, cost-to-goal, and cost-of-the-whole-path are calculated and an index of source cell is remembered for each neighbor around the active cell that is not an obstacle. If the neighbor is not already in the open or closed list, it is added to the open list. If the cell is already in the closed or open list and the new cost-to-here is lower, the cell is moved to the open list, and all costs and also the index of the source cell are updated. A more detailed description of the algorithm can be found in Section 4.4.5.

Exercise. Implement an algorithm that expands the search area according to the A* search algorithm. Define data structures for an open and closed list that also enable storing cost-to-here, cost-to-goal, cost-of-the-whole-path, and the index of the source cell. For the open list a priority queue is recommended, since the cells need to be sorted according to the cost-of-the-whole-path. Do not forget to include the conditions that terminate the search once the goal node has been reached or the open list contains no more cells.

If the goal cell has been reached, the optimum path from the start to the goal can be obtained by tracking back the source cells from the goal until the start cell is reached.

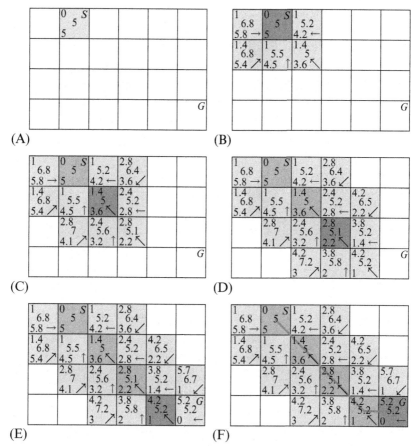

Fig. 8.32 Steps of the A* algorithm, from initial step (A) to final step (F), on a simple map without obstacles (Fig. 8.30A). The *number in the top left corner* of each cell is cost-to-here, the *number in the bottom left corner* is cost-to-goal, and the *number in the middle* is cost-of-the-whole-path. In the *right bottom corner* the direction of the source cell is indicated.

Exercise. Implement an algorithm for obtaining the path if the goal cell has been reached. Write a function that enables visualization of the path. Validate the algorithm with the visualization of all the steps shown in Figs. 8.32 and 8.33.

8.9.3 Optimum Path in the Map of the Environment

Although some automatic mapping approach can be used, in this case we will manually draw a map of the environment.

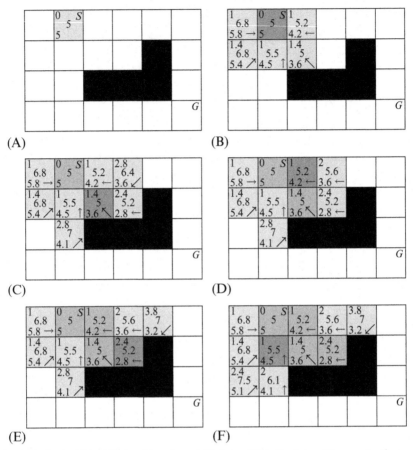

Fig. 8.33 Steps of the A* algorithm, from initial step (A) to final step (L), on a simple map with an L-shaped obstacle (Fig. 8.30B). See Fig. 8.32 for a description of the notation.

(Continued)

Exercise. Make a map of the environment in by using image manipulation software. Use a white color to mark free space and a black color to represent occupied space. Optionally different gray levels can be used in order to represent various types of obstacles or free space. In this case ensure that all grayscale levels above a certain threshold belong to free space. Save the image (map) in an appropriate lossless image format (e.g., BMP, PGM, or PBM).

Exercise. Write a simple program for loading the map of the environment in the form of an image file (e.g., PGM) into an appropriate data structure (e.g., matrix or simple data array) in a way that color levels of

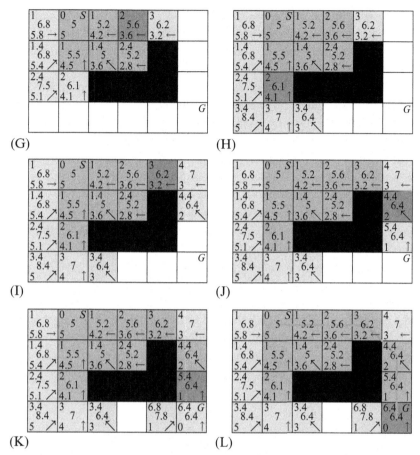

Fig. 8.33, cont'd.

each pixel can be read. Depending on the image format write a simple image processing program that enables determining if a particular pixel belongs to free space of occupied space. In the case of a grayscale image appropriate image thresholding needs to be implemented. If the image is already in binary format (i.e., every pixel has only two possible states) where one state represents free space and the other state represents occupied space, no additional image processing is required. In this case the image can be directly used as an input to the path searching algorithm.

Exercise. Write a program that enables definition of the start and goal locations. The start and goal locations can be given as program inputs or given interactively by user input during program execution. The most intuitive way of specifying the start and goal location is by the graphical user interface, where the user is given the ability to select the start and goal positions on the map with a mouse. In Matlab the function g i nput can be used to obtain the position of the mouse click in a figure. Enable manual determination of arbitrary start and goal locations within the map, although note that the start location is determined by a current mobile robot pose, which is normally given by a localization algorithm. The program should prevent the user from selecting locations that are not within the free space of the map. In such a case an appropriate notification should be raised and the nearest point in free space can optionally be selected, or the user is requested to repeat the selection process.

Exercise. Use the implemented path searching algorithm on the drawn map of the environment. Evaluate the performance of the algorithm with respect to different start and goal locations.

Fig. 8.34 shows an optimum path between the start and goal obtained by the A* path searching algorithm, where the searched cells (pixels) are also marked. The basic path searching algorithm requires some modifications in order to obtain the feasible path that the robot can follow. As it can be noted from Fig. 8.34, the optimum path is close to the obstacle boundaries. Since the robot dimensions are nonzero, the obtained path may not be feasible for the robot to follow.

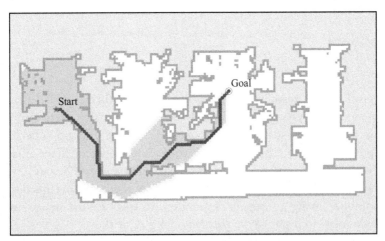

Fig. 8.34 Optimum path in the map shown in Fig. 8.29 between the start and goal location obtained by the A* path planning algorithm. Also, the searched cells (pixels) are marked.

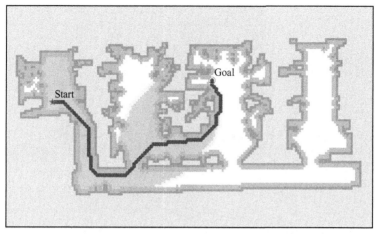

Fig. 8.35 Optimum path in a map with extended obstacle boundaries that take into account mobile robot width.

The problem of nonzero mobile robot dimensions can be solved in various ways. Assume that the robot can fit inside a circle with the radius $r = 40$ cm around the rotation center of the robot. In this case the map of the environment can be modified in a way that the free space in the vicinity of the obstacles, at distances below r, is considered as occupied space. The path searching can then be used without any other modifications, that is, the mobile robot can be assumed to be dimensionless. Fig. 8.35 shows the result of the path searching algorithm with extended obstacle boundaries.

Exercise. Implement a map (image) processing algorithm that enables extension of the occupied space into the free space for the distance of the robot radius. Evaluate the performance of the unmodified path searching algorithm on the modified map.

The approach of extending the obstacle boundaries to take the robot dimensions into account may lead to separation of free space into multiple unconnected areas. This may also happen to narrow passages that the mobile robot should be able to pass through but were blocked due to too wide obstacle boundaries, since the obstacles' boundary extension is normally larger than the robot's size. An alternative approach is to use *distance transform*. For every point the free space, a distance to the nearest obstacle is calculated. A visualization of distance transform is shown in Fig. 8.36. The distance inverse can then be added to the cost-to-goal (heuristic). In this way paths that are further away from the obstacles are

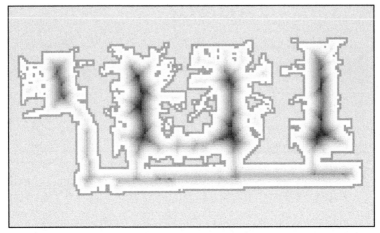

Fig. 8.36 Visualization of the distance transform in the free space (the *darker shade* represents greater distance from the nearest obstacle).

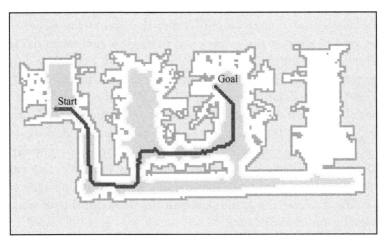

Fig. 8.37 Optimum path obtained with the use of the distance transform in the cost-to-goal.

preferred; therefore, better paths can be obtained that are optimally away from the nearest obstacles. In Fig. 8.37 it can be seen that the obtained path is along the middle of the corridors.

Exercise. Implement an algorithm that calculates the distance to the nearest obstacle for any point in the free space. Modify the cost-to-goal in the A* algorithm in order to take into account the distance transform.

Design an additional term in the cost-to-goal in a way that neglects long distances to the nearest obstacle, and limit the maximum value of the term. Make sure that the value of the added term is appropriately scaled with respect to the raw cost-to-goal.

Some additional aspects of the path planning algorithm can be found on a webpage [7], which includes many interactive examples.

REFERENCES

[1] L. Teslić, I. Škrjanc, G. Klančar, EKF-based localization of a wheeled mobile robot in structured environments, J. Intell. Robot. Syst. 62 (2010) 187–203.

[2] S. Thrun, W. Burgard, D. Fox, Probabilistic Robotics, vol. 4, MIT Press, Cambridge, MA, 2005.

[3] G. Klančar, D. Matko, S. Blažič, A control strategy for platoons of differential-drive wheeled mobile robot, Robot. Auton. Syst. 59 (2) (2011) 57–64.

[4] G. Klančar, M. Kristan, S. Kovačič, O. Orqueda, Robust and efficient vision system for group of cooperating mobile robots with application to soccer robots, ISA Trans. 43 (2004) 329–342.

[5] G. Klančar, D. Matko, Strategy control of mobile agents in soccer game, in: The 3rd International Conference on Computational Intelligence, Robotics and Autonomous Systems CIRAS 2005 & FIRA RoboWorld Congress Singapore 2005 FIRA 2005, CIRAS & FIRA Organising Committee, National University of Singapore, 2005, pp. 1–6.

[6] The OpenCV reference manual: release 2.4.3, 2012.

[7] Introduction to A*, 2016, http://www.redblobgames.com/pathfinding/a-star/introduction.html (accessed 07.06.16).

INDEX

Note: Page numbers followed by f indicate figures, t indicate tables and b indicate boxes.

G

Gaussian function, 293–294
Gaussian noise, 351
Global navigation satellite system (GNSS), 389, 390*f*, 392, 396, 396*f*, 433
GNSS. *See* Global navigation satellite system (GNSS)
GNSS signal receiver, 389, 390*f*, 391
Goalie motion control, 457–458, 457*f*
Graph-based path planning methods
    A★ algorithm, 199–201, 200*f*
    breadth-first search algorithm, 193–194, 193*f*
    depth-first search algorithm, 194, 195*f*
    Dijkstra's algorithm, 195–198, 198*f*
    greedy best-first search algorithm, 201–205, 202*f*
    iterative deepening depth-first search, 194–195, 196*f*
Greedy best-first search algorithm, 201–205, 202*f*
Ground mobile systems, 4

H

Heading measurement systems, 239–240
Hidden Markov process, 304, 304*f*, 305*f*
Histogram filter, 375
Holonomic constraints, 32–34
Hough transform, 265–268, 267*f*
Hue-saturation-lightness (HSL), 270–271
Hue-saturation-value (HSV), 270–271, 271*f*

I

Image-based visual servoing (IBVS), 465–466
Image features, camera, 268–282
Inertial navigation system (INS), 232–239
    motion sensors, 232–233
    pose error, 233
    rotation sensors, 232–233
    self-contained technique, 232–233
    signal-to-noise ratio, 233
    unit angular rate, 233–234
Information filter, 375
Informed algorithms, 193
Infrared light sensor, 420–421

line-following and crossroad-detection, 421, 422*f*
INS. *See* Inertial navigation system (INS)
Instantaneous center of rotation (ICR), 14
Internal camera model, 222–223
Internal kinematic model, 13–14, 231–232
Inverse kinematics, 14, 17–20
Iterative deepening depth-first search, 194–195, 196*f*

J

Jacobian matrix, 352–353, 436–439
Joint probability, 290

K

Kalman-Bucy filter, 375
Kalman filter, 234–235, 337–375, 339–340*b*
    algorithm, 343, 343*b*
    continuous variable, probability distribution of, 338*f*
    correction covariance matrix, 348
    correction step, 338, 343–344
    derivatives, 374–375
    extended, 351–374
    Gaussian function, 338
    Gaussian noise, 346
    for linear system, 346–347
    in matrix form, 345–351
    prediction step of, 342–344
    state estimation, 337
    updated state variance, 341–342
Kalman observability matrix, 303
Kinematic constraints, 32–33
Kinematic model, 13
    Ackermann steering principle, 24–26, 25*f*
    bicycle drive, 20–23, 21*f*
    differential drive, 15–20, 16*f*
    direct kinematics, 14, 17–18
    external kinematics, 14
    instantaneous center of rotation (ICR), 14
    internal kinematics, 13–14
    inverse kinematics, 14, 17–20
    motion constraints, 14
    omnidirectional drive, 27–31
    synchronous drive, 26–27, 26*f*

Printed in the United States
By Bookmasters